American Technological Sublime

American Technological Sublime

David E. Nye

The MIT Press
Cambridge, Massachusetts
London, England

First MIT Press paperback edition, 1996

© 1994 Massachusetts Institute of Technology

Set in New Baskerville by The MIT Press.
Printed and bound in the United States of America.

Library of Congress Cataloging-in-Publication Data

Nye, David E., 1946–
   American technological sublime/David E. Nye.
      p.   cm.
   Includes bibliographical references (p.   ) and index.
   ISBN 0-262-14056-X (HB), 0-262-64034-1 (PB)
   1. Technology—Social aspects—United States. 2. Technology—United States—History—19th century. 3. Technology—United States—History—20th century. 4. Sublime, The. I. Title.
T14.5.N93   1994
303.48′3—dc20   94-19273
                                                          CIP

to Leo Marx,
Sublime teacher

Contents

Acknowledgements

I had the good fortune to study with Leo Marx as an undergraduate at Amherst College, and this book is dedicated to him. Since I now spend most of my time on the other side of the Atlantic Ocean from Leo, we only spoke about this book a few times, and he can hardly be held accountable for my mistakes. But he did set my thoughts in motion. I first heard about the "technological sublime" in one of the many inspiring lectures he gave in 1965. Years later, when I finally realized how interesting this idea was, Leo was generous with encouragement and advice, as he has been on other projects during the intervening years.

In more specific terms, this book grew directly from a brief discussion of what I called the "electrical sublime" in chapter 2 of *Electrifying America*. This idea was developed further when Robert Shulman suggested that I write "something about republicanism and technology" for a special issue of the *American Transcendental Quarterly*. Fragments of that article resurface in many chapters here. Chapter 8 appeared in shorter and considerably different form in *Anthropology and History*, and I thank that journal for granting me permission to reclaim the material. During the process of writing I had many opportunities to test my ideas in lectures, particularly at the 1991 American Studies National Convention in Baltimore, at the Strong Museum in Rochester, New York, at Twente Technical University in the Netherlands, at Linköping University Center in Sweden, at Würzburg University's seminar on postmodernism, at Odense University's Center for Man and Nature, at the 1993 Norwegian American Studies Association meeting in Kristiansand, at the 1993 convention of the Italian American Studies Association in Venice, and at the Netherlands Institute for Advanced Study (NIAS). I was fortunate to be in residence as a fellow at NIAS during the 1991–92 academic year, during which time my early drafts were improved by comments from Hans Bertens, Mel van Eltern, Dunbar Ogden, and Bruce Kepferer. Others generously read portions of the manuscript and offered useful criticism: Mick Gidley, Rob Kroes, Robert Rydell, and Judith Kepferer. The Danish Research Council provided partial support for my year at NIAS, while the bulk of my grant was generously contributed by the Royal Academy of the Netherlands.

Many libraries made their resources available to me: the public libraries of Alexandria (Virginia), Baltimore, Boston, New York, and Rochester; the Library of the Congress; the libraries of the National Museum of American History, Amherst College, the University of California at Los Angeles, the Maryland Historical Society, Grand Canyon National Park, the University of Nevada at Las Vegas, Leiden University, and Odense University; the Danish Royal Library in Copenhagen; and the John F. Kennedy Library of the Free University of Berlin. Several journeys to the United States were crucial to the completion of my work, and I thank Odense University for partially funding two short research trips and the Electric Power Research Institute for inviting me from Denmark to lecture in Arizona (which, incidentally, made it possible for me to revisit the Grand Canyon). The Dibner Institute for the History of Science and Technology at MIT graciously provided office space during the final weeks of manuscript preparation, and Spanish-English Services gave me invaluable computer assistance.

All researchers know how much they depend on others, but the expatriate scholar has even more debts. Doing research on the United States from abroad is more difficult than anyone who has not tried it can imagine. I had to rely on colleagues and exchange students who were good enough to take time from their busy schedules to find materials for me or share the results of their own research. I particularly thank Ronald Johnson, Jeffery Stine, Jan Gretlund, Christen Kold Thomsen, Robert Baehr, Lene Brøndum, Michael Smith, Sarah Stebbins, Roland Marchand, and John O'Neill. I cannot thank them enough. Others whom I have never met were also generous with their time, notably Jerome Edwards, Barbara R. Kelley, John H. McCarthy, Vidar Pedersen, and Lisa McDonough. I thank my two formal reviewers (one anonymous, the other John Staudenmaier) for detailed and thoughtful readings. First and last, Larry Cohen guided me through this project with his usual editorial tact.

But most of all, I thank Helle Bugge Bertramsen.

Introduction

If any man-made object can be called sublime, surely the Golden Gate Bridge can. More than a mile long, it crosses the turbulent waters of San Francisco Bay between rocky headlands. Three times the height of the Brooklyn Bridge, it arches through crosswinds, clouds, and fog, slim and elegant, a triumph of engineering skill.[1] Icon of San Francisco and constantly featured on travel posters, postcards, and brochures, it has become an instantly recognizable landmark. Yet, like every sublime object, this magnificent piece of civil engineering cannot be comprehended through words and images alone. When visited, it outstrips expectations.

On May 24, 1987, the bridge was 50 years old, and to celebrate the occasion officials closed it to traffic for one day and allowed the public to walk across from both sides. On the day it was inaugurated,[2] 200,000 pedestrians paid a nickel each for a first look; in 1987 it seemed appropriate to ban cars once again for a day, abandoning the bridge's transportation function to emphasize its symbolic role. Since the novelty of the bridge had worn off, officials expected a smaller crowd than in 1937, and they were amazed to see an immense multitude gathering before dawn. The impatient crowd was as heterogeneous as the Bay Area's population and included every race, religion, and age group. A surprising number had been there 50 years earlier for the opening. Already at 5:30 A.M. some people flocked onto the bridge when a chain was dropped to let a truck drive across. By sunrise the deck was jammed, and the arched span flattened out under the weight of 250,000 people, with more than half a million more moving forward along the approach roads. Those trapped on the center of the bridge were pressed tightly together and could not move for hours. Police tried to get the crowd to back up, while engineers calculated whether the structure would buckle or collapse. The bridge began to sway in the wind, adding to

The fiftieth-anniversary celebration of the Golden Gate Bridge, May 24, 1987. Photo by Paul Kitagaki; courtesy of *San Francisco Examiner* and SABA Press Photos, Inc.

the tension. By late morning "footlock" had been overcome, but the majority never got to walk across. Yet most stayed through the day and into the evening for a massive fireworks display, a concert, and the spectacular lighting of the bridge's towers.[3] Despite the crush, the mood was surprisingly good. The people clearly loved their bridge.

San Francisco officials were unprepared for the massive turnout because they did not understand the American public's affection for spectacular technologies. Each day crowds visit the Kennedy Space Center, ascend St. Louis's Gateway Arch, and visit the observation decks of prominent skyscrapers in New York, Chicago, Boston, Minneapolis, and other major cities. The public response to the birthday of the Golden Gate Bridge was matched by the excitement at the centenary of the Brooklyn Bridge or the Statue of Liberty. For almost two centuries the American public has repeatedly paid homage to railways, bridges, skyscrapers, factories, dams, airplanes, and space vehicles.

The sublime underlies this enthusiasm for technology. One of the most powerful human emotions, when experienced by large groups the sublime can weld society together. In moments of sublimity, human beings temporarily disregard divisions among elements of the community. The sublime taps into fundamental hopes and fears. It is not a social residue, created by economic and political forces, though both can inflect its meaning. Rather, it is an essentially religious feeling, aroused by the confrontation with impressive objects, such as Niagara Falls, the Grand Canyon, the New York skyline, the Golden Gate Bridge, or the earth-shaking launch of a space shuttle. The technological sublime is an integral part of contemporary consciousness, and its emergence and exfoliation into several distinct forms during the past two centuries is inscribed within public life. In a physical world that is increasingly desacralized, the sublime represents a way to reinvest the landscape and the works of men with transcendent significance. As Émile Durkheim concluded: "The ideal society does not stand outside the real society: it is part of it. Far from being torn between two opposite poles, we cannot be part of the one without being part of the other. A society is not simply constituted by a mass of individuals who compose it, by the territory they occupy, by the things they use and the actions they perform, but above all by the idea it has about itself."[4] Since the early nineteenth century the technological sublime has been one of America's central "ideas about itself"—a defining ideal, helping to bind together a multicultural society. Americans have long found the sublime more necessary than Europeans, so much so that they have devised formations of the sublime appropriate to their pluralistic, technological society. Precisely because American society is so pluralistic, no single religion could perform that function. Instead, ever since the early national period the sublime has served as an element of social cohesion, an element that was already quite evident when the first canals were dug and steam engines were first harnessed to trains.

The members of a multicultural society need not agree on the precise meaning of a rite; it can create solidarity through participation. In David Kertzer's neo-Durkheimian view, "ritual can produce bonds of solidarity without requiring uniformity of belief."[5] The millions who travel to the Grand Canyon or Cape Canaveral can share an awed response to what they see without discussing or even articulating what their sublime encounter means. The crowd's infectious enthusiasm is an essential part of the atmosphere surrounding a world's fair, the celebration of a new technology, or an Independence Day celebration.[6] At such events organizers mediate the crowd's response through speeches, music, fireworks, and spectacular demonstrations, but unanimity is not necessary. The specific advantage of the sublime as a shared emotion is that it is beyond words.[7]

Yet the emotion, although ineffable, is not inevitable. Over time, the same objects cannot always be counted upon to evoke the sublime response. Their power often decays, and other alternatives are sought. Ultimately, the constant is not the technological object *per se*; it is the continual redeployment of the sublime itself, as a preferred American trope. Since the 1820s a number of interrelated American sublimes have emerged. Each of these articulates a distinct political and social relation to technology, and to some extent these coexist uneasily as alternative social constructions. Yet these contradictions are more latent than expressed, because the sublime encounter leaves observers too deeply moved to reflect on the historicity of their experience. Sublimity seems not a social construction but a unique and precious encounter with reality.

This book is organized historically, exploring forms of the sublime as they have emerged between 1820 and the present. Chapter 1 briefly examines the sublime as described by Edmund Burke and then sketches Immanuel Kant's synthesis and extension of previous theory. The purpose here is to provide background to readers unfamiliar with the sublime and to demonstrate the continuing relevance of the term in understanding present-day responses to natural objects. Chapter 2 examines the early-nineteenth-century emergence of the sublime in the United States as a distinctive form related to religion, gender differences, and politics. Chapters 3–5 describe the emergence of the technological, geometrical, and industrial sublimes in relation to railroads, bridges, skyscrapers, and factories. Chapters 6–8 turn to the electrical sublime, which emerged in the last decades of the nineteenth century in spectacular displays at world's fairs and pageants and soon became the basis for permanent night illuminations and the phantasmagoria of the Great White Way. These developments culminated in the integration of the technological, geometrical, and electrical sublimes in the stagecraft of the New York World's Fair of 1939. The final chapters examine three subsequent examples of the technological sublime: the atomic bomb, the first manned flight to the moon, and the rededication of the Statue of Liberty on July 4, 1986. Manifestations of the sublime on Independence Day will be a recurrent theme in each of the periods examined, marking the changing relations between technology and politics.

In view of these subjects it should be clear that, although this book deals with the history of technology, it is not a history of machines and structures from an engineering point of view. Rather, it is concerned with the social context of technology, with how new objects are interpreted and integrated into the fabric of social life.

The technological sublime was first discussed in Perry Miller's book *The Life of the Mind in America*. Miller appears to have coined the term, which was taken up and elaborated by Leo Marx in *The Machine in the Garden*. The most fully developed application of the term until now is to be found in John Kasson's *Civilizing the Machine*, an analysis of political, aesthetic, and utopian responses to mechanization in nineteenth-century America. Barbara Novak has also employed the term to discuss American painting. More recently, Roland Marchand, in his study of American advertising, wrote of the "power of man to manufacture the sublime," and John Sears devoted several pages to the technological sublime in his analysis of nineteenth-century tourism.[8] These authors have used the concept of the technological sublime in roughly comparable ways, and since they often have cited one another it is fair to say that the term has become common. Each has helped to define the concept, chiefly by example, yet to date it remains largely unexplored as a full-scale subject. Other research on the sublime in America has focused primarily on poetry and on nineteenth-century landscape painting, with little work on the years after 1890. The present volume systematizes and extends previous work to include both a wider range of phenomena and a longer time span.[9]

To trace such a broad topic through 200 years in one volume requires considerable selection. Expressions of the technological sublime are abundant, and I have adopted the strategy of examining a smaller number of examples in some depth rather than trying to survey the widest possible field. The Grand Canyon, Niagara Falls, and Virginia's Natural Bridge appear repeatedly in discussions of the natural sublime, for example, while most of the national parks are left out in the belief that their inclusion would lengthen the argument without changing it very much. One major area has been omitted: the experience of battle. The national anthem evokes "the rockets' red glare and bombs bursting in air," a reminder that the most powerful experiences of technology for many have long been encountered in warfare. This subject merits a study of its own.[10]

What are my criteria of selection? First, I have searched for the things that awed the public. Second, I have focused on phenomena that attracted maximum national attention: the Grand Canyon, Niagara Falls, the explosion of Mount St. Helens, the Erie Canal, the first transcontinental railroad, the Eads Bridge, the Brooklyn Bridge, the major international expositions, the Hudson-Fulton Celebration of 1909, the Empire State Building, Boulder (Hoover) Dam, the first atomic bomb, Apollo XI, and the rededication of the Statue of Liberty. Each of these was a national event that awed the multitude. In analyzing them I have not presupposed a trickling down of aesthetic theory to the masses; rather I have examined experiences which ordinary people have intensely valued. They have often, but not always, used the term 'sublime' to describe their experience. I want to emphasize that this study is not about the aesthetic education of the public and does not seek to trace the shifting definitions of 'sublime'. Rather, it is about repeated experiences of awe and wonder, often tinged with an element of terror, which people have had when confronted with particular natural sites, architectural forms, and technological achievements. This book is about the social construction of certain powerful experiences in industrial society, which is to say it is about the politics of perception. It does not primarily concern literature or the arts, but rather the public's experience of particular technologies.

Edmund Burke declared at the end of his *Philosophical Enquiry into the Origin of Our Ideas of the Sublime and Beautiful:* "It was not my design to enter into the criticism of the sublime and beautiful in any art, but to attempt to lay down such principles as may tend to ascertain, to distinguish, and to form a sort of standard for them; which purposes I thought might be best effected by an enquiry into the properties of such things in nature as raise love and astonishment in us. . . ."[11] Where Burke hoped to lay down immutable principles concerning both the sublime and the beautiful, I have a more modest goal: to sketch the emergence of new sublime objects that have "raised astonishment." I do not take the sublime to be immutable, and therefore its changing cultural and political meaning must form part of the subject. My first working title, *Varieties of Sublime Experience,* echoed William James's title *The Varieties of Religious Experience.* It suggested no historical development, but rather a range of possible experiences that coexist in time. However, I want to stress the historicity and the politics of sublime experiences, presenting them as emotional configurations that both emerge from and help to validate new social and technological conditions. This volume traces the emergence of new forms of the sublime, considering them not as absolute categories of aesthetic experience but as contingent categories within social and political systems.

Each new form of the sublime may undermine and partially displace older versions. Durkheim understood that "when . . . conflicts break out . . . they do not take place between the ideal and the real but between different ideals, between the ideal of yesterday and the ideal of today."[12] One person's sublime may be another's abomination. An environmentalist finds the Grand Canyon or Niagara Falls sublime and dislikes technological "improvements" such as bridges, canals, and dams. To an engineer, a bridge may simultaneously be both a work of art that is sublime in its scale and power and a technical feat that is legible to the trained eye; the same engineer may balk at the idea of trimming a bridge in ornamental lights, preferring the unadorned technological sublime to the electrical sublime. What seems sublime can vary from one individual to another. Longinus argued that sublimity was established by the consensus of people from different backgrounds; only a work that could arouse universal admiration qualified. But we live in a world splintered into interpretive communities, each claiming the right to establish its own aesthetic standards. Conservationists and ecologists disagree with civil engineers on the sublimity of dams; this results in conflicts such as the one over the Storm King Dam proposed by New York's

Consolidated Edison Company in the 1960s.[13] Were these differences merely private opinions without public consequences, the variable social construction of the sublime would be only a curiosity. But in the United States, where the sublime has increasingly become a group experience rather than a moment of private contemplation, these experiences often have overt political consequences, both as matters of public display and as issues of social policy. The questions of central concern in this study are these: What objects have Americans invested with sublimity? What responses have there been to these different objects? What is the larger ritual or political framework within which the sublime appears? What patterns emerge when the sublime is studied over time?

Americans were not the first to admire feats of engineering and architecture. Though the term 'technology' did not exist in antiquity, some classical authors did adapt the sublime to describe both man-made and natural landscapes. Statius was perhaps the first author "to devote whole poems to the praise of technological progress," and Pliny "successfully introduced this poetic topic to prose."[14] While earlier poets such as Horace and Lucretius extolled simplicity and a primitive life without luxury, the poets of the Roman empire regularly praised villas, baths, and aqueducts and the blending of nature with art. Statius even devoted an entire poem to "the praises of a good modern road . . . expressing joy at man's successful effort at levelling mountains, cutting down forests, building a firm surface across soft and shifting sands. . . ."[15] The ancient world likewise had established the notion of the "seven wonders of the world," all of which were man-made.

In eighteenth-century England the sublime also included architecture. Just as Roman literature adapted the sublime to its roads and other monumental public works, Burke took it for granted that two basic categories of the sublime, namely difficulty and magnificence, particularly applied to architecture, Nicholas Taylor points out that Burke's writing on this point was often less a theory than a codification of already-existing architectural achievements—"filing into appropriate categories for criticism the raw materials of a new sensibility which had already appeared among artists." In the following century Victorian cities were filled with structures that were not meant to be beautiful or picturesque, but rather awesome, astonishing, vast, powerful, and obscure, striking terror into the observer. The new railway stations, aqueducts, factories, and warehouses were rhetorical structures, demonstrating the power of the builders in what Taylor calls "a permanent harangue to the public." What later generations often came to perceive as Victorian ugliness had, Taylor writes, "a direct relationship to the permanence of the social hierarchy. It is, I believe, central to Sublimity, with its hugeness and massiveness and unashamed arrogance, that it was the aristocratic taste of the time."[16]

In the United States the sublime took a different turn, for a variety of political and economic reasons. Democratic principles were translated into a strong preference for Greek Revival architecture in public buildings. Furthermore, because the United States urbanized and industrialized considerably later than England, there were fewer impressive buildings in the private sector. With no royalty or aristocracy, architects had little opportunity to design massive, sublime structures for private use. Engineers, rather than architects, built the first man-made objects that Americans regarded as sublime, and what particularly distinguished their response from that of the classical age or the English Enlightenment was the focus on moving machines. Americans often favorably compared their technological achievements to those of the ancient world. Daniel Webster emphatically declared that "the hydraulic works of New York, Philadelphia, and Boston surpass in extent and importance those of ancient Rome."[17] Writing of the Philadelphia Exposition of 1876, Walt Whitman proclaimed:

Mightier than Egypt's tombs,
Fairer than Grecia's, Roma's temples.
Prouder than Milan's statued, spired cathedral,
More picturesque than Rhenish castle-keeps,
We plan even now to raise, beyond them all,
The great cathedral sacred industry, no tomb,
A keep for life of practical invention.[18]

"Sacred industry" rivaled the religious architecture of antiquity; in America technological achievements became measures of cultural value.

The two-century-long American project of the technological sublime is not identical to the currently fashionable postmodern "sublime." Jean-François Lyotard, who adapted Kant's theory of the sublime to his analysis of the postmodern condition, gave increasing emphasis to the sublime in his later work. He attacked the project of modernism, making little distinction between the sublime in the arts and the direct experience of the sublime. But the Grand Canyon or a rocket launch, unlike a book or a painting, is apprehended with all five senses. There is a very real difference between observing a volcanic eruption and, to use Lyotard's examples, looking at a Picasso or reading James Joyce. A volcano, unlike a painting, can kill the observer. An eruption can cause the terror that lies at the core of Burke's philosophy of the sublime and which later was an essential part of Kant's theory of the dynamic sublime. Take out terror and the mind is not transfixed; rather, it is free to engage in games of reference and to lose itself in an interior hall of mirrors.[19] Lyotard's early writing on the sublime celebrates avant-garde art and its continuous rule-breaking, which pushes the viewer toward the limits of perception and the intuition of the unrepresentable, producing an emotion that he calls "the sublime." Yet this emotion has nothing to do with fear. When Lyotard speaks of postmodern art, he is writing not about the sublime but about another form of the unspeakable, which might better be called the aesthetic of the strange.[20] Apparently unaware of the long tradition of the technological sublime, he sets his form of the sublime in opposition to what he (mis)conceives to be the rationality of the technicians.

In fact, the reemergence of the natural sublime in the eighteenth century soon led to technological versions of the sublime that have persisted down to the present.[21] Nineteenth-century engineers, architects, and inventors were hardly rational technicians, and they often embraced transcendental ideas. Along with clergymen, writers, and artists, they imbued technology with moral values. Likewise, ordinary Americans repeatedly demonstrated en masse their love of technological objects, from the Erie Canal and the first railroads to the space program of the 1960s and the 1987 celebration of the Golden Gate Bridge. The *San Francisco Examiner* editorialized that the bridge is "a gateway to the imagination," noting that "in its artful poise, slender there above the shimmering channel, it is more a state of the spirit than a fabricated road connection. It beckons us to dream and dare. First seen as an impossible dream, it became a moral regenerator in the 1930s for a nation devastated by depression." Like other forms of the technological sublime, the bridge seemed to confer not only economic benefits but "can do proof" that the nation's "inventive and productive genius" would prevail.[22] It was, and is, an outward and visible sign of an ideal America. This book will examine how such objects fuse practical goals with political and spiritual regeneration.

American Technological Sublime

A photo from John K. Hillers' 1879 geological survey of the Grand Canyon. Courtesy of National Archives.

1

The Sublime

The North American continent possesses every feature that a theory of the natural sublime might require, including mountains, deserts, frozen wastes, endless swamps, vast plains, the Great Lakes, and hundreds of unusual sights, notably Yellowstone, Mammoth Cave, Niagara Falls, and the Grand Canyon. Likewise, its tornadoes, hurricanes, floods, and other natural disasters are among the most terrifying phenomena one could encounter anywhere. It would be tempting to say that had no theory of the sublime existed, Americans would have been forced to invent one. In a sense, this is what happened, for by the middle of the nineteenth century the American sublime was no longer a copy of European theory; it had begun to develop in ways appropriate to a democratic society in the throes of rapid industrialization and geographic expansion.

The American sublime drew on European ideas in the fine arts, literature, and philosophy. In art history the concept of the sublime is often applied to paintings that are unreal, monstrous, nightmarish, or imaginary. In architecture a sublime building usually is vast and includes striking contrasts of light and darkness, designed to fill the observer with foreboding and fear.[1] Intellectual historians and literary critics have been particularly interested in eighteenth-century and early-nineteenth century texts on the sublime. David B. Morris writes:

The discovery of the sublime was one of the great adventures of eighteenth-century England: accompanying the establishment of a commercial empire, the growth of industrialism, the invention of the common reader, and the rise of the waltz, a taste developed among almost all classes of society for the qualities of wildness, grandeur, and overwhelming power which, in a flash of intensity could ravish the soul with a sudden transport of thought or feeling. . . . Sublimity liberated the eighteenth-century imagination from all that was little, pretty, rational, regular, and safe—although only for as long as the moment of intensity could be sustained.[2]

As Marjorie Hope Nicolson notes, English writers on the sublime (including Dennis, Shaftsbury, and Addison) agreed that the most important "stimulus to the Sublime lay in vast objects of Nature— mountains and oceans, stars and cosmic space—all reflecting the glory of Deity." To them, the experience in nature was primary; the "rhetorical" sublime was "only secondary." Despite this agreement on the primary stimulus of nature, however, it was difficult to classify the emotional content of this "moment of intensity." Shaftsbury argued that the sublime was the highest form of beauty. Addison saw the sublime and the beautiful as distinct categories. Burke agreed with Addison on this point, but he emphasized the terror of the sublime whereas Addison spoke of "pleasing astonishment" and "awe."[3] One can easily give too much weight to such differences, however. As Nicolson says, the important point is that "during the eighteenth century the English discovered a new world. In a way, they were like the imaginary cosmic voyagers who, from Lucian to writers of modern science-fiction, have traveled to the moon or planets to find worlds that puzzle, amaze, astound, enthrall by their very differences from our world."[4]

An actual new world had been discovered in the western hemisphere—one which, according to the first explorers, contained a wild profusion of monsters and previously unknown phenomena, including bullfrogs as large as dogs, mosquitoes the size of bats, mountains 50 miles high, strange winds that caused a living man's body to rot, earthquakes that toppled mountains, and enormous seagoing lions that seemed to glide over the water. Howard Mumford Jones surveyed the profusion of creatures and marvels described in early travelers' reports and concluded that "the New World was filled with monsters animal and monsters human; it was a region of terrifying natural forces, of gigantic catastrophes, of unbearable heat and cold, an area where the laws of nature tidily governing Europe were transmogrified into something new and strange."[5] Descriptions of such marvels continued unabated throughout the sixteenth and early seventeenth centuries, establishing a discourse about the Americas as an anti-Europe, a strange world that challenged every presupposition about nature.

Since this discourse was well established by 1700, it should not be surprising that the intellectual ferment over the sublime was transmitted to the American colonies, although, as is the case whenever a complex of ideas is carried from one culture to another, its content was transformed in the process. A modified form of the sublime emerged that was in harmony with American political, social, and religious conditions. Because it had originated in classical antiquity, the sublime was peculiarly suited to Americans as they increasingly sought to emulate the Roman Republic and the democracies of ancient Greece, after about 1750. As Jones notes, in the revolutionary years "classicism remained a powerful force, whether for propaganda, historical precedent, warning, or the theory of a republic."[6] In the years after the revolution, as Americans fashioned a discourse that identified the new nation with the landscape, their language gradually became permeated with classical ideas—not least the idea of the sublime. As Raymond O'Brien notes, "Pre-Romantic concepts of mountain gloom, Old World superstitions of the forest, and puritanically mundane views of nature were dissipated more slowly in the colonies; consequently there is a time lag apparent between the formulation of landscape theories of the sublime and picturesque and the adaptation of these ideas in America."[7] Such ideas reached a large audience only in the nineteenth century.

The history of the sublime from antiquity shows, if nothing else, that, although it refers to an immutable capacity of human psychology for astonishment, both the objects that arouse this feeling and their interpretations are socially constructed. The objects and interpretations vary not only from one epoch to another and from one culture to another but also from one discipline to another, and a large volume would be necessary to provide a history of the sublime from antiquity to the nineteenth century. Here a short summary must suffice.

As conceived in the first century, the sublime was defined as an attribute of oratory and fine writing. The anonymous author usually identified as Longinus wrote:

If an intelligent and well-read man can hear a passage several times, and it does not either touch his spirit with a sense of grandeur or leave more food for reflection in his mind than the mere words convey, but with long and careful examination loses more and more of its effectiveness, then it cannot be an example of true sublimity—certainly not unless it can outlive a single hearing. For a piece is truly great only if it can stand up to repeated examination, and if it is difficult, or rather impossible to resist its appeal, and it remains fairly and ineffaceably in the memory. As a generalization, you may take it that sublimity in all its truth and beauty exists in such works as please all men at all times. For when men who differ in their pursuits, their ways of life, their ambitions, their ages, and their languages all think in one and the same way about the same works, then the unanimous judgement, as it were, of men who have so little in common induces a strong and unshakable faith in the object of admiration.[8]

The sublime is identifiable by the repetition and the universality of its effect. In this definition, the sublime is not an esoteric quality. Rather, it is available to everyone, regardless of background.[9]

Discussions of the sublime usually begin with Longinus and then jump to early-eighteenth-century England, where the topic was taken up and elaborated by many authors—most notably Edmund Burke, whose *Philosophical Enquiry into the Origin of Our Ideas of the Sublime and Beautiful,* published in 1756, became the most influential work on the subject. In the United States it went through at least ten editions before the Civil War.[10] Most discussions treat 'sublime' as a noun, seldom noting that during the interval between Longinus and Burke it was also a verb meaning to act upon a substance so as to produce a refined product. Alchemists seeking to bring substances to higher states of perfection employed sublimation in their efforts to attain the philosopher's stone. Alchemy gave the term 'sublime' a special coloring that anticipated the later response to industrial objects. Before the eighteenth century it was not yet common to praise as sublime an object of natural grandeur, such as a vast forest seen from a mountaintop or a tempest raging over the sea, but it was common to call the process of converting a substance into a vapor by heating it and then cooling it down to a refined product 'sublimation'. Metaphorically, 'sublime' suggested pure realms of thought and attempts to obtain hidden knowledge.[11] Alchemy was not a failed proto-chemistry; its practitioners did not see themselves as objective scientists. Alchemists were not neutral observers, and what happened in their beakers, vials, and retorts were not objectified experiments. They believed that material transformations worked upon the spirit.

In contrast, the general tendency of the new science of the seventeenth century was, as Mulford Sibley puts it, "to despiritualize nature, to wipe out the distinction between animate and inanimate, and to create a sharp separation between the inner and outer worlds."[12] Seen in this perspective, the eighteenth-century form of the sublime is not only a rewriting of Longinus; it is part of the Enlightenment project of defining reason, a project that included not only the creation of the encyclopedia but also the definition of what was not reason. As Michel Foucault has argued, to define science it was necessary to define what it was not. The mystical relation between man and nature assumed by the alchemist was replaced by the ideal of scientific objectivity. The alchemical connotations of 'sublime' were largely forgotten. Burke and his contemporaries provided a checklist of the objective attributes in objects that could be expected to call forth sublime emotion, and Burke often speculated on how external objects affected the body.[13] The sublime of the eighteenth century was a permissible eruption of feeling that briefly overwhelmed reason only to be recontained by it.

Why did the sublime reemerge when it did, fastening attention on particular natural objects? The literature of the Middle Ages and of the Renaissance marginalized the sublime. Nicolson's *Mountain Gloom and Mountain Glory* investigates the dramatic revaluation of the natural landscape that occurred after the late seventeenth century in England. For centuries mountains were thought to be the deformities of a fallen world whose surface had been smooth at the creation. Until c. 1650, mountains were "warts, blisters, imposthumes, when they were not the rubbish of the earth, swept away by the careful housewife Nature—waste places of the world, with little meaning and less charm."[14] But this attitude began to change as astronomers demonstrated the existence of mountains on the moon and the planets and as geologists proposed theories that explained the formation of mountains through natural processes. Equally important, John Calvin argued that there had been mountains in Eden, and that they existed at the creation. No part of the natural world was inherently ugly or evil. Man's soul was deformed; the world was not. Calvin believed that God could be seen in the beauty of nature: "On all his works he hath inscribed his glory in characters so clear, unequivocal, and striking, that the most illiterate and stupid cannot exculpate themselves by the plea of ignorance." Protestants increasingly looked for God in "the mirror of his works."[15] Americans would later incorporate this view in a powerful version of the natural sublime. The central point is that the sublime was not part of a static view of the world, nor was it part of a proto-ecological sensibility that aimed at the preservation of wilderness. Rather, to experience the sublime was to awaken to a new vision of a changing universe. The reemergence of the sublime was part of a positive revaluation of the natural world that by the eighteenth century had become a potential source of inspiration and education.

This revaluation was well underway by the time of Burke. He established an absolute contrast between the beautiful, which inspired feelings of tenderness and affection, and the sublime, which grew out of an ecstasy of terror that filled the mind completely. The encounter with a sublime object was a healthy shock, a temporary dislocation of sensibilities that forced the observer into mental action. To seek out the sublime was not to seek the irrational but rather to seek the awakening of sensibilities to an inner power. Burke wrote to a friend after seeing a raging flood in Dublin: "It gives me pleasure to see nature in these great though terrible scenes. It fills the mind with grand ideas, and turns the soul in upon itself."[16] Burke's sublime was subjected to rational controls; he created a list of the attributes in objects that could arouse this passion: obscurity, power, darkness, vacuity, silence, vastness, magnitude, infinity, difficulty, and magnificence. Herder later argued that Burke had relied upon a Newtonian idea of attraction and repulsion according to which the beautiful attracted and the sublime repulsed. While this view is oversimplified, Burke's version of the sublime ultimately seems to rest on the view that human beings respond to certain terrible or vast objects in predictable ways. Similar usage of the term has continued since his time, and most textbook definitions of the sublime refer to powerful natural scenes that are universally available and that deepen and strengthen the mind of the observer. The Oxford English Dictionary notes this sense of the term as "Of things in Nature and Art, affecting the mind with a sense of overwhelming grandeur or irresistible power; calculated to inspire awe, deep reverence, or lofty emotion by reason of its beauty, vastness, or grandeur."

When Kant adapted Burke's theory to his own, he argued that because the sublime included pleasure as well as pain it was not the opposite of the beautiful. Kant linked the beautiful to quality and the sublime to quantity, and argued that

the beautiful brings with it a direct feeling of the expansion of life, and hence imagination; the feeling of the sublime is a pleasure, which arises only indirectly, being produced by the feeling of a momentary checking of the vital forces followed by a stronger outflow of them, and as involving emotional excitement it does not appear as the play, but as the serious exercise, of the imagination. Accordingly, it cannot be united with sensuous charm [the beautiful]; and as the mind is alternately attracted and repelled by the object, the satisfaction in the sublime implies not so much positive pleasure as wonder or reverential awe, and may be called a negative pleasure.[17]

The function of this negative pleasure was to unite aesthetics with moral experience. As John Goldthwait summarizes, for Kant "the sublime makes man conscious of his destination, that is, his moral worth. For the feeling of the sublime is really the feeling of our own inner powers, which can outreach in thought the external objects that overwhelm our senses."[18]

In the *Critique of Judgement* Kant divided sublime experience into two forms: the mathematical sublime (the encounter with extreme magnitude or vastness, such as the view from a mountain) and the dynamic sublime (the contemplation of scenes that arouse terror, such as a volcanic eruption or a tempest at sea, seen by a subject who is safe from immediate danger). The mathematical sublime concerns that which is incomparably and absolutely great. But since every phenomena in nature is measurable, and therefore great only in relation to other things, the infinity of the sublime ultimately is an idea, not a quality of the object itself. In the presence of this apparent infinity, Kant's subject experiences weakness and insignificance, but then recuperates a sense of superior self-worth, because the mind is able to conceive something larger and more powerful than the senses can grasp.[19] In this experience the subject passes through humiliation and awe to a heightened awareness of reason.

In the dynamic sublime, the individual confronts a powerful and terrifying natural force. Kant notes that "we can, however, view an object as fearful without being afraid of it." He gives the following examples:

Bold, overhanging and as it were threatening cliffs, masses of cloud piled up in the heavens and alive with lightning and peals of thunder, volcanoes in all their destructive force, hurricanes bearing destruction in their path, the boundless ocean in the fury of a tempest, the lofty waterfall of a mighty river; these by their tremendous force dwarf our power of resistance into insignificance. But we are all the more attracted by their aspect the more fearful they are, when we are in a state of security; and we at once pronounce them sublime, because they call out unwonted strength of soul and reveal in us a power of resistance of an entirely different kind, which gives us courage to measure ourselves against the apparent omnipotence of nature.[20]

Contemplating such dangers makes the subject realize that nature can threaten only his physical being, leading him to feel superior to nature by virtue of his superior reason. For Kant both the mathematical and the dynamic forms of the sublime are not attributes of objects; they are the results of a dialogue between the individual and the object, a dialogue in which the distinction between the senses and the ego is forcibly manifested. "Sublimity, therefore, does not reside in any of the things of nature, but only in our own mind, insofar as we may become conscious of our superiority over nature within, and thus also over nature without us. . . ."[21]

From Burke to Kant to later thinkers, the natural world plays a smaller and smaller role in definitions of the sublime, and the observer becomes central in defining the emotion as the mind projects its interior state onto the world. Burke insisted on the centrality of the natural scene in evoking the sublime. Kant emphasized that the mind was central in apprehending the sublime, thus shifting attention from physical nature to its perception.

Touristic practice came to somewhat the same conclusion as formal philosophy, arriving there by a different route. For at the same time that philosophers were deemphasizing the external object as the stimulus to sublime feelings, tourists were having more and more difficulty capturing the elusive emotion. Elizabeth McKinsey traces such a declension in *Niagara Falls: Icon of the American Sublime*. By the second half of the nineteenth century, she notes, the sublime was seldom an accessible emotion: "Changes in the image of Niagara Falls after about 1860 indicate a profound shift in attitude toward nature. Both the actual scenes at the Falls [marred by excessive tourism] and the aesthetic assumptions of artists who journeyed there reveal the eclipse of the sublime as a motive force in American culture."[22]

Since touristic experience and formal philosophy seem to point to the same conclusion, it would only seem necessary to illustrate the gradual disappearance of the sublime with extensive examples from the nineteenth century. But it will be the burden of this book to describe the popular sublime, a history of enthusiasms for both natural and technological objects that has lasted until our time and that answers to classic aesthetic theories only partially. This history will not trace the intellectual's sense of an attenuating connection to the world, nor will it be concerned with the sublime in literature and the fine arts. Rather, it will trace the continual discovery of new sources of popular wonder and amazement, from the railroad to the atomic bomb and the space program. Such a history requires a different definition of the sublime, one that treats it less as part of a self-conscious aesthetic theory than as the cultural practice of certain historical subjects. Even if the sublime is not a philosophical absolute but a historicized object of inquiry, I will argue, the sublime experience still retains a fundamental structure, regardless of the object that inspires it or the interpretation that is given to the experience.

At the core of any sublime experience is a passion that Burke defined: "The passion caused by the great and sublime in nature, when those causes operate most powerfully, is Astonishment; and astonishment is that state of the soul, in which all its motions are suspended, with some degree of horror. In this case the mind is so entirely filled with its object, that it cannot entertain any other, nor by consequence reason on that object which employs it. Hence arises the great power of the sublime, that far from being produced by them, it anticipates our reasonings, and hurries us on by an irresistible force." The Grand Canyon is a good example of such a natural object. William F. Cody, better known as Buffalo Bill, wrote in a visitor's book that the canyon was "too sublime for expression, too wonderful to behold without awe, and beyond all power of mortal description."[23]

The millions who travel to the Grand Canyon visit it in order to sense a magnificence that cannot be described or grasped through descriptions or images but must be experienced directly. The huge scale can produce an awareness of human insignificance, of natural power, of immensity, and of eternity. The experience of visiting it is akin to the classical definitions of the sublime, even if there is little reason to believe that ordinary Americans self-consciously visit such a site with either Burke or Kant on their lips. The canyon exemplifies much of what Burke said about the sublime, and it exercises a powerful hold on the imagination of most who visit it, particularly if they do more than merely park their cars and look at a small part of it from the rim. The first reaction to this sublime object is often incomprehension. Joseph Wood Krutch realized this in observing both his own reactions and those of others on first peering over the rim into the abyss: "At first glance the spectacle seems too strange to be real. Because one has never seen anything like it, because one has nothing to compare it with, it stuns the eye but cannot really hold the attention. For one thing, the scale is too large to be credited. . . . We cannot realize that the tremendous mesas and curiously shaped buttes which rise all around us are the grandiose objects that they are. For a time it is too much like a scale model or an optical illusion. One admires the peep show and that is all."[24] At first one stands outside the object as though one were looking through a frame at a peep show. It requires much more effort to "relate one's self to it somehow"; indeed, that may take days.

Krutch provides a specific example of Burkean astonishment, a state in which all internal reflection is suspended. The Grand Canyon opens up suddenly in the midst of a high plateau, and the Colorado River is so far away that it seems to be a small stream when it is in fact 300 feet wide. The canyon's sheer size is difficult to grasp. Its depth is so terrifying that many pull back in fear after their first glimpse. A late-nineteenth-century traveler reported one group's experience: "Our party were straggling up the hill: two or three had reached the edge. I looked up. The duchess threw up her arms and screamed. We were not fifteen paces behind, but we saw nothing. We took a few steps, and the whole magnificence broke upon us. No one could be prepared for it. The scene is one to strike dumb with awe, or to unstring the nerves; one might stand in silent astonishment, another would burst into tears. . . . It was a shock so novel that the mind, dazed, quite failed to comprehend it."[25]

But the Grand Canyon does more than suspend the mental faculties. Burke points out that "to make any thing very terrible, obscurity seems in general to be necessary. When we know the full extent of any danger, when we can accustom our eyes to it, a great deal of the apprehension vanishes."[26] The canyon meets this requirement admirably, because it is so large that proportion and scale are confusing for a long time after one first looks into it. Furthermore, because of weather conditions and because of the shadows cast by the walls, much of the canyon is obscured a good deal of the time. The Grand Canyon contains virtually all of the elements Burke associated with the sublime in natural landscapes, including power, vacuity, darkness, solitude, silence, vastness, infinity, magnificence, and color. It is 280 miles long and up to 18 miles wide. It seems infinite in both time and space, presenting 2 billion years of geology in 15,000 feet of tilted-up stone, carved down by the Colorado River. It offers so many intriguing views and so many vantage points that it can never be seen in its entirety. Burke noted "that height is less grand than depth; and that we are more struck at looking down from a precipice, than at looking up at an object of equal height."[27]

Who would deny that the mile-deep Grand Canyon is more impressive than a mile-high mountain range?

A well-traveled Welshman, Colin Fletcher, had these Burkean reactions when he first came up to the canyon's rim: "And there, defeating my senses, was the depth. The depth and the distances. Cliffs and buttes and hanging terraces, all sculptured on a scale beyond anything I had ever imagined." His initial reaction also included a strong sense of light, fusing colors, and a powerful silence. "In that first moment of shock, with my mind already exploding beyond old boundaries, I knew that something had happened to the way I looked at things."[28] The novelist and essayist Frank Waters noted that the Grand Canyon is unlike such landscapes as the prairies, the Rockies, or the bayous of Louisiana, which can be depicted reasonably well in photographs or paintings. In contrast, the Grand Canyon is a complex system of views which no single image can possibly convey. "It is the sum total of all the aspects of nature combined in one integrated whole. It is at once the smile and the frown upon the face of nature. In its heart is the savage, uncontrollable fury of all the inanimate universe, and at the same time the immeasurable serenity that succeeds it."[29] In these contradictions, the Grand Canyon contains most of the qualities Burke finds essential to sublimity, and it illustrates Kant's mathematical sublime.[30] The first geologist to survey the region, Clarence Dutton, recognized this exemplary quality and named one of the most impressive lookouts Point Sublime.[31]

In contrast, the volcanic eruption of Mount St. Helens in Washington State on May 18, 1980, exemplifies the dynamic sublime. An overwhelming force, it hurled millions of tons of pulverized rock into the air, creating a cloud that rose 60,000 feet. The volcano literally blew its top, reducing its height by 1500 feet. Dust fell to the thickness of half an inch 500 miles away, and nearby it covered fields with 8 tons of ash per acre.[32] The eruption evaporated a lake, melted a glacier, set innumerable forest fires, changed day into night, and unleashed mudslides that swept away every tree in a 120-square-mile area. The cloud of ash and rock moved so fast that drivers found they could not outrun it and were trapped in a blinding dry rain.[33] The Portland *Oregonian* noted: "Eclipsing Dante's horrible dreams of Hell, the mountain poured out burning pyroclastic clouds that incinerated everything they touched—animal, vegetable, mineral." One witness said: "When the mountain went, it looked like the end of the world."[34]

Yet more than one observer realized that the eruption's meaning could not be reduced to death and destruction. A pilot who saw the eruption from a safe distance recalled: "I consider it a great privilege to have seen it. It was just a beautiful show."[35] Many said that it was the most exciting thing they had ever seen. The *Rocky Mountain News* commented: "If it weren't for the loss of life and the devastation done to the environment, the eruption of Mount St. Helens in Washington might almost be enjoyed as one of the most awesome spectacles of unleashed energy that nature can display."[36] There were thousands of eyewitnesses to the blast, which had long been anticipated. Tourists were drawn to the site by tremors in the weeks before the eruption, and the governor had to cordon off the area and even evict people from their own land, creating much resentment among property owners. Despite the barricades and numerous public warnings, however, at least 77 people died, including an 84-year-old man named Harry Truman who had lived on Spirit Lake at the foot of the mountain for half a century.

Because the eruption had long been anticipated, one television reporter and many amateur photographers recorded the blast on film. Yet no medium could capture the totality of the event. For example, an amateur photographer who had his camera pointed at the mountain at the moment it erupted made ten images that recorded the event as well as photography could; but even these images do not record what it looked like entirely satisfactorily, because the cloud grew so quickly that during the sequence the photographer had to switch from a telephoto to a 50-millimeter to a wide-angle lens.

The witnesses to the eruption had an experience that was not only visual but also visceral. The ground shook. Lightning flashed out of the spreading cloud. There was thunder, and a sulfuric odor. One man recalled: "I saw a puff of steam come out, and then it looked like the whole mountain blew out sideways and just fanned out. The animals, the elk especially, were signaling and bugling. The whole forest was full of it."[37] Those who saw Mount St. Helens erupt did not subdivide and analyze their experience like philosophers, and each of them articulated only part of a sudden, overwhelming event. However, in accord with Thomas Weiskel's observations in *The Romantic Sublime*, they passed through a three-stage psychological experience. The first stage is that of normal perception, of a person with no immediate expectation of seeing anything extraordinary. On that serene Sunday morning the official observers decided not to make their routine flight over the volcano because there were so few tremors, and one man who had been watching the volcano for a week recalls being utterly bored just before the eruption. The second stage begins at the moment when the subject perceives a break in ordinary perception, and a gap opens up between the self and the object, as was the case when the mountain blew up. Weiskel explains that at this moment of astonishment "there is an immediate intuition of a disconcerting disproportion between inner and outer." In the third stage the subject recovers from the shock of the encounter and regains his equilibrium, creating a new relationship to the sublime object; now "the very indeterminacy which erupted in phase two is taken as symbolizing the mind's relation to a transcendent order."[38]

This psychological progression assumes, however, that the individual encounters the sublime object without warning and without expectations. Virtually everyone who sees an object that is considered sublime has heard of it first and comes to it with a set of expectations. Mount St. Helens was something of an exception, since even the scientists flying over the site admitted "This is our first volcano."[39] Ordinarily, however, the visitor does not see the Grand Canyon or any other site with innocent or unprepared eyes. Most sublime objects have become tourist sites. Their existence has been well advertised in advance, their appearance has been suggested by photographs, and their meaning has been overdetermined. As a result, in many cases tourists do not experience the sublime at all. And this is by no means a phenomenon of the late twentieth century. In 1818 one visitor to Niagara Falls remarked "the unbridled scope in which imagination delights to riot, magnifying what is small and exaggerating what is great" and noted "surely it will no longer be surprising that many, who take but a flying view of the wonders of Niagara, should depart utterly displeased that they are not still more wonderful."[40] Because the experience of standing before a sublime object is affected by expectations, it is often necessary to linger several days before preconceptions are overcome by direct experience.

McKinsey recounts the example of Nathaniel Hawthorne, who visited Niagara Falls and who later wrote a sketch about a man who came to the cataract with expectations so strong that they blocked the possibility of being astonished. Like many modern tourists, he experienced an inversion. The sublime object was not overwhelming. Instead, powerful expectations that exceeded the immediate appearance of the object created a disjunction between the man's consciousness and Niagara, leading to disappointment. (Weiskel considers this excess of expected meaning the determining quality of the "egotistical sublime.") Hawthorne's narrator, unlike most tourists, remains at the site for days and overcomes his initial disappointment by continually viewing the Falls until they overwhelm his inaccurate expectations.

Margaret Fuller's experience of Niagara Falls, recounted in *Summer on the Lakes, in 1843,* was somewhat different. She also had predetermined ideas of the site, but initially she had "a quiet satisfaction" when she "found that drawings, the panorama, etc. had given me a clear notion of the position and proportion of all the objects here; I knew where to look for everything, and everything looked as I thought it would." Though satisfied with the scene, she "thought only of comparing the effect on my mind with what I had read and heard. I looked for a short time, and then with almost a feeling of disappointment, turned to go to other points of view to see if I was not mistaken in not feeling any surpassing emotion at this site." Fuller, like Hawthorne, stayed for more than a week, viewing the falls from both sides, from Goat Island, and from a boat, and also taking in the whirlpool and the rapids below. During this time she found that every day Niagara's "proportions widened and towered more and more upon my sight, and I got, at last, a proper foreground for these sublime distances. Before coming away, I think I really saw the full wonder of the scene." After some days "it so drew me into itself as to inspire an undefined dread, such as I never knew before, such as may be felt when death is about to usher us into a new existence. The perpetual trampling of the waters seized my senses. I felt that no other sound, however near, could be heard." Nevertheless, her most powerful response was not to the falls, for which she was "prepared by descriptions and paintings," but to the rapids seen from Goat Island: "My emotions overpowered me, a choking sensation rose to my throat, a thrill rushed through my veins, 'my blood ran rippling to my fingers' ends.' This was the climax. . . ." Yet, far from being satisfied with this sublimity, Fuller was "provoked with [her] stupidity in feeling most moved in the wrong place."[41]

The examples of Hawthorne and Fuller are not unusual; the natural sublime is often blocked by powerful preconceptions. Americans of their generation had ample opportunities to envision Niagara before seeing it in person. Not only were many lithographs and other reproductions for sale; P. T. Barnum had a scale model of it, complete with running water, on display in his American Museum in New York. In 1840 a 200-square-foot moving diorama of the cataract went on display in Philadelphia. An even more spectacular canvas, by Godfrey N. Frankenstein, toured the United States between 1853 and 1859. It was 1000 feet long and took 90 minutes to unwind, accompanied by music and a commentary.[42]

Such familiarity with an object threatens to undermine its potential sublimity. Yet the mark of the truly sublime object, as Longinus emphasized, is that it grows in significance with repetition, as Margaret Fuller found in her week at Niagara and as many a tourist has realized after spending several days at the Grand Canyon. As early as 1882 Clarence Dutton warned travelers that it took time to apprehend the latter. He had observed that a visitor came "with a picture of it created by his own imagination. He reaches the spot, the conjured picture vanishes in an instant, and the place of it must be filled anew. Surely no imagination can construct out of its own material any picture having the remotest resemblance to the Grand Cañon." Ideally, it might seem that the sublime should be an unexpected encounter, a largely unmediated experience of discontinuity between the self and a startling natural object. Because the traveler is so prepared in advance, the sublime may seem to be swallowed up by representations in the mass media. Only a prolonged reexperiencing of the site can overcome the egotistical demands of the informed visitor. Dutton concluded that "those who have long and carefully studied the Grand Cañon of the Colorado do not hesitate to pronounce it by far the most sublime of all earthly spectacles." In contrast, a less spectacular but unexpected object of grandeur, such as the rapids below Niagara Falls, may more easily inspire a powerful response. Yet the sublime object cannot be extirpated by expectations. Indeed, even an "innocent observer" can only be certain of an object's sublimity by continually reexperiencing it to see if it gains rather than loses force through deeper acquaintance. The first view is almost never the "pure" experience of a hypothetical observer without preconceptions, but this by no means makes the sublime inaccessible to modern people.

The experience, when it occurs, has a basic structure. An object, natural or man-made, disrupts ordinary perception and astonishes the senses, forcing the observer to grapple mentally with its immensity and power. This amazement occurs most easily when the observer is not prepared for it; however, like religious conversion at a camp meeting, it can also occur over a period of days as internal resistance melts away. Kant distinguished between the mathematical and the dynamic sublime. In either case he expected that in the aftermath of the immediate experience the individual would become conscious of "our superiority over nature within, and thus also over nature without us." Yet this is not necessarily the conclusion everyone will draw from a sublime experience, particularly if the object is man-made rather than natural. The perception of what is immense and infinite changes over time and across cultures.[43] Although the interpretation developed in Weiskel's third stage may be more culturally determined than the astonishment at the core of the experience, even here previous perceptions shape responses. In short, American forms of the sublime are culturally inflected, including the awe bordering on terror of the second stage. The test for determining what is sublime is to observe whether or not an object strikes people dumb with amazement. The few experiences that meet this test have transcendent importance both in the lives of individuals and in the construction of culture.

2

The American Sublime

The classical definitions of the sublime were written by and for intellectual elites. Longinus, for example, begins with the words "If an intelligent and well-read man. . . ." Burke, Kant, Schiller, and later commentators, too, address an educated elite. While they assert that certain scenes will affect all minds in certain ways, they take no pains to demonstrate that this is the case, being content to let the reader experiment or to reflect on personal experience. And even if all readers of philosophy agreed, historians would still regard them as an interpretive community that might be labeled "readers of aesthetics," a group hardly representative of the whole population. This chapter will examine how non-philosophers developed their own understanding of the sublime, based on nationalism, gender divisions, religious convictions, new technologies, and political values. By the 1820s a distinctive American version of the sublime had emerged.

To trace this emergence, consider William Byrd, an eighteenth-century Virginian planter with an English education, whose extensive commentary on the wilderness of the Appalachian mountains, *History of the Dividing Line*, was imbued with European landscape ideas. Taken by itself, the book seems to prove that Americans of the day had already adopted the Enlightenment's ideas of nature, and that they no longer regarded the wilderness as a hostile region to be conquered. As an "intelligent and well-read man," however, Byrd did not represent the public of his day; rather, he was one of a wealthy group of tidewater planters who had the leisure to cultivate not only their fields but also their minds. Far from articulating widely held views, he was part of a tiny minority. The dominant view of nature was that of farmers and pioneers, who were determined to subdue the land and the Native Americans. They regarded both as obstacles to be overcome. To travelers, a landscape with no marks

Virginia's Natural Bridge. Courtesy of Collections of Library of Congress.

of settlement or agriculture upon it called up few pleasant associations; it was a forbidding solitude.[1] That point of view remained dominant until at least the 1820s, when such natural wonders as Virginia's Natural Bridge and Niagara Falls began to attract many visitors.

One can trace this changing response to natural scenery in the case of the Natural Bridge. For Jefferson's generation, scientific observation and the experience of the sublime were not thought to be incompatible, as his famous passage on the Natural Bridge in *Notes on the State of Virginia* demonstrates. He begins with a series of measurements, seeking to give a precise description of the site:

The Natural Bridge, the most sublime of nature's works . . . is on the ascent of a hill, which seems to have been cloven through its length by some great convulsion. The fissure, just at the bridge, is by some admeasurements, two hundred and seventy feet deep, by others only two hundred and five. It is about forty-five feet wide at the bottom and ninety feet at the top; this of course determined the length of the bridge, and its height from the water. Its breadth in the middle is about sixty feet, but more at the ends, and the thickness of the mass, at the summit of the arch, about forty feet. A part of this thickness is constituted by a coat of earth, which gives growth to many large trees. The residue, with the hill on both sides, is one solid rock of lime-stone. The arch approaches the semi-elliptical form; but the larger axis of the ellipsis, which would be the cord of the arch, is many times longer than the transverse.

Modern geology was being invented during the eighteenth century, and portions of *Notes on the State of Virginia* demonstrate that Jefferson was acquainted with the effects of water erosion and the ease with which limestone could be hollowed out into underground caverns. Yet, while aware of this gradualist notion of how a natural bridge might evolve, Jefferson showed a preference for the catastrophe theory, which held that violent cataclysms of earthquake and flood had destroyed many beautiful structures, of which only a few remained. When he suggests that the ridge "seems to have been cloven through its length by some great convulsion," he is reading the landscape of Virginia as evidence for the catastrophe theory. The imagination of such disasters heightens the pleasure of seeing the Natural Bridge. Likewise, Jefferson remarks that another landscape records "a war between rivers and mountains, which must have shaken the earth to its centre." Such observations were later systematized by Cuvier into a general theory of catastrophe, which would compete with uniformitarian theory until after Jefferson's death.[2]

Had Jefferson stopped his description of the Natural Bridge here, his account would remain an impersonal record of measurements placed in the context of contemporary scientific theories. But his text continues without a break to characterize the experience of viewing it from different positions, and here he introduces the language of the sublime. Simultaneously, his narration shifts from the third person to the second to the first:

Though the sides of this bridge are provided in some parts with a parapet of fixed rocks, yet few men have the resolution to walk to them, and look over into the abyss. You involuntarily fall on your hands and feet, creep to the parapet of fixed rocks, and peep over it. Looking down from this height about a minute, gave me a violent head-ache. If the view from the top be painful and intolerable, that from below is delightful in an equal extreme. It is impossible for the emotions arising from the sublime to be felt beyond what they are here, so beautiful an arch, so elevated, so light, and springing as it were up to heaven! the rapture of the spectator is really indescribable! The fissure continuing narrow, deep, and straight, for a considerable distance above and below the bridge, opens a short but very pleasing view of the North mountain on one side.[3]

Jefferson abandons the neutral scientific tone as he recalls powerful emotions and urges the reader to seek out, as a necessary counterpart to the view from below, the "painful and intolerable" view that gave him a headache. The natural bridge is sublime because it is terrifying, painful, almost intolerable, and yet at the same time delightful. It is enormous yet graceful, massive yet light. It induces both terror and rapture.

Jefferson's Natural Bridge was not yet a tourist site in the 1770s, when only a few had acquired a taste for the sublime. Viewing another striking landscape, Jefferson complained that "here, as in the neighborhood of the Natural Bridge, are people who have passed their lives within a half dozen miles, and have never been to survey these monuments." The taste for landscape was long limited to an educated minority, such as Byrd and Jefferson. In the 1840s Margaret Fuller still encountered people who lived unconsciously in the landscape, and to her they were part of it. Once, as she stood on a hill with a friend, looking at "one of the finest sunsets that ever enriched this world," a little cow-boy, trudging along, wondered what we could be gazing at. After spying about some time, he found it could only be the sunset, and looking too, a moment, he said approvingly, 'that sun look well enough.'"[4]

Fuller felt that the boy's speech was "worthy of Shakespeare's Cloten, or the infant Mercury." His uneducated eye pleased her, suggesting a native simplicity and good sense. But if his naiveté is a foil to her own superior sensibility and taste, his presence also underscores her separation from the land, which to her has become landscape.

   Fuller reveals once again the split in sensibility between those of European taste, educated to see the landscape in terms provided by Kant and Burke, and the majority of the population. The highly educated have received more scholarly attention, not least because they have left behind an abundant record of their feelings. John Quincy Adams, for example, reported his experience of seeing Niagara Falls in terms that clearly echo classical theory: "I have seen it in all its sublimity and glory—and I have never witnessed a scene its equal. . . . a feeling over-powering, and which takes away the power of speech by its grandeur and sublimity."[5] The popular response to the falls was less predictable. As early as 1822 a Scottish traveler noted: "There is a large tavern on each side of the river, and in the album kept at one of these, I observed that upwards of a hundred folio pages had been written with names within five months." Such customs continued, and in 1845 a book was published containing selections from what tourists had written in a similar album placed at Table Rock, above the falls.[6] A few made ironic remarks in this *Table Rock Album.* One man declared "It is only some water running over some rocks—that's all." Another complained "They ain't good for nothin' for manufacturin'; and they completely spile navigation—that's a fact."[7] Many of the writers simply declared that it was impossible to describe their feelings, but that they felt certain that seeing Niagara was a religious experience. For example, a visitor from Baltimore wrote: "It is utterly impossible for any man to give expression to the overwhelming feeling he experiences on beholding this display of the Great Creator's works. . . . This roar of Niagara is but a song of praise to he Almighty God." The majority adopted the language of the sublime, and they often wrote in verse. A man from Philadelphia declared Niagara the "Eternal-prototype of God!" and concluded his stanzas with "thou hast been / To me a lesson deep and ineffaceable / And I leave this spot, I trust, a better man." Another wrote: "What mind is not enlarged, what soul not filled with ennobling emotions, by the contemplation of such wonders? Let man behold with awe and admiration, and learn." A visitor from Michigan recorded:

when I saw Niagara, I stood dumb, "lost in wonder, love and praise." Can it be, that the mighty God who has cleft these rocks with a stroke of his power, who has bid these waters roll on to the end of time, foaming, dashing, thundering in their course; can it be, that this mighty Being has said to insignificant mortals, "I will be thy God and thou shalt be my people?". . . Roll on! thou great Niagara, roll on! and by thy ceaseless roaring, lead the minds of mortals from Nature's contemplation up to Nature's God.[8]

As these responses indicate, the rhetoric of the religious sublime became the standard way of understanding the meaning of the falls. Moreover, this rhetoric was woven together with the nationalistic language of exceptionalism, so that Niagara became a sign of a special relationship, or a covenant, between America and the Almighty.[9]

Such feelings lay behind much of the public outcry over the commercialization of the falls later in the century. The adjacent lands were privately owned, and, as at the Natural Bridge, the owners charged admission. Tourists complained about the number of vendors and guides who impeded their progress, and about paying an entry fee each time they approached an outlook or crossed a footbridge.[10] Many small entrepreneurs set up businesses and amusements, choking the area around the falls with unsightly structures, signs, and carnival crowds. In the 1860s many tourists spent more time enjoying these commercial pleasures than they did gazing at the falls.[11] By the 1880s a reaction set in, and sentiment to preserve the falls from commerce was strong enough to force the legislature into action. The lands around the falls were purchased and put under the control of the state of New York.

At the dedication of the Niagara State Reservation, in 1885, James C. Carter gave an oration that expressed what had become the orthodox view of Niagara Falls. He took it for granted that the spectacle of the falls was sublime, and went on to declare:

There is in man a supernatural element, in virtue of which he aspires to lay hold of the Infinities by which he is surrounded. In all ages men have sought to find, or to create, the scenes or the objects which move it to activity. It was this spirit which consecrated the oracle at Delphi and the oaks of Dordona; reared the marvel of Eleusis, and hung in the heavens the dome of St. Peter. It is the highest, the profoundest, element of man's nature. Its possession is what most distinguishes him from other creatures, and what most distinguishes the best among his own ranks from their brethren.[12]

By conflating the man-made and the natural, Carter suggests that the technological sublime is identical with the natural sublime. Here is that typical American amalgamation of natural, technological, classical, and religious elements into a single aesthetic. In it, natural wonders, such as Yosemite, the Grand Canyon, Niagara Falls, and Yellowstone, became emblems of divinity comparable to the wonders of the ancient world and the greatest architectural achievements of modern times.

The convergence of popular and refined responses to the falls suggests that, as Kant proposed, the experience of sublimity is based on a universal capacity for a certain kind of emotion. But Americans nevertheless shaped this emotion to their own situation and needs. The sublime object is by definition something one is not accustomed to, something extraordinary. It virtually requires that one be an outsider. It does not require that one be comfortable, however. For example, a California-bound migrant passing through Wyoming could not look at the landscape as a tourist might. He admitted that "the scenery might have been judged 'sublime' by others, but he asked 'what charm had barren prairies to us who wanted grass and water?'"[13] Yet many early travelers strove to suggest the bleak magnificence of the western landscape, and, as Lewis Saum discovered, the "surprising thing about the common people is that they flirted as much as they did with the terror-laden sublime." For settlers confronting tornadoes, violent thunderstorms, blizzards, mountains, and other unaccustomed rigors of the west, Saum notes, "the category of the sublime not only did much to render the wilderness, vastness, and chaos of the American landscape acceptable, but it also did somewhat in rendering other, hitherto objectionable things less objectionable."[14] Those undergoing hardships were hardly immune to the sublime response. On the contrary, catastrophe seemed to induce it. When the Ohio River flooded Cincinnati in 1832, carrying away much of the town and forcing thousands to leave their homes, a correspondent wrote to the *New England Magazine*: "All is tumult, hurry, excitement, distress. . . .

Access to the city from every quarter is cut off." Yet, in the midst of adversity he found "no sad countenances." "The universal expression," he continued, "is that of amazement. Men forget their own petty grievances, in the universal sublimity of the scene. I have surveyed it from an eminence, which commands the whole view. I can only say it is a miniature of the Deluge."[15] This reaction is reminiscent of Burke's response to a flood in Dublin, but the differences are also important. Whereas Burke wrote of his individual feelings, Americans in the 1830s already responded more as a group and often saw sublime events in biblical terms. The Cincinnati flood was a miniature of the deluge, and later the Chicago fire of 1871 and the San Francisco earthquake and fire of 1906 would be called "American apocalypses."

Surveying the world from an eminence was the special province of landscape paintings, the most popular native genre, which explored not only scenes of tranquility but also what Thomas Cole termed "the more terrible objects of nature."[16] Charles Sanford has noted that both Cole and the poet William Cullen Bryant "had need of the sublime to celebrate what they felt was peculiar and unique about American scenery, which the concept of the beautiful was incapable of expressing." They found a way to "salvage the sublime for patriotic service" by adapting Burke's ideas "to the Scotch moralists—chiefly Kames, Blair, and Alison, which united sentiments of the sublime to a great moral idea assumed to exist in and behind nature." In this way, "the passions released into art by the sublime" would "be harnessed to lofty spiritual ends."[17] This modified philosophy also informed the travel books that guided early tourists through the wide range of American landscapes, and it was this philosophy that seemed self-evident to the visitor to the Natural Bridge or Niagara Falls.

As Americans became tourists in their own country, interest in sublime landscapes became not an idle diversion but an act of self-definition. John Sears found that "tourism had deeper cultural sources than the need for diversion. Tourism played a powerful role in America's invention of itself as a culture. . . . It was inevitable when they set out to establish a national culture in the 1820s and 1830s, that they would turn to the landscape of America as the basis of that culture."[18] Lacking the usual rallying points (a royal family, a national church, a long history memorialized at the sites of important events), Americans turned to the landscape as the source of national character. So marked was this tendency that few public monuments were built before the Civil War. Wilbur Zelinsky notes that "the post-Revolutionary American landscape was remarkably bereft of monuments of any sort for decades, and exceedingly few of the rare pre-1850 items have survived."[19] Bunker Hill Monument, one of the earliest large monuments, was completed in 1843, and George Washington's home, Mount Vernon, was only rescued and restored in the 1850s. Until 1856, when a statue of Washington was erected, New York City contained no monuments of any kind to the Revolution. In short, during the antebellum period natural monuments such as Niagara became repositories and representations of the national spirit.

Journeys to natural wonders began to take on the character of pilgrimages in the Jacksonian period, broadening to include the trans-Mississippi west after 1865. Yosemite, Yellowstone, and the Grand Canyon were added to the ideal itinerary. Alfred Runte concluded that by the time of the Civil War Americans had "embraced the wonderlands of the West" as "replacements for [the] man-made marks of achievement" celebrated in European cultures. This attitude certainly continues today. The anthropologists Victor and Edith Turner note that "every year millions visit the national parks and forests (the precincts of 'Old Faithful' in Yellowstone Park irresistibly recall the cultural landscape of a major religious shrine), mostly, no doubt, for recreational purposes, but partly to renew love of land and country, as expressed in 'secular psalms' like 'America the Beautiful.'"[20] Just as European pilgrims once traveled to Santiago de Compostela or to Jerusalem, Americans seek out Niagara Falls, Yosemite, and the Grand Canyon. In doing so, they break with the daily round of habit, seeking renewal through a transcendent experience of natural power. The anthropologist of religion Mircea Eliade has argued that all cultures set aside certain places as sacred precincts. These sites preserve an original relation to the deity, offering the chance to step out of ordinary time and into the sacred time of an eternal present. To seek the sublime is, in effect, to step out of historical time into the eternal now.

Such journeys to the natural sublime began in the early nineteenth century, when the Natural Bridge, largely ignored in Jefferson's youth, began to attract a steady stream of tourists. Samuel Kercheval visited the spot in 1819. In his often reprinted account, he writes that he "was so struck with the grandeur and majesty of the scene as to become for several minutes terrified and nailed to the spot, and incapable to move forward." He continues:

"After recovering in some degree from this, I may truly say, agonizing mental state of excitement, the author approached the arch with trembling and trepidation."[21] Many such descriptions were published. The Natural Bridge became nationally famous and supported a small tourist industry. Nearby, a hotel and farmers rented out lodgings. The bridge was privately owned, and visitors were charged admission. Virginians were fond of pointing out that it was 55 feet higher than Niagara Falls.

The Natural Bridge had long been an obligatory stop on any Southern tour in 1904, when Clifton Johnson found that it still carried a road over the valley but remained part of a rustic setting. The gradualist theory of evolution rather than catastrophe theory had become the accepted explanation of the Natural Bridge, but the resolution of this conflict had hardly detracted from its appeal as a tourist attraction, and it still supported several hotels nearby. Johnson wrote:

Its immensity quite took my breath away. Nothing one has read or imagined can wholly prepare the visitor for this herculean span of rock across that abysmal chasm. Viewed from below it seems lifted into the very sky. Trees and bushes grow on its tip as on a mountain summit, and the swallows dart under it so far above the spectator as to make the arch appear like another firmament. The grace and regularity of the bridge suggest human handiwork, but doubtless in ages past the stream hollowed out a cavern in the valley, the roof of which all fell in long, long ago, save for this sturdy fragment.[22]

Johnson was also quick to point out that for the past century tourists had been writing their names on the underside of the arch. "George Washington" was carved on the rock 25 feet above the ground, and Washington's reputed presence at the site had stirred others to surpass him by climbing ever higher up the walls to place their names. The inevitable result was that one young man "early in the last century [before 1820], after out-rivaling all his predecessors in the height to which he attained, found he was placed in such a situation that it was impossible to descend." He escaped death by climbing the rest of the way to the top of the arch by cutting hand grips in the soft limestone with his knife.[23]

Thus were many sublime objects defaced and conquered. Visitors to Niagara Falls and the Grand Canyon painted their names on accessible rocks. Yet this defacement was not necessarily meant as disrespect for the object. Rather, the tourist identified with the site, wishing to share in its importance, to declare himself present in it, and to put his autograph on it. Writing on the natural object underlines the fact that sublimity is not inherent but a social construction.

The development of Niagara Falls and the Natural Bridge as tourist sites suggests some of the ways in which Americans adapted the sublime to their own society. Although in philosophical discussion the sublime is usually treated as an emotion enjoyed in solitude, in America it has quite often been experienced in a crowd. Natural wonders are usually surrounded by tourists, and virtually every technological demonstration, such as a world's fair or a rocket launch, provides a sublime experience for a multitude. A crowd requires many facilities—toilets, first aid, food, public transportation, supervision, and accommodations—that seem vulgar and demeaning to those in search of the classical sublime experience of solitary meditation in unspoiled surroundings. Thus, there are constant discussions about the obtrusive presence of tourist facilities and complaints about the crowds at Niagara Falls, Yosemite, the Grand Canyon, and other natural wonders. Such discussions can miss a larger point: the presence of a crowd can enhance the interest in an object, confirming its importance. The psychology of the crowd creates additional meanings. From the organizers' point of view, exciting and pleasing the crowd became a matter of techniques to be learned and refined. The sublime soon became not the result of serendipity but rather a scheduled part of travel. The sublime was considered to be practically guaranteed during a tourist's trip to Niagara Falls, the Grand Canyon, or Yosemite. It became a matter of promotion and public relations. Specialists emerged who could provide the best access to sublime experience, or who had discovered ways to enhance it or make it more predictable. In the case of Niagara Falls, these enhancements were mostly in the form of new vantage points—for example, taking crowds on a boat ride to the base of the falls and onto ledges behind them. A series of bridges, paths, and ladders made it possible to see Niagara anew. At the Grand Canyon, promoters provided trails, outlooks, mules, guides, rafting, camping, and hotel facilities. By 1993 the annual number of visitors was up to 5 million—13,000 people a day, demanding parking, food services, and so on. National shrines attract a national audience.[24]

In addition to being an event organized for the crowd, the popular sublime differs from that of the philosophers in that it is less clearly articulated. Many visitors' experiences are nearly inchoate, described in words such as 'amazing', 'fabulous', 'astounding', and 'overwhelming'. This failure of public speech to describe the sublime is, of course, one of the hallmarks of the emotion itself. People are awed and virtually dumbstruck, and most of the discourse only points at emotions without giving many details. And for most people the sublime retains linkages to the magical and the marvelous that have been removed from formal philosophy. While the highly educated classified their reactions to natural events and felt gratified that their reasoning made them superior to nature, the vast majority of people were ever ready to pay homage to the spectacular with wonder and delight (mixed with irreverence and ironic complaints if something failed to satisfy as advertised). Yet the popular sublime was never wholly commercialized. Though the sublime object might be part of a tour package, the experience itself could not be guaranteed to occur. One had to do more than just pay the price of admission; like a religious penitent intent on grace, one had to be receptive and patient as well.

Because of its highly emotional nature, the popular sublime was intimately connected to religious feelings—particularly in the nineteenth century, when revivals periodically swept communities into a frenzy. The sublime could hardly avoid becoming intimately interwoven with popular religion. Indeed, one of the most intense revivals took place not far from Niagara Falls, in western New York State, and that region came to be known as the "burned over" district—a metaphor expressing the belief that the region had been swept by waves of conversions, as if ravaged by the fire of God. Little wonder that Niagara was a site much frequented by revivalists, and that visits to Chautauqua in the later nineteenth century were often accompanied by visits to the cataracts. Likewise, the Baptists constructed a church near the Natural Bridge and used a pool directly beneath its arch for baptisms. One observer noted that these ceremonies were especially impressive, "in that place, always tinged in ordinary times with a semi-religious atmosphere." He went on:

In the many times that this writer attended this religious drama beneath the mighty arch of the Bridge, he does not recall that anything ever occurred to strike a false note, no unforeseen incident to mar the solemnity of the scene. The assemblage of people, whether drawn thither by idle curiosity or by a truly religious motive, were never other than orderly and quiet and attentive. It was as if the concrete evidence right at hand of the Almighty's handiwork in this gigantic struc-

ture, touched a chord in the heart of even the most thoughtless, causing him to conform, outwardly at least, to the spirit of the occasion.[25]

This description lays bare both how organized religion appropriated the natural sublime for its own purposes and how such ceremonies influenced the "thoughtless" and enforced conformity on
witnesses. Revivalist religion denounced intellectual approaches to
salvation and insisted on the need to move the sinner's heart. The
sublime was easily reconciled with such a doctrine, and the use of
the Natural Bridge as a site for baptism is but one of many examples of the religious appropriation of the natural sublime.
Revivalists often sought out dramatic backdrops for their efforts,
and visitors to Niagara, Yosemite, or the Natural Bridge wrote many
poems that professed to see God manifested in these natural works.
In 1819, for example, the publisher of the *Lexington Gazette* composed these lines on the Natural Bridge:

Beneath this noble arch, men wondering stand;
'Twas fashioned here, by an Almighty hand,
An Architect Divine, whose voice can call
Worlds into being—or, decree their fall.

If I had never known Jehovah's law,
This scene had taught me reverential awe;
Inspired my soul its feeble powers to raise,
Admiring nature–nature's God to praise.

Atheist! contemplate this grand scene, one hour
And thou shalt own there is a God of Power.[26]

In view of Calvinism's propensity to read nature as a second scripture, it seemed perfectly appropriate, as Perry Miller observed, to
fuse the sublime and religion in the Jacksonian period. "The
Sublime and the Heart! That they should, so to speak, find each
other out and become, in the passions of the Revival, partners—this
is a basic condition of the mass civilization of the nation." The
revivalist cast of mind explicitly rejected "the cultivation of the intellect," and camp meetings instead emphasized spontaneity and practical results.[27]

Because of the centrality of women in the religious life of the
United States, the sublime was also inflected by gender. In the sexual politics of the nineteenth century, women were generally regarded as the repositories of feeling, religion, and sensitivity, and they

were considered peculiarly sensitive to natural wonders. Elizabeth McKinsey gives the example of a man who visited Niagara several times in the 1820s, both with and without female companions, and concluded, "the view of anything grand or sublime in nature or art is not worth two pence in selfish solitude, or rude male companionship, unembellished by the sex."[28] Women, whom the cataracts moved to tears, liberated his own emotional response.

Women's responses were not limited to those that men expected of them. Their attitudes toward nature, examined by Annette Kolodny, often did not emphasize the sublime. Characteristically, their domestic fantasies focused on the great plains and concerned themselves with the beautiful.[29] Margaret Fuller found the Middle West of the 1840s parklike, a vast garden of spring wild flowers and new farms. She and other women writers of the time typically described the frontier and the middle border not in the language of the sublime but in terms of the beautiful and the picturesque. Such literary travelers found, however, that the women who lived on the middle border had so many domestic burdens that they seldom could leave home to enjoy the natural scenes around them. Their aesthetic concerns were usually limited to the immediate surroundings, especially their gardens and their efforts at landscaping.

Fuller and other early feminists experienced an institutionalized bifurcation in sensibilities: the sublime was for men; the beautiful and the picturesque were suited to effeminate men, women, and preachers. This division was sanctioned by the highest authority. Kant himself declared in his early essay *Observations on the Feeling of the Beautiful and Sublime* that men had an innate appreciation of the sublime, whereas women were more disposed to the beautiful:

. . . all the other merits of a woman should unite solely to enhance the character of the beautiful, which is the proper reference point; and on the other hand, among the masculine qualities the sublime clearly stands out as the criterion of his kind. All judgements of the two sexes must refer to these criteria. . . .

The fair sex has just as much understanding as the male, but it is a beautiful understanding, whereas ours [the male] should be a deep understanding, an expression that signifies identity with the sublime. [30]

Kant's work, translated into English in 1799, crystallized widespread assumptions about how gender entered into the definition of appropriate emotions. Fuller, both in the act of traveling and in her writing, was demanding for women the right to appreciate not only the beautiful but also the "masculine" emotions associated with the sublime.

Yet if women made the sublime their own, it was as observers. Works of the technological sublime were decidedly male creations. Throughout the nineteenth century, women's intellectual faculties were judged to be different in kind from men's. Even more than the professions of medicine, architecture, and law, engineering was over- whelmingly male, and women were thought to be intellectually inca- pable of higher mathematics. Women were deemed inferior, and their education was usually restricted to the humanities. Kant argued: "Deep meditation and a long-sustained reflection are noble but difficult, and do not well befit a person in whom unconstrained charms should show nothing else than a beautiful nature. Laborious learning or painful pondering, even if a woman should greatly suc- ceed in it, destroy the merits that are proper to her sex. . . . A woman who has a head full of Greek, like Mme. Dacier, or carries on fundamental controversies about mechanics, like the Marquise de Chatelet, might as well even have a beard. . . . A woman will there- fore learn no geometry." Educational institutions reflected this sexu- al division of knowledge. From the founding of Harvard University in 1636, for 200 years universities were for men only. Even after col- leges were created for women, technical education was generally denied them, and most engineers were male throughout the nine- teenth century. In addition, it was an extremely mobile profession; engineers constantly moved from one large assignment to the next one. In an age when women were expected to remain at home and preserve the domestic sphere as a haven of repose, such a nomadic profession was so unorthodox for a female as to be almost unthinkable.

Some men did not want women present when they contemplated natural wonders. In 1864 this male response to nature was formal- ized by Clarence King, who spoke for many geologists when he complained of the women and "literary travelers" who stood at Inspiration Point in Yosemite Valley, inflated at the sight, and used "cheap adjectives" to describe it.[31] The "literary traveler" was femi- nized and, by implication, incapable of responding accurately to the landscape. Only a man such as King, who was both a Yale- educated reader of Ruskin and a geologist hardened by years of field work, could achieve a balance of aesthetics, scientific knowl- edge, and practical experience, which together made possible a masculine sublime response to the grandiose nature of the far west.

King could satirize landscape painters, and yet he could comment so perceptively on art that John Ruskin sought his acquaintance after overhearing his comments in a London gallery. In a public lecture he once placed himself in "that resolute band of nature-workers who both propel and guide the great plowshare of science on through the virgin sod of the unknown." Yet, as Michael Smith points out, in his private notebooks King explored the possibility that the scientist ought to combine the sensibilities that his age assigned to men and women. If in his public statements he often posed as a macho mountaineer, in the notebooks he admitted the value of "feminine" receptivity.[32]

Mark Twain wrestled with the same problem when, in *Life on the Mississippi*, he expressed disdain for an educated but inexperienced steamboat passenger's response to a Mississippi sunset. As Leo Marx has argued, the pilot observes the river in terms of its dangers, such as sandbars, snags, and currents, while the passenger sees it in terms of landscape conventions. This difference, Marx notes, is coded by Twain in terms of gender: "On the beautiful surface nature has obvious feminine characteristics (softness, dimples, graceful curves), but the subsurface is represented by objects with strongly masculine overtones (logs, bluff reefs, menacing snags)." When it came time to write *Huckleberry Finn*, says Marx, "It would have been absurd to have had Huck Finn describe the Mississippi as a sublime landscape painting." Rather, Twain's achievement was that Huck spoke neither the analytical language of the pilot nor the stilted clichés of landscape convention, but a vernacular language.[33]

King and Twain were exceptional observers, equally comfortable in the lecture hall and the mining camp and able to speak to most of their contemporaries. But few were able to achieve such a balance. Many women, like Fuller, struggled to overcome the limitations on their responses imposed by a gender-inflected language so that the sublime could be democratically enjoyed by both men and women.

The popular sublime became part of the emergent cultural nationalism of the United States in the nineteenth century. The American public celebrated the fact that a spectacular sight was the biggest waterfall, the longest railway bridge, or the grandest canyon, and they did so with a touch of pride that Europe boasted no such wonders. Natural places and great public works became icons of America's greatness.

By the 1830s, sublime technological objects were assumed to be active forces working for democracy. This is exemplified by the popular enthusiasm for the Erie Canal. Built in only eight years over the 300 miles from Albany to Buffalo, this canal more than doubled the volume of west-east trade in its first year of operation, stimulating urban development along its banks, speeding up westward migration, and making western agricultural products available to eastern cities. In addition, the canal was immediately perceived as a powerful political link binding the Great Lakes region to the East. Just as important, the canal became a tourist site, a demonstration

A depiction of the Grand Canal Celebration. Courtesy of Collections of Library of Congress.

of American engineering skill, and one of the first icons of the technological sublime. Its numerous locks, aqueducts, and cuts through solid rock demonstrated control over natural forces. Several sites became famous landmarks, including the 802-foot stone aqueduct that carried the canal over the Genessee River at Rochester and the ladder of five locks in Lockport that carried boats up a giant staircase over the Niagara Escarpment.

When the Erie Canal was officially inaugurated in October 1825, for the first time the whole nation celebrated a technological project. Months beforehand, committees of citizens in New York and Albany organized the celebration, issuing suggestions and requesting cooperation between state and local authorities. They called for cannon to be placed at intervals between Buffalo and Albany, and that "the entrance of the first boat from Lake Erie into the canal be announced by a discharge from the cannon near at the lake, and that it be followed by successive discharges from the cannon on the line to the city of Albany."[34] A procession of canal boats wended its way from west to east, carrying a selection of products from the western states and territories that the canal would serve. The boats halted in every village of consequence for speeches, dinners, and fireworks that had been arranged by local committees. In Utica, for example, the day was celebrated by an excursion on the canal, the boats decorated with flags and "accompanied by a band of music." A "cold collation" of food had been prepared for each boat. In the evening, all the bridges over the canal and their adjoining buildings were illuminated, and fireworks were ignited. An illuminated barge with an orchestra on board proceeded along the canal, leading the citizens to a "Grand Ball and concert." Afterwards the local paper noted the general enthusiasm: ". . . the whole proceeding was emphatically the work of the citizens—some classes were particularly distinguished and none felt the generous impulse more forcibly than the mechanics and military: —their zeal was unbounded." The overall effect of the event was that "political animosity has for a time been banished by the generous burst of popular enthusiasm; and a whole people have poured forth the overflowing of their gratitude and mingled in a general acclamation of joy."[35] A recent history of the canal notes that "crowds gathered everywhere. Visitors in holiday spirit flocked to the canal from miles around. Horses and carriages filled the roads. People covered the towpath, crowded the bridges, and pressed forward to hear speeches of welcome, speeches of response, and speeches of farewell. Arches garlanded with evergreens and flowers spanned the canal at nearly every village and supported banners praising Clinton, republicanism, or internal improvements. Bands played and cannon thundered in salute. Local militia companies escorted the guests to feast at the best hotel. . . . "[36] Governor De Witt Clinton's orations en route repeatedly linked internal improvements with prosperity, with "the duration of the Union," and with the "holy cause of Republican Government."[37] As the procession moved east, more boats joined the flotilla. When they reached the Hudson at Albany, a week after leaving Buffalo, a vast parade was followed by a banquet at the capitol building. Later, as the flotilla moved down the Hudson, a contingent of steamboats from New York City joined the procession to New York Harbor, where seagoing vessels joined the "Grand Aquatic Display." The flotilla then sailed out to Sandy Hook, where water from Lake Erie was ceremoniously poured into the Atlantic Ocean. Clinton declared in his dedication that "the first arrival of vessels from Lake Erie, is intended to indicate and commemorate the navigable communication, which has been accomplished between our Mediterranean seas and the Atlantic Ocean." After the flotilla returned to New York, there was a "Grand Procession" of some 5000 marchers, including many workingmen. "A surfeit of processions, speeches, banquets, bell-ringing, illuminations, and fireworks" lasted the rest of the day and long into the night.[38]

The Erie Canal ceremonies provided the opportunity to celebrate more than the completion of the canal. They united the militia, political leaders, merchants, mechanics, and the general public. Aside from creating a general sense of well-being, the speeches and editorials stressed the political values of republicanism.

American republicanism was less a creed than a language that could be used to explain the citizen's relation to the state. It was rooted in Machiavelli's civic humanism and in the founding fathers' reading of classical authors. Its key concepts were independence, virtue, popular sovereignty, citizenship, and commonwealth.[39] Since politics was expected to inspire vigorous debate and continual self-examination rather than automatic patriotism, another realm of unquestioned allegiance was needed to unite the citizenry. Hence the centrality of the natural and technological sublimes. While voters might disagree on the issues of the day, they could agree on the uplifting sublimity of Niagara Falls, the Natural Bridge, or the Erie Canal at Lockport. Such sights, it was felt, purified and uplifted the mind and helped individuals see themselves as members of a larger community. The canal was understood as a product of democracy. As the Utica paper stressed, "the successful completion of this great work must have an important bearing upon the destinies of this nation, and eventually upon the whole world," because of the "proof which it will present to all mankind of the capabilities of a free people, whose energies, undirected by absolute authority, have accomplished, with a sum insufficient to support regal pomp for a single year, a work of greater public utility, than the congregated forces of Kings have effected since the foundations of the earth." The Erie Canal showed how a free people could "complete without a burthen" a work "surpassing in its benefits to the human race the most splendid monuments of ancient or modern history."[40] The citizen who contemplated such public improvements became aware of the power of democracy and saw himself as part of the moral vanguard, leading the world toward universal democracy. These manmade objects became national symbols. Traveling to America's natural wonders and great public works became the act of a good citizen, just as a pilgrimage to Jerusalem was the sign of a good Christian. American democratic virtue could not be based on a state religion—that was forbidden by the constitution, and in any case there were many different sects. Nor could it be based on adherence to ancient traditions, since there were none. But democratic virtue could be invigorated by the powerful experience of sublimity. Hence the central importance of visiting Niagara Falls, the Hudson River Valley, and the Erie Canal, which could be done all in one trip. Contemplation of these sites or of the Natural Bridge taught the individual his place in the world, lifting him out of himself.

Yet the experience of the natural sublime was not intended to justify preserving the wilderness or halting development. The Jeffersonian ideal was not the wild but the agrarian, not the frontier retreat but the rural township with a free public school in the middle.[41] Thomas Jefferson's democratic landscape was the rural grid of roads and section lines. In 1785 the National Survey began to lay out the grid pattern that is still visible in the checkerboard of farms that can be seen from any airplane passing over the Middle West. It applied science to the division of the continent into freeholds for small farmers, whom Jefferson believed would be innately good citizens. As Jefferson put it in *Notes on the State of Virginia*, "The country produces more virtuous citizens."[42] Jefferson cultivated a taste for sublime landscape, yet he also focused on the settled agrarian state and on the steady habits and good morals which he believed it encouraged. He saw the land as a commodity to be divided into uniform squares and then made available to independent farmers, yet he turned to portions of the landscape as sources of spiritual inspiration.

The sublime was inseparable from a peculiar double action of the imagination by which the land was appropriated as a natural symbol of the nation while, at the same time, it was being transformed into a man-made landscape. One appeal of the technological sublime in America was that it conflated the preservation and the transformation of the natural world. This conflation was in part characterized by what Bryan Wolf has called "an astonishing capacity of mind, an ability to consume the world as nothing more than a plenum of nutrients in that characteristically American project of self-making."[43] "At its most audacious," Wolf continues, "the sublime entailed a virtual substitution of self for world: it was an egotistical affair conceived in pride and consummated in an incestuous twining of nature back into the self, the NOT ME into the ME." Wolf wrote these words in an essay on Herman Melville. As literary scholars so often do, he imperialistically swallows all reality into his own discipline as he goes on to say: "The key to this project was language, the uncanny talent of words to usurp the place of things." In fact, however, most Americans of the Jacksonian period had read little philosophy, and they were swallowing the world not through language but through direct action. They were assaulting the natural world with axes, shovels, plows, and railroads, literally reworking the landscape, usurping the place of natural things with man-made objects. They were vigorously projecting themselves into the world, mixing their labor with it, and building internal improvements.[44]

The transformation of the land did not abate. Saum notes in his study of diaries and letters from the period that "the common man's pleasure in the pastoral . . . derived not from his comparative proximity to nature but from his recognition that, in agricultural and pastoral settings, nature had been subdued and rendered orderly."[45] Andrew Jackson echoed these convictions in his second annual message to Congress: "What good man would prefer a country covered with forests and ranged by a few thousand savages to our extensive Republic, studded with cities, towns and prosperous farms, embellished with all the improvements which art can devise or industry execute?"[46] Jackson's rhetorical question articulated what, after two centuries of pioneering and settlement, was the dominant view. He began by evoking the contrast between savages in the forest and a prosperous, settled landscape. Had he stopped there, his description would have been close to Jefferson's vision. But the end of his statement introduces the characteristic note of his generation, which wanted not a balanced, pastoral state but "all the improvements which art can devise or industry execute." Much as Jacksonians wanted these changes, however, they were suspicious of a centralized government or a permanent civil service. Indeed, they abolished the nation's central bank on the grounds that it was a threat to democracy. They wanted technological progress, but they rejected the political and economic structures needed to ensure it. As Marx notes, they refused to recognize "a root contradiction between industrial progress and the older, chaste image of a green Republic."[47] Instead, they believed hopefully that mechanical improvements would be harmonious with nature. Many asserted that industrial development was not merely compatible with democracy but a direct outgrowth of it. This idea had already reached full expression in the 1825 celebration of the Erie Canal. In 1841 a journalist underlined the apparently indissoluble relationship between democracy and advancing technology:

In exact proportion to the extension of political freedom and the diffusion of popular intelligence, has been the advance of invention in the useful arts. . . . As political power has been diffused among the great mass of men, the human mind has been directed to those inventions that were calculated to confer solid benefits upon the mass. Among the most important of these useful inventions is the discovery of the mariner's compass, the arts of printing and cotton spinning, and last of all, the science of navigation by steam, everywhere displaying its triumphs upon the rivers, the lakes, and the oceans of the world, the crowning victory of the mechanical philosophy of this nineteenth century.[48]

The Jacksonian version of the sublime focused as easily on the "victory of the mechanical philosophy" as on nature, and enfolded both in the larger scenario of Manifest Destiny. As Donald Pease puts it, "in the ideological American rendition, the sublime was not man's but Nature's discourse. . . . Some order beyond Nature seemed to command man to get in touch with Nature's higher will and to obey the implicit command to move beyond Nature."[49] Nature was understood to have authored the script sanctioning its own transformation in the service of an inevitable destiny. History was to be President Jackson's story of the creative subjugation of "a country covered with forests" to produce "cities . . . embellished with all the improvements which art can devise or industry execute." In this scenario both natural and man-made objects became part of the discourse of Manifest Destiny. Those who praised Niagara Falls and a new railroad did not see any inconsistency in embracing both. Each could be comprehended within what Albert Boime has called "the magisterial gaze"—a "diagonal line of sight . . . taking us rapidly from an elevated geographical zone to another below and from one temporal zone to another, locating progress synchronically in time and space."[50]

Landscape paintings, in their repeated use of the "magisterial gaze," embodied Jackson's ideology. Like Clarence King, who loved the wilderness but nevertheless sought to "propel and guide the great plowshare of science on through the virgin sod of the unknown,"[51] nineteenth-century Americans saw no irreconcilable contradiction between nature and industry; rather, they enjoyed contemplating the dramatic contrasts created by rapid progress.[52]

In the political realm, Kasson concludes, republicanism— despite early suspicions of capitalism—"developed into a dynamic ideology consonant with rapid technological innovation and expansion." In particular, "the promise of labor-saving devices strongly appealed to a nation concerned with establishing economic independence, safeguarding moral purity, and promoting industry and thrift among her people."[53] Many observers justified internal improvements as a spur to labor. George S. White, an apologist for industrialization, wrote in 1836: "Let our legislators be assured, that while they are extending towards its completion that system of improvement planned and hitherto carried forward with so much wisdom, they are putting into operation a moral machine which, in proportion as it facilitates a constant and rapid communication between all parts of our land, tends most effectually to perfect the civilization, and elevate the moral character of the people. . . . An idle population is ever vicious and degraded."[54] In this vision, the Erie Canal and other internal improvements were components of a "moral machine." They ensured not only prosperity but also democracy and the moral health of the nation.

Both natural wonders and mechanical triumphs like the Erie Canal were said to elevate the moral character of the people. To safeguard democracy, the ideology of republicanism demanded virtuous and

A view of the Upper Village of Lockport. Courtesy of Collections of Library of Congress.

active citizens who voted wisely and participated in their society. The need for active citizens in public life was itself relatively new, having emerged after the breakdown of royal authority in seventeenth-century England. This need became acute in the United States after the revolution. J. G. A. Pocock has argued that the revolutionary generation had to invent new forms of civic virtue and a citizenry motivated by non-pecuniary motives.[55] Republicanism was suspicious of capitalism because the entrepreneur was motivated by short-term self-interest. Republicanism was not entirely comfortable with a laissez-faire economic system, but it could embrace the useful arts. As Joyce Appleby has noted, "classical republicanism denied liberals their forward-moving, freedom-loving makers of history." At the same time, republicanism spoke "a language which wrote capitalists out of the political script."[56] The inventor of a new device, in contrast, rendered humanity a service in perpetuity. If the initial republican heroes were revolutionary statesmen and generals, by the 1820s inventors—Benjamin Franklin, Eli Whitney, Samuel Morse, Robert Fulton—seemed ideal republican heroes, because their new machines benefitted all of society. At the banquet of the Erie Canal celebration at Albany, three portraits had been emphasized: Governor De Witt Clinton, George Clinton (vice-president of the United States from 1805 to 1812), and George Washington.[57] In the Jacksonian age, however, engineers, builders, and inventors began to occupy important places in pantheon of republicanism. These new heroes and the machines they created were often celebrated on Independence Day, technology and republicanism thus being merged in one event. The participation of ordinary citizens in these occasions, an innovation of the late eighteenth century, helped to articulate the role of the individual in the new democratic state.

Not for another generation was the Fourth of July—celebrated in Philadelphia in 1777, only a year after independence was declared—widely recognized as a holiday.[58] Commemorations in the 1790s were often quite partisan, with separate ceremonies sponsored by competing parties. By the 1820s, however, the Fourth had become broadly popular, having shed most of the partisan spirit. It became an occasion for bombastic rhetoric not only celebrating the revolution but also tracing the progress of the nation and predicting its future greatness. By the time of the Civil War the celebration had become a fixture in most communities. Shortly before his death, Jefferson interpreted the meaning of the holiday in a famous letter:

All eyes are opened or opening to the rights of man. The general spread of the light of science has already laid open to every view the palpable truth, that the mass of mankind has not been born with saddles on their backs, nor a favored few, booted and spurred, ready to ride them legitimately, by the grace of God. There are grounds of hope for others; for ourselves, let the annual return of this day forever refresh our recollections of these rights, and an undiminished devotion to them.[59]

Jefferson placed the Revolution in an international context, as part of a worldwide movement toward democracy. He conceived of science as the enemy of superstition and blind tradition; the handmaiden of democracy, it found expression in public improvements. In Jefferson's vision, ritual observance of the Fourth of July inculcated civic virtue, giving it international and historical dimensions as the ritual rejection of undemocratic European models of government. Jefferson wished "this day" to "forever refresh our recollections" of the rights won in the past, with the hope that they would soon be won by others.

Independence Day exemplifies how a culture invents rites in which, to use Durkheim's words, "the social group reaffirms itself periodically."[60] John Adams understood this need at the very moment that the founding fathers declared America's independence. He felt that the event "ought to be solemnized with pomp and parades, with shows, games, sports, guns, bells, bonfires, and illuminations, from one end of the continent to the other, from this day forward, forever more."[61] These were in fact the chief elements of the republican ceremony in the 1820s, notably at the inauguration of the Erie Canal. Politicians organized these elements into an effective presentation. Parades and other displays were important preliminaries, military salutes excited the crowd, and fireworks and illuminations were suitable for the finale. But speeches were the central element—an Independence Day oration was the start of many a political career. Fourth of July ceremonies were quite distinct from the disorderly revels of the lower classes, and they provided the wealthy and the political elite with an opportunity to display their power. Local organizers invariably orchestrated the participation of many groups—including skilled workers, who usually marched in parades arranged by trade and profession.

During the nineteenth century new technologies became increasingly important in Fourth of July events, which are perhaps best understood as a form of street theater. Gradually, rather than commemorate the past, the Fourth began to emphasize the social effects of technology and to compare America to other civilizations. For example, Daniel Webster, speaking from the steps of the new addition to the Capitol Building in Washington on July 4, 1851, said: "The network of railroads and telegraphic lines by which this vast country is reticulated have not only developed its resources, but united emphatically, in metallic bands, all parts of the Union. The hydraulic works of New York, Philadelphia, and Boston surpass in extent and importance those of ancient Rome."[62] Similar sentiments marked the inaugurations of new canals, railroads, bridges, and buildings. The importance of technology in public ceremonies was in evidence as early as July 4, 1817, when Governor De Witt Clinton thrust a spade into the ground to officially begin the digging of the Erie Canal. Later Independence Days saw the official opening to navigation of the first section of the Erie Canal (1820); the start of the construction of the Pennsylvania Grand Canal (1826), the Baltimore and Ohio Railroad (1828), and the Baltimore and Ohio Canal (1828); the inauguration of the Boston and Worcester Railroad (1835); the dedication of the Eads Bridge in St. Louis (1874); and the rededication of the Statue of Liberty (1986). President John Quincy Adams presided over the 1828 canal ceremony. Not to be outdone, the Baltimore and Ohio had the last living signer of the Declaration of Independence, Charles Carroll, inaugurate the building of the first American railroad. Carroll solemnly declared "I consider this among the most important acts of my life, second only to that of signing the Declaration of Independence, if, indeed, second to that."[63] The railroad promised to be a moral machine that would promote public virtue and so preserve the nation he had helped create in 1776.

As technological achievements became central to July Fourth, the American sublime fused with religion, nationalism, and technology, diverging in practice significantly from European theory. It ceased to be a philosophical idea and became submerged in practice. In keeping with democratic tradition, the American sublime was for all—women as well as men. Rather than the result of solitary communion with nature, the sublime became an experience organized for crowds of tourists. Rather than treat the sublime as part of a transcendental philosophy, Americans merged it with revivalism. Not limited to nature, the American sublime embraced technology. Where Kant had reasoned that the awe inspired by a sublime object made men aware of their moral worth, the American sublime transformed the individual's experience of immensity and awe into a belief in national greatness.

The Railroad: The Dynamic Sublime

In the Jacksonian era the United States was still overwhelmingly a rural nation. The startling introduction of railroads into this agricultural society provoked a discussion that soon arrived at the enthusiastic consensus that railways were sublime and that they would help to unify, dignify, expand, and enrich the nation. They became part of the public celebrations of republicanism. The rhetoric, the form, and the central figures of civic ceremonies changed to accommodate the intrusion of this technology, as one can see by examining events staged between 1828 (when Baltimore celebrated the construction of the first railway) and 1869 (when most of the nation celebrated the completion of the first transcontinental railway). In the four decades between these two events, Americans integrated the railroad into the national economy and enfolded it within the sublime.

The word 'technology' was formulated by the Harvard professor of medicine Jacob Bigelow in 1828, the year Andrew Jackson was elected president. The term gradually became common during the years of the railroad's development, but it was unknown to the first railroad promoters. In his book *Elements of Technology* Bigelow encouraged the fusion of science and art, which he felt was characteristic of industrial society. He made what are now familiar claims for technology: "The labor of a hundred artificers is now performed by the operations of a single machine. . . . We accomplish what the ancients only dreamt of in their fables; we ascend above the clouds, and penetrate into the abysses of the ocean."[1] Only gradually did 'technology' acquire the all-encompassing sense it commonly has today. In the middle of the nineteenth century, technical universities began to call themselves institutes of technology, and the word began to take on the connotation of utilitarianism (as distinct from 'science'). Technology had to do with machines, patents, and systems of production; science was the province of "pure" research into fundamentals and did not concern itself with applications. Although technology lacked some of science's prestige and its claims to be an objective search for the great truths of the universe, it was a term of approbation, and by the later decades of the nineteenth century the engineer was a formidable social hero.[2] Particular machines and structures were often called sublime during the nineteenth century, but no one linked 'technological' and 'sublime' until after the Second World War. Just as Bigelow fused science and the arts, Americans of Jackson's era subsumed new machines within the framework of the natural sublime. They saw the construction of a railroad or a bridge as a triumph of "art," and they assumed that such advances could not fail to increase human happiness. In the early nineteenth century the machine was, as Leo Marx puts it, "a token of that liberation of the human spirit to be realized by the young American Republic."[3] Most Americans rejected English critiques of new machines, and there were virtually no Luddites among American workers. In a characteristic reply to Thomas Carlyle's attacks on "mechanism," Timothy Walker declared in the *North American Review* that mechanization had "emancipated the mind, in the most glorious sense." He continued: "From a ministering servant of matter, mind has become the powerful lord of matter. Having put myriads of wheels in motion by laws of its own discovery, it rests, like the Omnipotent Mind, of which it is the image, from its work of creation, and calls it good."[4] It was with such overblown rhetoric that many people sanctified acts of invention.

The railroad seemed the most obvious example of a liberating machine. The orator Edward Everett summarized a widespread attitude when he declared that the railway locomotive was "a miracle of science, art, and capital, a magic power . . . by which the forest is thrown open, the lakes and rivers are bridged, and all Nature yields to man."[5] In the 1830s the French traveler Michel Chevalier noted "a perfect mania in [the United States] on the subject of railroads" and concluded, after compiling a vast list of examples of railways in mines, shipyards, penitentiaries, and tobacco factories, that "the Americans have railroads in the water, in the bowels of the earth, and in the air. . . . When they cannot construct a real, profitable railway from river to river, from city to city, or from State to State, they get one up, at least as a plaything or until they can accomplish something better. . . ."[6] In the six years after 1830, when the first rail line went into operation, more than 1000 miles of track were laid, mostly in smaller sections, and "at least two hundred railroads were being operated, built, projected, planned, or merely talked about."[7] This intense interest had begun well before then. Frederick Gamst concludes that "American agitation for railroads began not in 1830 but at least twenty years earlier."[8]

Agitation finally led to action in Baltimore, where the construction of a railroad was begun on July 4, 1828.[9] Baltimore had grown to be slightly larger than Philadelphia, and with 80,600 inhabitants it was second only to New York in size among American cities.[10] The participants in the Baltimore ceremony understood the railroad in political and economic terms. Its construction was to increase Baltimore's prosperity, make Baltimore the focal point for western trade, and bind the union together. These typical Jacksonian sentiments were accompanied by belief in the inspirational value and the moral influence of new machines.[11]

July 4 dawned clear and cool for the season. Between 50,000 and 70,000 spectators lined the streets and filled every available window at 8 A.M., when the procession began. Although in 1828 it was quite unusual to see more than a few hundred people at one time, virtually the whole body politic of Baltimore was on display to itself, either in the parade or along streets; the sheer mass of people had a sublime quality.[12] The city had already been packed and bustling with activity the day before. A local chronicler noted that "on the afternoon and evening immediately preceding, all the roads to town were thronged with passengers, while in the city itself, the livery and incessant crowds in Baltimore's streets; the movements of various cars, banners, and other decorations of trades, to their several points of destination; the erection of scaffolds, and the removal of window sashes, gave many 'notes of preparation' for the ensuing fete."[13] This gradual filling of the city did not turn the populace into a passive crowd of spectators—a considerable percentage of them were directly involved in the celebration.

Besides the usual elements, the patriotic observances were infused with oratory and symbolism appropriate to the beginning of the first steam railroad in the United States.[14] It would link Baltimore to the Ohio Valley, opening it to the western trade, just as New York had tapped the Great Lakes region with the Erie Canal (completed 3 years earlier). When choosing July 4 for their celebration, local leaders were imitating canal builders, who had a decided preference for that date. The Erie Canal had been started on July 4,

1817, and on July 4, 1828, the competing Baltimore and Ohio canal was being inaugurated in Washington.

The repeated choice of July 4 as the date to commence large projects might appear to be a disingenuous use of patriotism by promoters intent on winning public support. Yet the Baltimore celebration was not entirely a private venture, as the financing demonstrates. Some of the backing came from the legislature, some from the City of Baltimore, and some from private investors. Furthermore, most citizens, whether they advocated free trade or protectionism, valued improvements to the transportation system. Workingmen's associations, expecting to benefit directly both from the construction of the line and from the increase in trade that it promised, gladly participated in the ceremonies. *Niles' Weekly Register* declared the following day: "The most splendid civic procession, perhaps, ever exhibited in America, took place in this city, yesterday, the 4th of July on the occasion of laying the first corner stone of the Baltimore and Ohio railroad. . . . Between fifty and sixty associations appeared with their banners, cars, and insignia. . . . The concourse of spectators, citizens, and strangers was exceedingly great and a glorious tribute was paid to one arm of the triumphant AMERICAN SYSTEM of internal improvements."[15]

Like other public works, Baltimore's new railroad was intended to transform the local economy; yet it was understood to be more than a profit-making venture. The parchment scroll deposited in its foundation stone announced that it would confer three "important benefits upon this nation" by "facilitating its commerce, diffusing and extending its social intercourse and perpetuating the happy Union of the Confederated States."[16] Such claims, which were common in the rhetoric surrounding public works in the Jacksonian period, reveal a continual anxiety that the divisive force of regionalism might undermine the union. Promoters continually played upon fears of disunion when arguing for the advantages of their projects. A generation weaned on the Missouri Compromise of 1820 and on sectional tensions embraced railroads with enthusiasm.[17] Railways were expected to expand the nation, even though many feared it was already too large and too heterogeneous. Public speeches revealed "a deep fear that the nation will not endure."[18] Thus, it was entirely characteristic that in Baltimore one of the central symbols was "the good ship 'Union' completely rigged on Fells Point."[19] A local historian noted that "as the various bands of music, trades, and other bodies in the procession passed before it, it was

evident from their greetings that they regarded this combined symbol of our confederacy and navy with especial approbation."[20] In the
midst of the parade, when Charles Carroll, a director of the railroad and the last surviving signer of the Declaration of
Independence, reached the *Union*, the following dialogue was sung
out for the crowd:

Ship ahoy! What is the name of that ship and by whom commanded?
The Union; Captain Gardiner.
From whence came you and where bound?
From Baltimore, bound to the Ohio.
How will you get over the mountains?
We've engaged passage by the railroad.[21]

The crowd expected this promise to be literally true: the ship of
state would pass from the seaboard to the western states through
the construction of the railroad. To reemphasize the point, the
crew of the *Union* sang, to the tune of "Hail to the Chief," the following words: "Hail to the road which triumphant commences, /
Shall closer t'unite the east and the west. . . ." Thus the railroad was
understood as a means to several ends: prosperity for Baltimore,
access to western markets, and the preservation of the union, tying
it together with bands of iron. All three themes would persist
throughout the antebellum period.[22]

The celebration of 1828 was not an example of the technological
sublime. Rather, it established the context within which it would
appear. Not only was the word 'technological' unknown; the vast
majority of the spectators had never even seen a railroad. The idea
that it was appropriate to lay a cornerstone and consecrate it by
"pouring wine and oil and scattering corn upon the stone," as
though for a building, immediately suggests some confusion.[23] The
cornerstone was not part of the rail line or of an administrative
building; it lay in an open field along the original right-of-way. Not
only did it bear no useful relation to the railroad; soon it was not
even near the tracks, because the line was relocated. For some years
the stone was entirely forgotten and inadvertently buried. Though
it has since been recovered and placed behind an iron fence to protect it from vandalism, the stone has never served as a very potent
symbol of the railroad. Rather, it is a reminder of another conception of the railroad: as a guarantor of stability and local prosperity.
The crowd that came to see the consecration of this piece of
dressed granite and to celebrate Independence Day had not yet

embraced the railroad as a machine. They understood it as an idea to be added to an older cluster of ideas about the useful arts, prosperity, political unity, and patriotism. The event provides a starting point from which to trace the development of the technological sublime, which became indissolubly connected to railways and which achieved ascendancy in the following decades.

There was a gap between the rhetorical flourishes of the inauguration and actual construction. Despite all the speechifying, parading, and public enthusiasm of 1828, the B&O took much longer to build and cost much more than originally anticipated. In 1830 regular service began on the first 13 miles, from Baltimore to Elliot's Upper Mills. *Niles' Weekly Register* announced: "Tens of thousands will embrace the opportunity of seeing the novelest work yet attempted in the United States, of travelling 26 miles in two and one half hours, without danger or fatigue . . . every minute passing something new." And what was this wonder? "Each Wagon (drawn by one horse) will carry from 25 to 30 passengers," with a change of horses after 13 miles. At the time this was a remarkable novelty, and

A Baltimore and Ohio passenger car, 1830. Courtesy of Collections of Library of Congress.

the newspaper reported that, owing to strong public interest, "passengers will be under the necessity of going and returning in the same coach, until a sufficient number of additional carriages can be furnished." Several weeks later the public interest was still not satiated, and *Niles' Weekly* reported: ". . . no one can rightly estimate this work without seeing it, and travelling upon it. The best written accounts of the undertaking afford no more than an idea of it." The wonder had to be "realized by personal experience."[24]

In 1836, when it had been hoped the whole line would be in service, steam engines had been introduced, but only 85 miles of road were in operation, and it had not reached far enough to make a dramatic difference in regional trade. It had not even reached a coal mine. It did carry about 100,000 passengers a year,[25] and it was profitable but only at a rate of 3%. An estimated $8 million still had to be raised in order to complete the line, but investment enthusiasm was weak. In 1836 "a man of the times" wrote a pamphlet to the citizens of Baltimore, attempting to rouse them to action in response to the opening of a Pennsylvania railroad, which would serve many of the same markets originally sought by the Baltimore and Ohio. He warned: "Your day of prosperity is gliding by, and the streams of your power are stealing from you." For the moment, he declared, "The West has gone to Philadelphia."[26] In fact, the B&O's charter from the Pennsylvania legislature, which permitted it to serve Pittsburgh, expired in 1843, before the tracks reached there, and instead the line had to be built to Wheeling in 1851.[27]

As the mandatory round trips required of the first riders on the B&O suggest, railroads owed much of their early financial success to pleasure seekers who wanted to find out what it was like to ride on rails and to see how the world looked when shooting by at 15 or 20 miles an hour. One of the earliest and most popular tourist attractions was Mauch Chunk, a transport center at the edge of the Pennsylvania anthracite coal fields. Tourists began to frequent the site in the 1820s to see the 9-mile gravity railroad that carried coal down to a navigable waterway. Gravity railways on a smaller scale were common inside coal mines, and in Quincy, Massachusetts, an outdoor line had been built from a granite quarry to the harbor. Mauch Chunk's line was built on a larger scale than anything previously attempted.[28] Since no one had ever seen a steam railway, the sight of coal cars apparently under their own power rumbling down hills seemed a wonderful innovation. Soon tourists were not only looking but paying for the privilege of traveling in the cars. In 1846 the attraction was further enhanced by "another ingenious mechanism called the 'switchback,'" which "began returning empty coal cars to the new mines in Panther Creek Valley. . . . This railroad zigzagged back and forth as it descended into the valley, the direction of the cars reversed by self-operating switches."[29] As John Sears notes, the system foreshadowed the later development of the roller coaster but ran over a much larger route than any amusement park ride.[30] Not only did the gravity railroad offer visitors the opportunity to travel at what was then great speed; it also provided a glimpse of rugged mountain landscapes from the rushing cars. The panoramic views at the tops of crests were succeeded by the pell-mell rush down into valleys, or by the series of reversals accomplished by an intricate self-regulating system of switches. There was no need to steer or control the movement of the cars; the passengers were completely passive observers, alternately terrified and delighted by the trip. Mauch Chunk offered the first visual experience of riding at high speed, and it began to accustom the public to the new experience of riding on rails.

On Christmas Day of 1830, the citizens of Charleston, South Carolina, were the first people to ride on a steam train in the United States. About 140 people, including public officials, a brass band, soldiers firing a cannon, and stockholders, were on the official run over the 6-mile line.[31] This cast of characters was prophetic: politicians, Charleston investors, and military men rode on the new device, while the general public stood along the tracks and cheered. The sight of a line of vehicles moving without horses or oxen to pull them was startling enough. Even more startling was the size and number of the vehicles, for people were only accustomed to seeing one wagon pulled by a team. But most amazing of all was the speed. For most of human history, sailing ships had been the

The "Best Friend," the first locomotive built in the United States for actual service on a railroad. Courtesy of Collections of Library of Congress.

fastest form of transport over any distance. A galloping horse, the fastest thing on land, could be ridden only a few miles before needing a rest. The "iron horse" was far more powerful, and tireless, and on its first appearance it drew enthusiastic crowds. Passengers were surprised "that the motion at the rate of 24 miles an hour is not attended with any inconvenience, except in looking at stationary objects—that persons read and converse perfectly at their ease. When two engines, travelling at this rate, meet—the effect is described as astonishing; it is like a rushing together of the 'wings of the wind,' if the wind has wings."[32] A passenger on the first steam train to leave Charleston used the same image: ". . . we flew with the wings of the wind at the varied speed of fifteen to twenty-five miles an hour."[33] This velocity was "annihilating space and time." Riders also found that the railway changed the appearance of the local landscape. The slow unwinding view seen from a wagon or a horse was transformed into a sliding world that seemed to move by while the passenger sat immobile. The eye was not prepared to see these hurtling objects glimpsed in a rush, and had to learn to focus on the distant panorama. Ralph Waldo Emerson expressed the passenger's excitement at seeing the world defamiliarized by speed: "What new thoughts are suggested by seeing a face of country quite familiar in the rapid movement of the railroad car."[34]

Aside from the novel sensation of speed, the thunder of the engine, and the transformation of the landscape, however, passengers often found their first railroad journey uncomfortable. Sparks from the engine burned their clothing, the seats were hard, and the movement was often jerky because the connections between cars were slack. Nevertheless, passengers were usually exhilarated. In 1837 Samuel J. Parker listed the usual complaints in an account of his first journey: "The time after midnight was excessively wearisome, as we 'enjoyed' the English style cars, with eight on a seat riding backward and eight more facing these backward riders, with feet interlocked and one lantern as a lamp to two such satanic English style compartments, and the glass sliding rack, rattling as the springless cars rattled and thumped over the strap-iron rails spiked to the long sleeper logs that made the track." "Yet," he continued, "to me and to most of us this first night and ride in the cars was sublime, as an excitement and a novelty," despite the fact that "the sparks of the locomotive flew over us in a perpetual shower often burning holes in exposed clothes."[35] After observing American railroads, Chevalier asked: "Is there anything which gives a nobler idea of the power of man than the steam engine in the form in which it is applied to produce motion on railroads? It is more than a machine, it is almost a living being."[36]

Despite the general enthusiasm, however, there was opposition to the railroad. Self-interested parties such as turnpike operators, canal boat owners, and innkeepers along stage lines attacked railroads as dangerous and tried to impede their construction. Many people were afraid of the new machine. One observer described a passing train as "frightening the horses attached to all sorts of vehicles filled with people from the surrounding country, congregated all along at every available position near the road . . . causing thus innumerable capsizes and smash-ups of the vehicles and the tumbling of the spectators in every direction."[37] A clergyman in Connecticut declared that seeing moving locomotives had the potential to drive some people insane.[38] By 1840 there had already been enough accidents with steam engines that a book was devoted to the subject.[39] How, then, did people come to terms with these new technologies that overturned the existing order?

Ambivalence toward technology was more prevalent in England than in the United States. When the Liverpool and Manchester Railway opened in 1830, it drew a crowd estimated at 400,000, which lined the tracks for most of the distance. But whereas at Baltimore two years earlier the public had eagerly gathered to celebrate technological advancement, many in the English crowd protested the new railway, and some spattered mud on the first passengers. Many faces wore scowls. On a prominent spot above the crowd, a poorly dressed hand-loom weaver sat in silent protest; he was understood by Fanny Kemble to be "a representative man, to protest against this triumph of machinery, and the gain and glory which the wealthy Liverpool and Manchester men were likely to derive from it."[40] The English were prone to view industrialization in terms of satanic mills, frankensteinian monsters, and class strife; the Americans emphasized the moral influence of steam, and often sought to harmonize nature and industrialization.[41]

When Charles Dickens wrote about American railroads he adopted the imagery of monsters. Of one journey he wrote:

On it whirls headlong, dives through the woods again, emerges in the light, clatters over frail arches, rumbles upon the heavy ground, shoots beneath a wooden bridge . . . dashes on hap-hazard, pell-mell, neck or nothing . . . and unaccustomed horses plunging and rearing, close to the very rails—there—on, on, on, tears the mad dragon of an engine with its train of cars; scattering in all directions a shower of burning sparks from its wood fire; screeching, hissing, yelling, panting; until at last the thirsty monster stops beneath a covered way to drink, the people cluster round, and you have time to breathe again.[42]

At first a Dickensian nervousness was not unknown in the United States, but then fear is a central element of the sublime response. At virtually the same moment as Dickens was touring America, in 1839, a New Yorker saw a train at night and wrote in his diary: "Just imagine such a concern rushing unexpectedly by a stranger to the invention on a dark night, whizzing and rattling and panting, with its fiery furnace gleaming in front, its chimney vomiting fiery smoke above, and its long train of cars rushing along behind like the body and tail of a gigantic dragon—or the d—l himself—and all darting forward at the rate of twenty miles an hour. Whew!"[43] In 1835 Emerson reported hearing a joke about a farmer who first saw a train at night and believed it was "like hell in harness," and the same year Davy Crockett published a southern version of this story.[44] Jacksonian-era Americans clearly liked to imagine the terrifying impact of a railway train on the senses of someone unprepared for the encounter, evoking Dickens's imagery only to mock it. Class was not a factor in these imaginative encounters, which in effect divided society into two groups. On the one side were innocents who would feel astonishment and terror at the sudden irruption of the railroad into their experience. Unlike someone viewing the Grand Canyon or Niagara Falls for the first time, they could not decode the experience into familiar categories of perception, and therefore they turned to the supernatural: dragons, monsters, or visions of hell. On the other side, separated by laughter, were those who knew they were looking at a railway, and who could feel the sublimity of a powerful machine under human control. Not only did they possess the knowledge necessary to understand the railroad, but, since they knew it did not immediately threaten them, the power and danger of the train became an essential part of the frisson of this new form of the sublime. Their laughter expressed superior knowledge, declaring the difference between those who understood the new machine and were moved by its sublimities and those who did not understand it and ignorantly interpreted it in supernatural terms. At the same time, such jokes record the persistence of some unease with the railroad and the desire to dismiss this fear by projecting that response onto others. The joke suggests that the machine is monstrous only to the ignorant.[45]

The initial nervousness soon gave way to exhilaration. In 1841 an American orator began his address by asking "What varieties of elocution are not blended to make railroads sublime!" He meant this not as a reproach but as a preliminary to his own eulogy to "a Titanic colossus of iron and of brass, instinct with elemental life and power, with a glowing furnace for his lungs, and streams of fire and smoke for the breath of his nostrils." As Perry Miller notes, the "public speech of the period everywhere resounds with these addresses to the railroad—the steam horse that mounts the Alleghenies. . . ."[46] The enthusiasm permeated all levels of society. The Swedish traveler Frederika Bremer noticed on her visit to the United States between 1849 and 1851 that when American boys drew on their school slates they almost invariably chose to depict engines, steamboats, and railways.[47] Walt Whitman was so enthralled with the locomotive that he addressed a poem "To a Locomotive in Winter," lovingly evoking each of its mechanical parts in turn and calling the engine "Type of the modern—emblem of motion and power—pulse of the continent." Another excerpt:

Fierce-throated beauty!
Roll through my chant with all thy lawless music, thy swinging lamps at
    night,
Thy madly-whistled laughter, echoing, rumbling like an earthquake,
    rousing all,
Law of thyself complete, thine own track firmly holding,
(No sweetness debonair of tearful harp or glib piano thine,)
Thy trills of shrieks by rocks and hills return'd,
Launch'd o'er the prairies wide, across the lakes,
To the free skies unpent and glad and strong.[48]

Kant mentions only natural forces in his account of the dynamic sublime, but he notes that this experience reveals "a power of resistance of an entirely different kind, which gives us courage to measure ourselves against the apparent omnipotence of nature."[49] What better way to measure oneself against nature than through great works of manufacturing and engineering? Jacksonian-era Americans experienced both the dynamic and the arithmetical sublime in their response to new machines. The arithmetical reformulation (examined in subsequent chapters) emerged when engineers began to build massive objects whose scale and permanence made them appear to be triumphs over the physical powers of nature. Great bridges overcame natural obstacles; tall buildings surmounted the force of gravity; dams restrained the largest rivers of the continent. In contrast, the dynamic technological sublime focused on the triumph of machines—particularly the railroad, but also the telegraph and the steamboat—over space and time.

The steamboat was popular not least because its invention was usually attributed to Americans, notably Robert Fulton.[50] It hastened communications, knitted the union together, and stimulated the economy, but to the first generation to know it the steamboat was also a mechanical marvel. In 1818 the *American Journal of Science* declared it sublime and wondered over "how completely such a floating palace transcends the wildest dream of which the builder of the gigantic Pyramids or even Archimedes himself might be supposed capable."[51] Henry Clay later declared: "Nature herself seems to survey, with astonishment, the passing wonder, and in silent submission, reluctantly to own the magnificent triumph, in her own vast domain, of Fulton's immortal genius!"[52] Nature had become an astonished, passive witness to man's achievements.

Similarly, a lecturer in Charleston, inspired by local railway construction, declaimed on "The Moral Influence of Steam."[53] Promoters of the Western Railroad of Massachusetts sent a letter to ministers explaining "the moral effect of Rail-Roads," and urged them to "take an early opportunity to deliver a Discourse on the Moral effect of Rail-Roads in our wide extended country."[54] By appealing to clergymen as arbiters of public opinion, the Western Railroad demonstrated clearly how inseparable technology and religion were in the United States. Such efforts may have been reactions to adverse comment from the pulpit. In Ohio, where no railroads had yet been built in 1828, a local school board declared: "If God had designed that His intelligent creatures should travel at the frightful speed of 15 miles an hour by steam, He would have foretold it through His holy prophets. It is a device of Satan to lead immortal souls down to Hell."[55] Nathaniel Hawthorne later made a similar notion the kernel of his satirical tale "The Celestial Railroad," in which Bunyan's pilgrims hurry toward the celestial city on a railroad only to be diverted into a tunnel to hell at the last moment.[56] But these were the fears of a minority.

In contrast to these early responses, most contemporaries found the speed of the locomotive and its sense of danger essential ingredients of its sublimity. A *Scientific American* reporter called the sight of a rushing train at night "sublime and terrific."[57] Charles Caldwell expressed what became the dominant view in "Thoughts on the Moral and Other Indirect Influences of Rail-Roads," which appeared in 1832 in the *New England Magazine*. Adopting the rhetoric of the natural sublime, he declared: "Objects of exalted power and grandeur elevate the mind that seriously dwells on them, and impart to it greater compass and strength. Alpine scenery and an embattled ocean deepen contemplation, and give their own sublimity to the conceptions of the beholders. The same will be true of our system of Rail-roads. Its vastness and magnificence will prove communicable, and add to the standard of the intellect of our country."[58] Leo Marx has noted that in such passages "the awe and reverence once reserved for the Deity and later bestowed upon the visible landscape is directed toward technology, or rather the technological conquest of matter."[59] Jacksonians did not require a secularized sublime; the conquest of space and time also could be seen as God's handiwork. In 1851, on the occasion of the opening of the Cleveland and Columbus Railroad, S. C. Aiken declaimed in a sermon: "In a moral and religious point of view, as well as social and commercial, to me there is something interesting, solemn, and grand in the opening of a great thoroughfare. There is sublimity about it, indicating not only march of mind and a higher type of society, but the evolution of divine purposes, infinite, eternal—connecting social revolutions with the progress of Christianity and the coming reign of Christ."[60] Whether the railroad was understood to be a direct expression of the Almighty or a secular force, it was seen as a sublime engine of moral development.

The railroad was regarded as an engine of progress and western expansion. Like the Erie Canal and its imitators, railways were expected to stimulate economic development wherever they reached. In "The Young American" Emerson wrote: "An unlooked for consequence of the railroad is the increased acquaintance it has given the American people with the boundless resources of their own soil. . . . It has given a new celerity to time, or anticipated by fifty years the planting of tracts of land, the choice of water privileges, the working of mines, and other natural advantages. Railroad iron is a magician's rod, in its power to evoke sleeping energies of land and water."[61] Horace Greeley emphasized that "railroads in Europe are built to connect centers of population; but in the west the railroad itself builds cities. Pushing boldly out into the wilderness, along its iron track villages, towns, and cities spring into existence, and are strung together into a consistent whole by its lines of rails, as beads are upon a silken thread."[62] The dynamism of the railroad made it possible to merge westward expansion and Manifest Destiny with the sublime.

As Barbara Novak points out, Americans of the Jacksonian period worked out their version of the sublime at the very moment when they were beginning to industrialize: ". . . the rhetoric of the technological sublime developed concurrently with the nature rhetoric."[63] The machine—whether locomotive, steamboat, or telegraph—was considered to be part of a sublime landscape, and at first it was included in pastoral paintings as a harmonious part. This confusion of the natural and the technological sublime was encouraged by the owners and managers of the railroads, who not only offered clergymen tips on sermons but also commissioned paintings of the landscapes visible along their lines. The public was exhorted to use the railways not only because of their speed and convenience but also as a means of enjoying scenery. As early as 1831 potential passengers on the Baltimore and Ohio were told that its "scenery partakes of the 'sublime and the beautiful'" and that "in the spring and summer it is much like that on the Green Mountains." Equally impressive were the "magnificent" Carrollton viaduct (which was 300 feet long and 65 feet high), the "deep cut" that was three-fourths of a mile long, and a high embankment "57 feet above the natural surface of the ground."[64] In short, railway scenery was a mixture of the natural and the man-made sublime. As this notion became widespread, engineering projects took increasingly prominent places in landscape paintings[65] and became featured parts of tourism. The Main Line portage between Pittsburgh and Philadelphia, which carried canal boats 1400 feet up over the Allegheny Mountains, was rightly considered "an engineering marvel and a scenic wonder."[66] Later the great "horseshoe curve" of the Pennsylvania Railroad became a tourist site, both for those who rode the train and for those who came to watch. And the impressive railway stations that were constructed in every major city were ornate and self-conscious symbols of progress.

These improvements seemed equivalent to the wonders of the ancient world. In 1840 an essayist noted proudly that Americans had "dug through plains, hills, and solid rocks, in our long lines of canals and railroads, works that have stamped upon the soil a lasting impression, which, if the republic were swept away, and all records of its existence blotted out forever, would be viewed by posterity with the same wonder with which we now gaze upon the mouldering ruins of Rome, the marble temples of the Acropolis, the pyramids of Egypt, and the track of the Appian way."[67] Technological developments were thus understood as continuations of the great architectural and engineering works of the ancient world, which ancient authors often had proclaimed to be sublime. An article extolling the advantages of modern machinery in the *North American Review* cited Cicero's *De Natura Deorum* and concluded: "Vanquished Nature yields! Her secrets are extorted. Art prevails! What monuments of genius, spirit, power!"[68]

The attribution of sublimity to human creations radically modified the psychological process that the sublime involved. Whereas in a sublime encounter in nature human reason intervenes and triumphs when the imagination finds itself overwhelmed, in the technological sublime reason had a new meaning. Because human beings had created the awe-inspiring steamboats, railroads, bridges, and dams, the sublime object itself was a manifestation of reason. Because the overwhelming power displayed was human rather than natural, the "dialogue" was now not between man and nature but between man and the man-made. The awe induced by seeing an immense or dynamic technological object became a celebration of the power of human reason, and this awe granted special privilege to engineers and inventors.[69] The sense of weakness and humiliation before the superior power of nature was thus redirected, because the power displayed was not that of God or nature but that of particular human beings.

Thomas Weiskel wrote of the natural sublime that "there can be no sublime moment without the implicit, dialectical endorsement of human limitations."[70] The technological sublime does not endorse human limitations; rather, it manifests a split between those who understand and control machines and those who do not. In Kant's theory of the natural sublime, every human being's imagination falters before the immensity of the absolutely great. In contrast, a sublime based on mechanical improvements is made possible by the superior imagination of an engineer or a technician, who creates an object that overwhelms the imagination of ordinary men. Yet this inspiring effect is only temporary. Machines that arouse awe and admiration in one generation soon cease to seem remarkable, and the next generation demands something larger, faster, or more complex. By implication, this form of the sublime undermines all notions of limitation, instead presupposing the ability to innovate continually and to transform the world. The technological sublime proposes the idea of reason in constant evolution.

While the natural sublime is related to eternity, the technological sublime aims at the future and is often embodied in instruments of speed, such as the railway, the airplane, and the rocket, that annihilate time and distance.

Emerson understood the implications of the technological sublime when he proclaimed: "Nature is thoroughly mediate. It is made to serve. . . . It offers its kingdoms to man as the raw material which he may mould into what is useful. Man is never weary of working it up. . . . One after another his victorious thought comes up with and reduces all things, until the world becomes at last only a realized will, —the double of man."[71] In his formulation, the machine does not dominate the spirit; rather, man's mental processes use the machine to create a new and original relation to nature. At the end of the essay "Art," Emerson recognized the baser uses of technology but confidently brushed them aside:

Is not the selfish and even cruel aspect which belongs to our great mechanical works, to mills, railways, and machinery, the effect of the mercenary impulses which these works obey? When its errands are noble and adequate, a steamboat bridging the Atlantic between Old and New England and arriving at its ports with the punctuality of a planet, is a step into harmony with nature. The boat at St. Petersburg which plies along the Lena by magnetism, needs little to make it sublime. When science is learned in love, and its powers are wielded by love, they will appear the supplements and continuations of the material creation.[72]

Here Emerson was rejecting the common distinction between the useful and the fine arts. Technology, as a form of art, could be a central part of a merging of man and nature, in which the world would become the spiritualized will, or double, of man. In "The Poet" he chided his contemporaries for thinking that industrialization was not in harmony with the landscape: "Readers of poetry see the factory-village and the railway, and fancy that the poetry of the landscape is broken up by these; for these works of art are not yet consecrated by their reading; but the poet sees them fall within the great Order not less than the beehive or the spider's geometrical web. Nature adopts them very fast into her vital circles, and the gliding train of cars she loves like her own."[73]

Like Emerson, many at first preferred to see technological achievements not as violations of nature but as extensions and imitations of it. As Miller put it, "man's conquest of the mountains was not a violation of Nature but an embrace." This line of thought had it that, whereas Europeans created artificial-looking engineering works based on Roman models, "the American imitates nature, with whose great works he is in constant communication, and, like a spider, constructs a bridge light in appearance, but sufficiently strong." It followed that "only an appreciation of the grandeur of such a fall as that of Niagara could fit a man to construct the bridge that spans its river."[74] The natural sublime would inspire the engineer to produce works in harmony with it. The *New England Magazine* declared: ". . . the railroad is but one step in the ascending staircase on which the races are mounting, guided and cheered by heavenly voices. . . . We only mark the beginning."[75] Note the characteristic use of the plural here. This new version of the sublime was not based on solitary communion with nature. Rather, it was a communion, through the machine, of man with man. This new sublime implicitly proposed a group experience of its own potential greatness, rather than an individual experience of the power of the non-human.

The collective realization of the power of human intellect was the hallmark of celebrations of another new machine of the day: the telegraph. Conceived, patented, and constructed by Samuel F. B. Morse, the telegraph astounded observers in 1838 at its first public demonstrations. Wonder and incredulity brought excited crowds to the first telegraph offices, which often provided seating for spectators.[76] Instantaneous communication was literally dislocating, violating the sense of the possible. To prove that the new machine worked required at least three people: a sender, a receiver, and a witness. In practice, there were many witnesses, and they could scarcely believe that it was possible to sever language from human presence. Twenty years later the telegraph still excited general admiration and wonder. On August 9, 1858, the *New York Times* declared: "The Telegraph undoubtedly ranks foremost among that series of mighty discoveries that have gone to subjugate matter under the domain of mind." Whereas in the natural dynamic sublime the mind had to recuperate a sense of its superiority after an initial exposure to overwhelming power, in the technological sublime the overwhelming force itself was man-made. The sublimity lay in realizing that man had directly "subjugated" matter, and this realization was a collective experience.

Overcoming isolation was a general theme of the new version of the sublime. At the Boston Railroad Jubilee of 1851, the mayor exulted that the city had invested "to relieve us of the natural isolation of our position." He continued: "Hills have been cut in sunder, vallies [*sic*] have been filled up, and running waters have been spanned, to facilitate our communication with every section of the land. Our iron pathways are our rivers."[77] The machine enables society to unite and realize its common needs. The featured speaker of the day, Edward Everett, expatiated on this theme, first describing the area between Boston and Canada as a "horrible wilderness, rivers and lakes unspanned by human art, pathless swamps, dismal forests that it made the flesh creep to enter, threaded by nothing more practicable than the Indian's trail, echoing with no sound more inviting than the yell of the wolf and the warwhoop of the savage: these it was that filled the space between us and Canada." Having depicted nature in this rhetoric, he went on triumphantly: "By the magic power of these works of art, the forest is thrown open—the rivers and lakes are bridged—the valleys rise, the mountains bow their everlasting heads; and the Governor-General of Canada takes his breakfast in Montreal, and his dinner in Boston;— reading a newspaper leisurely by the way which was printed a fortnight ago in London." With this framework established, Everett turned to an explicit critique of William Wordsworth, who had opposed the construction of a railway into the Lake District and who had published a sonnet, "On the Projected Kendall and Windermere Railway," beginning with the lines "Is there no nook of English ground secure/From rash assault?"[78] Everett scoffed at such sentiments. While acknowledging that Wordsworth was "a most distinguished poet," he found him "entirely mistaken." "The quiet of a few spots may be disturbed," Everett said, " but a hundred quiet spots are rendered accessible. The bustle of the station house may take the place of the Druidical silence of some shady dell; but Gracious Heavens? Sir, how many of those verdant cathedral arches, entwined by the hand of God in our pathless woods, are opened to the grateful worship of man by these means of communication!"[79] Everett further implied that Wordsworth's position was elitist. The railroad put the seashore and the mountains "within the reach, not of a score of luxurious, sauntering tourists, but of the great mass of the population, who have senses and tastes as keen as the keenest. You throw it open, with all its soothing and humanizing influences, to thousands who, but for your railways and steamers, would have lived and died without having breathed the life-giving air of the mountains." (This argument was reported to have met with

"Immense Cheering.") Everett asserted that there was no contradiction between nature and industrialization. Rather, the railway would make the unspoiled natural world available to everyone. The *New York Daily Times* published a similar argument in 1851, asserting that the railroad made it easier for more people to see the beauty of the Hudson Valley.[80]

In such arguments, nature paradoxically had a "humanizing influence" as a sublime landscape and yet remained distinctly non-human. The wilderness would no longer belong to "savage" Indians, whom it somehow had not humanized; it would belong to what Everett unabashedly described as "this great Anglo Saxon race."[81] In these and other references to race, Everett inadvertently laid bare a fundamental contradiction within the technological sublime. On the one hand, like the natural sublime, it posited the existence of a universal human nature. Because all men and women were presumed to be the same when it came to sense impressions, all could benefit equally from a visit to the mountains or the sea. They would all be uplifted by the "moral influence" of the landscape. What then of those natives who had always lived in the landscape? They were presumed to be aesthetically underdeveloped, blighted beings offering a suggestive contrast to the white man. Native Americans recur persistently in evocations of the railroad in the landscape. Chevalier, for example, described seeing a train after sundown on the Petersburg and Roanoke Railroad in Virginia as follows: "The engine came on at its usual rate of speed through a narrow clearing cut for the road in one of the primitive forests, formerly the domain of the great king Powhatan and his copper-colored warriors."[82] The rhetorical function of these Indians is to suggest the difference between savagery and civilization, and to place every reader in a superior position.

But this superiority was illusory. When nature ceased to be the only source of sublimity, the technician became a creator of experiences. Instead of searching for fundamental relations between man and nature, the inventor found ways to dominate and control nature. Implicit in this shift was a diminished role for the farmer-citizen in the new industrial polity. This marginalization was apparent in the landscape, which was no longer dominated by the abstract geometry of the national survey—transportation lines had reshaped the contours of the land and created new nodal points. Cities, which Jefferson had hoped would never dominate America, mushroomed wherever rivers and railways met. Steamship routes and railway tracks created new commercial value in land. At every level the technological sublime displaced agrarianism.

As Americans began their long romance with the railroad, they celebrated its advance across the continent in spectacles and parades, many of them on Independence Day. Besides demonstrating the relative cultural power of the various groups that took part, these spectacles suggest, through the size and composition of the audience, the degree of public interest. The parades not only mirrored the society; they also built social relations and expressed the self-perceptions of various groups.[83] They were dramatic enactments that created a bond of shared perceptions among the participants. As early as the fifteenth century the architect and poet Leone Battista Alberti recognized the community-building function of civic events. In his *Ten Books of Architecture* he discusses "Public shows," which he believed served "wonderfully to revive and keep up the Vigour and Fire of the Mind," encouraging participants to "grow more humane, and be the closer linked in Friendship one with another."[84] "Public shows" do not merely reproduce the social order; they animate and revive it. This was particularly the case with a ceremony celebrating the opening of a technological achievement, for it marked a moment of dramatic transition. Rather than preserve an existing order, the advent of a new technology heralded social transformations.

Celebrations of major technological projects changed in form over time. Workers were prominent at first, but in later years they were marginalized. When the Erie Canal was opened in 1825, Alan Trachtenberg notes, "firemen, carpenters, millwrights, merchants, militiamen, cabinetmakers, and so on, all subscribed as groups to the gala celebration, they built floats for the parade, and marched together." In such events, "civic identity was expressed through the participating voluntary organizations."[85] Likewise, workers actively took part in the Baltimore and Ohio celebration of 1828, swelling the parade into an immense procession of more than 5000, divided into four divisions. The first contained state and national dignitaries in carriages, Masonic fraternities, and the directors and the engineers of the railroad. Significantly, even at this early date the engineers walked with the directors and not with workingmen such as the steam engine makers. The second and by far the largest division comprised workingmen's associations, in the following order: groups of farmers, gardeners, plough makers, millers, bakers, victuallers, brewers, tailors, blacksmiths, steam engine makers, weavers and dyers, carpenters and joiners, stone cutters, masons and brick-layers, painters, glaziers, and twenty more. Note that unskilled workers were omitted. Third came juvenile associations, and last came the mayor and other local officials. These groups of working-men and artisans did not merely march; they also demonstrated their crafts as they moved along the streets. The printers, aboard their car, printed an ode celebrating the day; the farmers threshed grain on one float and milked a cow on another; one of the painters completed a portrait while borne along on a car; the machine makers operated a lathe. The sumptuously dressed tailors made a coat, which they presented to railroad director Charles Carroll later in the day. The carpenters made a scale model of a Greek temple. In short, the workingmen were not merely present in the parade; they demonstrated their skills before the public, appearing before them as large groups, with their own distinct dress, badges, and insignia.[86] It was a parade that enacted social identity and announced the composition of significant actors in the formation of the body politic.

The history of parades and inaugurations traces the devolution of workers' social standing. Susan G. Davis emphasizes in *Parades and Power* that in the early nineteenth century tradesmen and arti-sans as well as elected officials took prominent parts in celebrating the Fourth of July, but that the working poor increasingly held sepa-rate ceremonies, marking a breakdown in the sense of community. Not only did working people begin to disappear from parades; their dramatic roles changed as well. When 75,000 Whigs gathered for a great public celebration in Baltimore in 1844, the workingmen in the procession appeared less as independent artisans than as dependent wage earners.[87] By the Great Railway Celebration of 1857, many working groups were absent, owing to bad labor rela-tions with the Baltimore and Ohio Railroad.[88] The declension con-tinued after the Civil War. By the last third of the nineteenth centu-ry, civic identity was expressed through a smaller contingent of elected officials, entrepreneurs, builders, and engineers. In effect, the workingmen, who constituted most of the 5000 members of the 1828 procession, had been eliminated. As the technological sub-lime reached its apotheosis and cast up the engineer as the heroic genius responsible for its creation, workers were gradually eliminat-ed from the rituals celebrating industrial achievements. At the opening ceremony for the Brooklyn Bridge in 1883, "the only visi-ble institutions were the militia and government; the only official parade was an escort for President Arthur from his hotel to the bridge."[89] In contrast to New York in 1825 or Baltimore in 1828, only the first and last sections of the parade remained. Diminished worker presence in the railroad jubilee at Boston in 1851 and that at Baltimore in 1857 adumbrated their total exclusion by 1883.

Defined out of the procession, workers held their own Fourth of July celebrations, often demonstrations of ethnic solidarity. In 1829 the Philadelphia *Mechanics' Free Press* noted: "No class of the community has so generally and constantly manifested a sense of hilarity on the Fourth of July as the Working People. While more of the wealthy classes have gradually withdrawn themselves from all public display on this national holiday as ungenteel, the toil worn artisan has continued to set it apart with mirth and jollity." These events contested the power of civic elites, at times in burlesques of the fine clothing of official parades. Davis notes that "the parades of the working poor challenged and denied respectability and the social order based on property."[90] Scoffing humor diminished the social distance between workers and their "betters." It is essential to note that the use of sublime technological display could virtually never play a part in such a parade or demonstration. The sublime use of technology was, rather, part of a formal parade of politicians, military heroes, and national leaders.

The deemphasis of the working class can be clearly seen in the Boston Railroad Jubilee of 1851, a three-day hymn to human industry. The occasion was the completion of the new rail line to Canada, which opened the prospect of increased trade (particularly during the winter months, when the St. Lawrence River was impassable). New England was understood to be a region that had been improved far beyond its natural state by industrialization. In 1849 the mayor of Boston declared:

The sterile soil, the rugged surface, the stern climate, and the want of navigable streams in New England, would have seemed to render it improbable that it would ever be considerably peopled or that any great commercial mart should arise within its borders. . . . But the resolution and intelligent industry of our fathers surmounted every obstacle. . . . What may we not expect of the future destiny of Boston now that her iron highways, extending in all directions, bring her into convenient proximity with every section of the land?. . . . The long winter of New England isolation is broken; —she now warms and flourishes in friendly and thrifty intercourse with the luxuriant West.[91]

As Baltimoreans had in 1828, Bostonians in 1851 saw the railroad as the harbinger of prosperity and the enemy of sectionalism. They lost no opportunity to connect it with patriotic sentiments. President Millard Fillmore was received on the outskirts, "on the spot where our fathers gathered to hail the coming of Washington." He was then escorted past 20,000 people to the center of the city. To emphasize the connection to the nation's founding, the reception had been scheduled on "the anniversary of the completion of the Federal Constitution."[92] President Fillmore noted in his address that General Washington had traveled to Boston to take command of the revolutionary army. "Why is it," he asked, "that the distance which it took him eleven days to travel over, and that, too, when a most critical state of affairs called for the utmost speed, has now been passed over by me, as a matter of pleasure, in almost as many hours?" The answer to this rhetorical question was obvious: internal improvements had knitted the union closer together. This theme would be repeated everywhere. A large banner stretched across the street along the parade route bore the single word 'Union', and private homes and shops all along the route displayed the flag.

At the State House a vast crowd overflowed the galleries to welcome the president and to hear Secretary of State Daniel Webster. He did not disappoint them; there was continual applause. At the heart of his speech, Webster declared: " . . . the bitterest, ablest, and most anti-American press in all Europe, within a fortnight has stated that, 'in everything valuable, in everything that is for human improvement, exhibited at the World's Fair, the United States go so far ahead of every body else as to leave nobody else in sight.' "[93] With this reference to the Crystal Palace Exposition, the first world's fair, Webster placed American technological achievements in an international framework and identified them as emblems of national identity. The remaining speeches of the day did not take up this theme, however, they repeatedly addressed the sectional crisis. Speaker after speaker declared the indivisibility of the union—a sentiment that was particularly applauded when expressed by visiting southerners, such as Secretary of the Interior Alexander Stuart.

The second day of the jubilee was devoted to an excursion and boat races in Boston Harbor and to the reception of the governor general of British North America, with fireworks in the evening. Friday, September 19, the third and final day of the jubilee, was a public holiday. The streets were decked with innumerable flags, primarily American but also many British and some French.

Intermingled with them were patriotic banners welcoming the president and the governor general and reiterating the themes of peace, union, and prosperity.[94] The city was packed with people from dawn to midnight. The procession was 3 miles long and took 2 hours to pass. It had eleven divisions, including a vast number of local and state elected officials, members of the Massachusetts legislature, Canadian officials, invited guests, Army and Navy officers, fraternal associations, 1000 members of the Mercantile Library Association, Harvard University students, and a contingent of school children in 32 cars decorated to represent the four seasons.

"Trades and mechanical pursuits" were in the seventh division, which had no fewer than 50 sections. As in Baltimore, some of the artisans (including hatters, turners, granite cutters, and printers) were engaged in their vocations as they rode through the crowds, though actual production was less in evidence than in 1828. In addition, many of the displays were of luxury goods, such as ornamental furniture and silver tea services. Others functioned as advertisements, such as the gigantic painting that promoted a newspaper and the Boston Museum's carload of novelty items. Another significant difference between the Baltimore and Boston parades was the absence in Boston of bakers, victualers, and bricklayers and the lesser importance given to masons, carpenters, and house painters. Whereas ordinary trades had been deemphasized, considerable space was given to specialized trades—copper plate printers, school furniture makers, fire engine manufacturers, locomotive works, a foundry, and so on. The Boston and Chelsea Paper Hangings Manufacturing Company had three cars, the first displaying a printing machine operated by "men and boys in the act of printing paper, which, as soon as dry, was rolled up by another machine, managed by women."[95] The second car carried a large variety of paper, and the third represented a parlor which a paperhanger and his apprentice were decorating. In such displays one can unmistakably see the emergence of a new industrial order in which artisans were becoming wage earners in factories, producing goods that they could not always afford for themselves.

Yet perhaps the most telling feature of this display of work in Boston is its lack of internal coherence. In the 1828 Baltimore parade, the trades had been deployed as a coherent system of occupations: painters marched next to glaziers, tailors next to cloth makers, and so forth. In the Boston parade this image of organic relation was shattered; positions in the procession were determined by lot. This procedure resulted in meaningless sequences, such as this: sewing machines/fireworks/wooden wares/railway cars and this: plumbers/carpet makers/occupants of Faneuil Hall Market/paper hangers.[96] The parade could no longer be "read" as a coherent representation of labor. Instead, it presented a jumble of specialized activities, each pursued for profit only, with no connection to other trades. In this shift, labor ceased to be a calling and became a commodity.

In his later years Emerson observed that "a scientific engineer, with instruments and steam, is worth many hundred men, many thousands."[97] This remark suggests the differences between the Baltimore celebration of 1828 and the Boston Railroad Jubilee of 1851. In the Boston procession highly skilled workers, politicians, and military figures bulked larger than before while basic tradesmen were displaced by specialists. Less actual production took place on the floats, luxury goods began to take precedence over ordinary goods, and independent artisans were replaced by factory workers tending machines. Symptomatically, the random order in which the trades appeared announced the failure to conceive of work as an integral part of the community. In practice, technical advances undercut the rhetoric of the technological sublime, which asserted an organic unity between the machine and nature. Parades that celebrated technological advancement increasingly deleted from the script the skilled worker, who was expected to join the other spectators: the unskilled, women, and perhaps an occasional Native American or African-American. Those viewing a majestic parade were not likely to think much about the groups who did not march in it. Technological achievements were becoming part of a hegemonic system in which the nation merged with its most impressive technological and natural works.

Although the first railways may have been greeted enthusiastically by artisans, they did not prove to be egalitarian employers. In their reliance on a hierarchical command structure and a centralized authority inspired by the military they provided models for later corporations. As geographically extended enterprises, railroads were controlled by telegraphic communication to central offices, and they did not tolerate the self-directed artisan who wished to control the tempo of his own work.[98] Wage cuts during economic downturns, loss of control in the workplace, and frequent accidents soured labor-management relations. According to Walter Licht, "pioneer railwaymen protested in concert; they received support from other members of their communities [, and] destruction of private property was a tactic frequently employed."[99] In the 1850s conflicts and strikes became frequent and railway promoters became known as notorious Washington lobbyists who obtained land grants and federal subsidies on a massive scale.[100] Nor did railroads unite the regions, as had been hoped. Instead, they accelerated industrialization in the Northeast and helped make the South a dependent, agricultural region. Just as the unity of the trades had been replaced by specialization, the railroad had encouraged regional economic specialization. The South's cash crops—tobacco, sugar, cotton—were sold to mills in the North. The national unity so fervently hoped for in the republican rhetoric of public virtue was gradually redefined as economic unity animated by self-interest. The railroad had not brought prosperity to all classes, nor had it been equally beneficial to all regions.

Such changes were, however, difficult to visualize, whereas a direct encounter with the technological object had sensory power. The essence of America seemed to emerge as railways triumphed over every obstacle. In his last years Whitman observed: "I am not sure but the most typical and representative things in the United States are what are involved in the vast network of Interstate Railroad Lines—our Electric Telegraphs—our Mails (post-office)—and the whole of the mighty, ceaseless, complicated (and quite perfect already, tremendous as they are) systems of transportation everywhere of passengers and intelligence. No works, no painting, can too strongly depict the fullness and grandeur of these—the smallest minutiae attended to, and in their totality incomparably magnificent."[101] Americans found the railway thrilling, and proved passionately fond of everything connected to it. In the second half of the nineteenth century, John Stilgoe notes, "the train, and particularly the fast express, struck few observers as a monstrous machine soiling a virginal garden. Instead it seemed a powerful romantic creature inhabiting an environment created especially for it."[102] Public attention focused not on the conditions of labor but on the excitement of technological experience. The early tourists who rode the gravity railroad at Mauch Chunk apparently took little interest in the coal mines that had built the railroad, nor did they inquire into the lives and working conditions of the miners. Likewise, few who rode the railroad concerned themselves with the human cost of building it, or inquired deeply into the social effects of railroads beyond their immediate promise of prosperity and speed. Travelers learned to focus on the immediate experience of seeing the mechanical perfection and the power of the locomotives. They enjoyed riding in new forms of transportation and seeing new landscapes, concentrating on novel physical sensations and new vistas. These dislocations of sensibility, in turn, became talismanic moments of the dynamic sublime, distilling the railroad's meaning. (A similar process recurred in the later responses to the automobile, the airplane, and the rocket.)

For a century after the first appearance of steam trains the railroad remained a central part of the technological sublime, celebrated in songs and stories and manifested in enormous terminals, roundhouses, and freightyards, longer tunnels and bridges, higher speeds, and more luxurious cars. Perhaps the high point of this romance occurred with the building of the first transcontinental railroad. The idea of a line from the Atlantic to the Pacific was almost as old as the railroad itself, having been proposed in Congress in the 1840s and crystallized into detailed plans by the 1850s. Yet the project seemed so formidable that few expected to see the continent spanned in the next decade, and initially no private investors saw profits in the largely barren, dry, and unpopulated Rocky Mountain region. The Baltimore and Ohio had required a quarter century to reach Wheeling on the Ohio River, a much shorter distance through less formidable and more densely populated country.

Yet the transcontinental railroad suddenly surged across the continent. In 1865, with land grants from Congress, railroads in California and Nebraska started to build toward each other, each trying to maximize its share of the federal subsidies given for every mile of track laid. Rapid progress provided newspapers with a riveting story of hardship, ingenuity, and engineering skill. By 1866 reporters spoke of the railroad as "the greatest undertaking of Western civilization," linking New York not only with San Francisco but also with the Far East.[103] Whitman later spiritualized that theme in "Passage to India," presenting the railroad as the fulfillment of Columbus's dream. He called 1869 the "year of the marriage of continents, climates and oceans."[104] The climax to 4 years' frantic work came at Promontory Point, Utah, on May 10, 1869, when the first transcontinental line was completed with the driving of the "golden spike." A. J. Russell's carefully composed photograph showed the meeting of a locomotive from the east and one from the west, with workmen and representatives from each line clasping hands in the center. Above the workers, posed on the cowcatchers of the two locomotives, one man stretched to pour Central Pacific champagne into a waiting Union Pacific glass.[105] Omitted from this image were the Chinese laborers who had chiseled and blasted a roadbed across the Sierras at wages of less than $2 a day and who had set a record by laying 10 miles of track in 12 hours; however, some of the Irish workers who had laid track across Nebraska and Wyoming into Utah were included.[106] The photograph also could not show that many of the workers had not been paid in weeks. A few days earlier, 300 men had blocked the line at Piedmont Station to prevent the vice-president of the Union Pacific from reaching the ceremony until they received back wages.[107]

Russell's photograph was first widely seen as a wood engraving a month later, in *Frank Leslie's Illustrated Newspaper*.[108] However, the achievement itself was immediately made known by telegraph. A wire attached to the "golden spike" permitted crowds around the country to "hear" each stroke of the hammer. The last blow sent the

A. J. Russell's photograph of the "golden spike" ceremony. Courtesy of Collections of Library of Congress.

news instantaneously across the nation and triggered cannonades in Sacramento, San Francisco, and New York. Church bells rang in virtually every city. Three days of largely spontaneous jubilation began in San Francisco. The Chicago celebration "was entirely impromptu" and "almost every man, woman, and child in the city did their part" as an enormous procession formed, "unique in appearance and immense in length, which at the lowest estimate was seven miles."[109] Vice-President Schuyler Colfax and other dignitaries "addressed an overflow crowd at Library Hall, in which they spoke eloquently of the great era which this day marks in the history of our country."[110] The news likewise brought huge crowds into the streets of Buffalo, St. Louis, and other western cities. The driving of the golden spike sparked a celebration of the technological sublime from coast to coast, linking the nation together not only in

Preparing the telegraph line for the golden spike. Courtesy of Collections of Library of Congress.

space but also in time. Durkheim argued that in such "great movements of enthusiasm" the feelings expressed "do not originate in any one of the particular individual consciousnesses" but "come to each one of us from without and can carry us away in spite of ourselves."[111] In 1869 each local crowd was conscious of being part of a larger celebration, stretching from coast to coast. People literally felt united by the railway. A three-column front-page story in the May 11 New *York Times* noted that in Philadelphia "the bells on Independence Hall were rung, drawing a crowd into the streets, thinking a general alarm or fire was being rung. The people soon ascertained the reason of the ringing of the bells, and flags were immediately hoisted everywhere. A large number of steam fire engines were ranged in front of Independence Hall, with screeching whistles, hose carriages, bells ringing, etc. Joy was expressed in every face. . . ."

In a telegram to San Francisco, the mayor of New York compared the transcontinental line with the "completion of the Erie Canal," which had "riveted the silver chain of western inland seas" to the eastern seaboard half a century before. On Wall Street and in Printing House Square a holiday mood prevailed. "Flags were displayed on City Hall, on all the newspaper offices, and on the prominent hotels. Every countenance seemed to bear a look of supreme satisfaction," and once the spike had been driven "there was booming of cannon, peals from Trinity chimes, and general rejoicing."[112] A special service at Trinity Church was fully reported in the city papers, including long excerpts from the sermon. Preaching on Zachariah 14:9—"And the Lord shall be king over all the earth, in that day there shall be one Lord, and his name one"—Dr. Vinton counted the blessings secured by the new railroad. It would stimulate commerce, populate an empty region, and unite a far-flung nation. Like the ancient Roman roads, used by the apostles to spread the gospel, the railway was "a means, under Divine Providence, for propagating the Church and the Gospel from this, the youngest Christian nation, to the oldest lands in the orient, now sunk in Paganism and idolatry."[113] Church, government, and business reiterated the now-established catechism of railways. The New York Chamber of Commerce telegraphed its counterpart in San Francisco "in grateful thanksgiving to Almighty God, the Supreme Ruler of the Universe, on the completion of the continental line of railway." The telegraph went on: "The new highway thus opened to man will not only develop the resources, extend the commerce,

increase the power, exalt the dignity and perpetuate the unity of
our Republic, but in its broader relations, as the segment of a
world-embracing circle, directly connecting the nations of Europe
with those of Asia, will materially facilitate the enlightened and
advancing civilization of our age." The unity of man and nature,
assumed in the first years of the nineteenth century, had been redefined as technological mastery that marched in hand with religion.
Yet most people comprehended these changes as extensions of natural forces. Emerson wrote: "The chain of Western railroads from
Chicago to the Pacific has planted cities and civilization in less time
than it costs to bring an orchard into bearing."[114] The metaphors of
planting and cultivation still seemed appropriate for the railroad.

In 1828 there was no steam railroad anywhere in the United
States, and the telegraph had not been invented. Four decades later
the trip from Baltimore to San Francisco required only 7 days. It
opened the wonders of Yosemite and the great sequoias to tourism,
and it spurred economic development. As the Baltimore and Ohio
had done before the Civil War, the Union Pacific commissioned
painters and photographers, including Thomas Moran and Albert
Bierstadt, to depict the scenery along its routes.[115] Less than a year
after the line opened, Russell's photographs were being projected
as lantern slides by lecturers on the lyceum circuit[116] and reproduced in national magazines. As Susan Danly notes, the natural sublime intertwined with technological conquest, creating tensions
within the imagery: ". . . the vastness of the American West literally
put into perspective by the railroad track; the geometry of new
cities juxtaposed with the panoramic vista of mountains; man
dwarfed by the monuments of nature, while technology stands triumphant in the wilderness."[117] For nineteenth-century Americans
these were not ironic contradictions but dramatic contrasts; the
rugged western landscape and the transcontinental railroad were
complementary forms of the sublime that dramatized an unfolding
national destiny.

Bridges and Skyscrapers: The Geometrical Sublime

The dynamic technological sublime was embodied in the telegraph, the steamboat, and the railroad, which conquered space and time. Equally important was the conquest of natural obstacles and forces, accomplished by bridges and skyscrapers. These structures were assimilated into a new version of the mathematical sublime that began to emerge as the public attempted to explain the feelings induced by seeing a vast panorama of man-made objects.

This new form, the geometrical sublime, had to do with triumphs over nature more emphatic than those of the antebellum period. Whereas the dynamic form of the technological sublime had emphasized the movement of information over wires and railways across the natural landscape, transforming it into a mere backdrop, the geometrical sublime was static and appeared to dominate nature through elegant design and sheer bulk. It found expression first in bridges and soon afterward in skyscrapers. All these structures expressed the triumph of reason in concrete form, proving that the world was becoming, in Emerson's words, "a realized will"—"the double of man."[1]

Although public enthusiasm for railways remained high after the Civil War, by that time they had lost their novelty, and their financing seemed more problematic. Indeed, the rise of corporate control over what had been seen as public works was a lively and acrimonious issue in politics as early as the 1840s.[2] And the desirability of working for the railroads, once deemed unquestionable, now seemed doubtful, at least to the navvies who laid the rails and the thousands of railway employees who often went on strike.

In 1877, in response to wage cuts, striking railway workers paralyzed traffic in most of the United States. This industrial action began on the Baltimore and Ohio but quickly spread to other lines. State and national troops were sent in to quell the disturbances by firing on the crowds and taking prisoners. As the railroad lost its novelty and became a symbol of class divisions, Americans looked elsewhere for examples of technological achievement. In the decades immediately after the Civil War, bridges proved particularly attractive. In a sense, this was only a shift in interest from one part of the railroad to another, since many of the important bridges built in America stemmed directly from railroad development.

In the early years of the railroads a few impressive stone bridges were built on the Baltimore and Ohio line to Washington, but in general railroads found stonework too costly and too time-consuming.[3] Of the exceptions to this generalization, none was more striking than the Great Stone Bridge that carried the Great Northern Railway into Minneapolis over 23 arches, each almost 100 feet high, which described a curve as they marched across the Mississippi River.[4] Although such structures had considerable appeal as symbols of progress, they were built using Roman techniques and materials. The technological sublime required novelty.

The suspension bridge was first patented in the United States in 1796. By 1810 about 50 such bridges had been built, using hand-forged iron chains. The largest of the early suspension bridges was a 306-foot span across the Schuylkill River in Philadelphia.[5] In the 1830s, John A. Roebling, an engineering graduate of the University of Berlin who had emigrated to the United States, developed wire rope, which proved stronger, more flexible, and more consistent than chains. Roebling had developed this wire rope for pulling canal boats, but he soon adapted it for use in a bridge over the Allegheny River in Pittsburgh. The success of that bridge led to Roebling's design for the Niagara Suspension Bridge (1855), which had a main span of 821 feet and which was the first suspension bridge in the world to carry a railroad. Roebling consciously planned that it should have "a very graceful, simple, but at the same time substantial appearance." Held up by "four massive cables" carried by large supporting columns, it was intended to be "unique and striking in its effect and quite in keeping with the surrounding scenery."[6] It became almost as great a tourist attraction as the falls themselves, and a favorite subject of painters and lithographers.[7] By the middle of the nineteenth century this vision of the man-made bestriding nature had a powerful hold on the imagination. Roebling was soon given commissions for other impressive bridges. By 1860 he had completed the two-span Allegheny River Bridge in Pittsburgh. His 1058-foot Cincinnati Bridge over the Ohio River was regarded as so important to the future of Cincinnati that work on it continued in 1864 and 1865, when the Civil War had halted most large construction projects. The longest bridge in the world when put in service in 1867, it aroused great admiration.[8]

Though John Roebling was the most prominent American bridge builder and the acknowledged master of the suspension form, thousands of other bridges were built in America during the nineteenth century in many different forms and styles. Probably the most important of these was the Eads Bridge in St. Louis, which became a national symbol as it was built across the Mississippi in the years immediately after the Civil War. Like many of the railways, this bridge was considered vital to the continuing economic health of a city. St. Louis had rapidly lost ground to Chicago during the Civil War, when the Mississippi was closed to navigation. Chicago already had spanned the river and had become the most important railway center in the country. Construction of a bridge, long discussed, had become absolutely essential to postwar economic recovery in St. Louis. Yet the Mississippi River below its junctions with the Missouri and the Ohio was a daunting adversary for any builder. A flood could quadruple its normal rate of flow of 1.6 million gallons per second. The extreme variation between low and high water was over 40 feet, and a 3-foot-deep current of sand at the bottom hindered construction. Wind stress was also a serious problem—the region was subject to violent thunderstorms and tornadoes. Worst of all, the bedrock lay well below the surface, particularly on the eastern side.[9]

The bridge that Colonel James Eads designed to meet these challenges was a bold break with previous designs. Its construction would be followed avidly in the press, becoming a national story. The first large cantilever bridge, and one of the first such bridges to be built with steel, it would require the construction of two huge piers in the unruly river. It was to have only three spans, each over 500 feet in length—more than 200 feet longer than any previously built. Even more controversial was Eads' decision to use tubular steel, then an untried material. The bridge's ambitious size and design and the proposed use of new materials all drew attention to it. Many builders—including 27 consulting engineers employed by a rival company intent on constructing a different bridge from the Illinois side—declared that the structure would collapse. (One in four of the iron bridges constructed during this time did indeed fall down.) Nevertheless, Eads overcame his critics and set to work in August 1867.

The first problem was to build the piers. A coffer dam was sufficient to sink them down to bedrock on the western side, where it
was not far below the mud; but on the eastern side the bedrock was
about 100 feet lower. Eads soon learned of a new device, the pneumatic caisson, that required high atmospheric pressure. As David
Plowdon neatly summarizes: "A pneumatic caisson is essentially a
box, open on the bottom, upon which masonry for the bridge towers or piers, is built. It is filled with compressed air of a high
enough atmospheric pressure to counterbalance the water pressure
and to enable men to work inside it directly on the bottom of the
river. As the material is excavated and more and more masonry is
built on top of the caisson, it sinks further and further into the
riverbed."[10] The danger of this new technique for workers was that
when oxygen enters the blood at a high atmospheric pressure it
must exit slowly at a gradually decreasing pressure. When a worker
suddenly went from the bottom of the river to the surface, the gases
in his blood expanded rapidly and he suffered from terrible muscu-
lar pains, called "the bends." This then-inexplicable illness crippled
many caisson workers and killed 14.

Once the piers were completed, arches of hollow steel tubing
were erected, beginning from each pier and working outward in
two directions at once to maintain a balanced load. As a contemporary writer explained, "the weight of one side act[ed] as a counter-
poise for the construction on the other side of the pier. They were
thus gradually and systematically projected over the river, without
support from below, til they met at the middle of the span, when
the last central connecting tube was put in place by an ingenious
mechanical arrangement, and the arch became self-supporting."[11]
The bridge seemed deceptively low because of the great length of
the spans, but it was high enough to permit steamboats to pass
underneath. It carried two railway tracks and a roadway above them,
and it was constructed so solidly that it still remains in service.

The young architect Louis Sullivan "followed every detail of the
design, every measurement, every operation." As he put it: "Here
was Romance, here again was man, the great adventurer, daring to
think, daring to do. Here again was to be set forth to view man in
his power to create beneficently."[12] And, like Sullivan, the public as
a whole felt that the bridge represented an epic confrontation
between man and a powerful natural obstacle. The bridge was a
tourist site even before its completion. Many ventured down the
stairways of the piers, briefly savoring the danger of going below

The Eads Bridge under construction.

the water level to observe the workmen sinking the caissons to the bedrock. The excitement reached its climax in a series of events that marked the completion of the bridge. On May 24, 1874, over 15,000 pedestrians came to the western side of the bridge and paid admission to walk across and back. Then it was tested with progressively heavier loads, including the first train on June 9. General William Tecumseh Sherman drove the last spike in the line that linked St. Louis to the East. On July 2 much more weight was placed on the bridge in a public test. A contemporary noted that "thousands of citizens flocked to witness the interesting spectacle," and they filled the upper roadway of the bridge. In addition, "the levee north and south was crowded with people" waiting to see if the bridge would hold. "Fourteen locomotives, with their tenders full of coal and water, were loaned for the occasion. All were heavy machines and all were crowded with spectators. Seven were coupled together, and crossed and recrossed the Bridge on each track, stopping on the middle of each span. Then the fourteen locomotives, in two divisions of seven each, moved out on the two tracks, and stopped on the center of each arch, side by side. Finally, the four-teen locomotives, all in a line on one track, slowly moved across the Bridge." During all these demonstrations, the "excitement was intense, as with almost breathless interest they watched the immense load move. . . . To the unskilled it seemed incredible that the slender arches could support such a burden. Men swarmed over the engines, and to the crowd below a thousand lives seemed hanging by delicate threads." As it became clear that the bridge held the fourteen locomotives with ease, "a conviction that the Bridge was immeasurably strong settled into every heart, and a joy-ful shout of triumph went up from the vast throng."[13]

The Eads Bridge was dedicated with much fanfare two days later, on the Fourth of July. The ceremonies included a 15-mile-long parade that retained the structure of a Jacksonian procession and included all the major trades and occupations. One newspaper's list of the marchers required seventeen full columns of type, and it did not pretend to be complete. As at Baltimore in 1828, many work-men displayed their crafts on the horse-drawn floats as they crossed the river to Illinois and returned. One could see plasterers complet-ing a small house, upholsterers covering chairs, printers running off copies of the Declaration of Independence, and stove manufac-turers baking cookies. Many free samples were distributed to the crowds, "including 2000 hand-made tin cups." A whole section of the parade was devoted to the men and machines that had made the bridge, including models of derricks, a model of one of the caissons, a 30-foot model of the western arch, specimens of granite and limestone, and "a huge painting, mounted on a frame on wheels, of the laying of the first stone for the first pier."[14]

The sheer mass of spectators was an even more impressive sight. A new train pulled 15 palace cars carrying 500 dignitaries. "The spectacle of the thronged boats and the crowds which completely covered the banks, lighted up by brilliant colored lights on the Bridge, was exceedingly fine." As a finale, "pyrotechnics were dis-played from the upper road-way, while a vast concourse of people beheld them from the levees and from the decks of a numerous flotilla of steamboats." But some of these fireworks failed to go off, including a "Phantom Train" that was to run the length of the bridge. The fireworks proved inadequate to the magnitude of the object they celebrated, "dwarfed into comparative insignificance by the height and length of the Bridge and by their distance from the spectators." "The effect of the mottoes and other elaborate designs was less impressive than was expected."[15] An object of such magnitude was its own best advertisement. The plaque fastened over one arch read: "The Mississippi, discovered by Marquette, 1673; spanned by Captain Eads, 1874." This pairing summarized the historical narrative that was emerging with the technological sublime: from discovery to conquest, the explorer giving way to the engineer. The ritual observances in St. Louis, like those that celebrated the Erie Canal, embraced the engineer as the prototype of the good citizen.[16]

The completed Eads Bridge became a favorite of printmakers and photographers. Currier and Ives made several images of it, and stereoscopic views were also sold. During a visit, Walt Whitman "haunted the river every night lately, where I could get a look at the bridge by moonlight. It is indeed a structure of perfection and beauty unsurpassable, and I never tire of it. . . . I am out pretty late: it is so fascinating, dreamy."[17] A journalist noted that "in summer the bridge is a breathing place. . . . Lovers and others come out upon it, and sit on seats conveniently left at several points along its extent."[18] Yet, for all its popularity, the bridge did not carry enough traffic in its first year. In April 1875 it went bankrupt.

By the time the Eads Bridge was dedicated, preparations were underway to build the Brooklyn Bridge. Here, as David McCullough chronicles, the Roebling family erected its finest public work, overcoming personal illness, political corruption, "caisson disease," faulty materials, and a host of technical problems.[19] By the time the Brooklyn Bridge was completed, in 1883, it had already captured the public imagination. Photographs taken from its unfinished towers had been circulated widely, and the general population enthusiastically awaited the time when they could walk from one side of the East River to the other.[20] On opening day, sightseers packed the railways, steamers, and roads into the city and both banks of the river and a vast flotilla of ships and small craft. A contingent of battleships had arrived for the occasion, and they were anchored downstream. As usually seems to be the case when a crowd awaits a technological display, the spectators were well-behaved despite being crammed together. Aside from the bridge itself, the spectacle they awaited was the biggest fireworks display yet held in America, designed to overcome the problems encountered in St. Louis. The great piers of the bridge served as launching platforms, and more rockets were set off from boats, from the shore, and even from gas balloons far overhead.

The Brooklyn Bridge.

Earlier in the day only a fraction of the spectators heard the speeches or even had a clear look at President Chester A. Arthur as he walked from the Fifth Avenue Hotel to the bridge, accompanied by a regiment of soldiers, a military band, the governor, the mayor, and a few other officials. They made a short parade in comparison to the 5000 who had marched in Baltimore in 1828, the multitudes who had celebrated the completion of the Erie Canal in 1825, or the St. Louis procession of 1874. Those who had worked on the bridge, those who had supplied its materials, and even those who had supervised its construction were not present. At the dedication ceremonies, the speakers agreed that the Brooklyn Bridge signified a tremendous human achievement. As Mayor Seth Low of Brooklyn put it: "The impression upon the visitor is one of astonishment that grows with every visit. No one who has been upon it can ever forget it. This great structure cannot be confined to the limits of local pride. The glory of it belongs to the race. Not one shall see it and not feel prouder to be a man. . . . It is distinctly an American triumph. American genius designed it, American skill built it, and American workshops made it."[21] The new bridge seemed to prove the superiority of American civilization over all that had gone before. Another orator of the day declared the bridge was "a trophy of triumph over an obstacle of Nature."[22]

Abram Hewitt, the mayor of New York, contrasted the pyramids and the new bridge, noting that the workers of ancient Egypt had received the equivalent of two cents a day whereas those who worked under Roebling got $2.50 a day. Moreover, the Brooklyn Bridge was useful; the pyramids were not. Furthermore, the bridge was "built for peace, not war; for free commerce between cities, not for celebration of the dead." It "broke down barriers, and thus stood for 'the solidarity of the human race.'" And it was "a lesson to those few in the nation who would obstruct free trade with the artificial barriers of high tariffs." Most important of all, Hewitt declared, the bridge had helped to reveal the corruption of the Tweed Ring, which had stolen upwards of $20 million of public funds during the years of its construction.[23]

While these matters were all concerns of the day, and undoubtedly formed a part of the Brooklyn Bridge's meaning for those who saw it in 1883, over time it has acquired a wide range of symbolic associations. It has been painted, written about, and celebrated in museum retrospectives.[24] Why is this bridge capable of calling forth so many literary and artistic interpretations? It remains a sublime piece of architecture, as may be established by the test Longinus proposed for the sublime. Astonishment seems to grow with subsequent visits, and the consensus of people from different backgrounds ever since its dedication has been that it is a moving spectacle. Lewis Kachur has noted that critics have repeatedly praised its union of monumental stone piers and ethereal spun-steel cables.[25] Yet the appreciation has gone well beyond such observations, registering that powerful surge of emotions that is the hallmark of the sublime. Joseph Stella, who painted and sketched the bridge many times, found himself drawn to it when he returned to the United States from Europe just after World War I. "Many nights," he recalled, "I stood on the bridge. . . . I felt deeply moved, as if on the threshold of a new religion or in the presence of a new DIVINITY."[26]

Granting that a wide spectrum of people find the bridge sublime, it remains to specify the qualities that create this effect. Hewitt and the other orators on dedication day made little attempt to explain the bridge's architectural attraction, except to dwell on the difficulty of its execution. This is not to be overlooked, of course. As Burke noted: "When any work seems to have required immense force and labour to effect it, the idea is grand. Stonehenge, neither for disposition nor ornament, has any thing admirable; but those huge rude masses of stone, set on end, and piled each on other,

turn the mind on the immense force necessary for such a work."
Likewise, the two massive towers of the Brooklyn Bridge force one
to reflect on the tremendous, coordinated effort necessary to sink
them into the riverbed. The bridge overwhelms the observer not
only with its size but also with the rhythmic repetition of its cables
and with its soaring length, which suggests the infinite. Burke wrote
that "hardly anything can strike the mind with its greatness which
does not make some sort of approach towards infinity," and this the
Brooklyn Bridge does extremely well—especially if the weather happens to be overcast or foggy, so that it becomes difficult to discern
the other end of the span with precision. But size alone does not
explain its effect on the imagination. Burke also noted that architectural designs that are "vast only by their dimensions" are "always
the sign of a common and low imagination." The proportions of
the Brooklyn Bridge—particularly its two large towers and its gently
arching roadway, which rises to meet the downward slope to the
cables—avoid the banality of a merely horizontal span.
Furthermore, there is a pleasing contrast between the wire cables
that carry the roadway and the heavy towers. From a distance the
cables appear light, almost gossamer compared to what they support. Nor should the somber colors of the bridge be overlooked.
Burke argued that dark and "fuscuous" [dingy] colors add to the
sublime effect, while bright or gay colors detract from it.[27]

Like the Natural Bridge, the Brooklyn Bridge is sublime when
seen from many different positions. From below, it offers a variety
of stunning perspectives: it reaches out over smaller buildings into
the air, it hangs high above the ships that pass continually under it,
and from either end it seems to stretch an almost impossible distance across the water, effortlessly carrying a tremendous load of
traffic. Seen from either shore at a distance, it makes all the structures near it seem poorly designed by comparison. One of the best
perspectives is from the promenade on Brooklyn Heights, where
one is elevated enough to see the traffic on the bridge and far
enough removed to measure its monumentality against
Manhattan's skyscrapers. Contemporary architects agree: "Whoever
has walked the bridge at sunset and seen Manhattan catch fire
through an intricate set of abstract lines, knows that Roebling had
created a work of art. The network of cables satisfies the static
requirements of the bridge, but in addition creates a dynamic
vision for the stroller, as moiré's effects enhance the picture in the
eye of the moving viewer."[28] The bridge affords a majestic view of

the skyline and the harbor, and a dizzying look down to the waters of the East River. It is a transitional object not only because it is part of a transportation network but also because it gives the eye new vantage points. The almost endless multiplication of views and perspectives it offers up to contemplation continually insist on the primacy not of nature but of a man-made landscape of steel and stone.

The Eads Bridge and the Brooklyn Bridge pioneered the geometrical sublime, which was still appreciated when San Franciscans celebrated their Golden Gate in 1987. The new perspectives these bridges offered intrigued many artists—including the photographer Walker Evans, who on the basis of his work for the Farm Security Administration and in *Let Us Now Praise Famous Men* might appear to be a critic of technology. Evans' photographs often visualize the meeting point between a local culture and industrial civilization, as in a view of a railway station and tracks cutting through a small town or of a utility pole that both links the local community to society and signifies the town's increasing isolation from nature. But Evans also did a portfolio of images of the Brooklyn Bridge which testify to a fascination with the multiple points of view created by a massive technological object. As Trachtenberg notes, these images are "not merely recognitions of the bridge, but constructive visions, unexpected organizations of objects in space: not simply representations of Brooklyn Bridge, but also of the act of seeing, of the photographic discovery that form follows point of view."[29] Evans not only celebrated the bridge; he also celebrated the camera as an instrument of knowledge capable of expressing the multiplicity of a monumental technological form. This too was in keeping with the geometrical sublime, which breaks down the man-made object into a series of perspectives, forcing the viewer to reconstruct the experience of the object through interior triangulation.

Some cities—Bergen, Budapest, Salzburg, Rio de Janeiro, Montreal—possess natural vantage points that provide panoramic views. But in the United States natural elevations near major cities were comparatively rare, and one could hypothesize that Americans built skyscrapers to compensate for a lack of such advantages. Yet this observation does not account for the rise of skyscrapers in Chicago and New York in the late nineteenth century and their virtual absence from major European capitals such as London and Berlin, both of which were in flat areas and were larger and more densely populated than their American counterparts.[30] Another potential explanation of the prevalence of skyscrapers in the United States is that they were a logical economic development, forced into existence by rapid growth. On this view, Chicago built skyscrapers to solve the problem of limited land in the central city, which was caught between Lake Michigan and the railway yards. Likewise, the land available on Manhattan was limited. This explanation sounds plausible, but it is not as easily substantiated as one might suppose. Mark Girouard concluded, after examining real estate prices in New York and Chicago, that, "the land-shortage theory is superficially attractive. . . . But it does not, in fact, bear looking into."[31] The high costs of building higher were not justified by rents, and many areas of Manhattan and Chicago were comparatively undeveloped when the first skyscrapers went up. There was still abundant space for more modest structures. In short, traditional economic factors do not account for a penchant for tall buildings. London, after all, was much larger than any American city throughout the nineteenth century, but no skyscrapers appeared there, and well into the twentieth century their construction was resisted. One of the first tall buildings in London was an American-owned Hilton Hotel.

While functional explanations are suggestive but not convincing, the availability of new materials and technologies does partially explain the development of skyscrapers at the end of the nineteenth century. Tall buildings entailed high risks for the builders and imposed demanding standards for safety, wind shear, ventilation, and internal transportation. In practical terms, such buildings were not feasible until the last decades of the nineteenth century. Most obviously a safe and reliable elevator was necessary—a condition met in the 1860s, when one came into regular use at New York's Fifth Avenue Hotel.[32] Just as important were techniques of steel construction that reduced the volume of the building needed for supporting walls. The architects of early tall buildings in Chicago, notably Louis Sullivan and William LeBaron Jenney, learned the potentials of structural steel from the Eads Bridge and similar engineering projects.[33] Chicago pioneered the use of steel, and the Chicagoan John Root was one of the first to develop a "floating raft" foundation—"a solid mass of cement webbed with a mesh or grill of steel, giving it very great transverse as well as crushing strength."[34] Yet, as attractive as the technical explanation for the advent of skyscrapers may seem, in practice the first skyscrapers in New York did not use structural steel, which was not approved in the local building code until 1889.[35] The desire to build upward forced a change in the code. It is essential to note, in this connection, that structural steel was cheaper to use than iron, making possible more stories for the same investment. Equally important, new electrical technologies made it easier to light and ventilate large structures. Although European architects had access to the same materials, only in the United States did the tall building become a characteristic feature of the city. In short, technological innovations did not call the skyscraper into being, even if they did facilitate upward expansion.

Economic and technical factors alone are not sufficient to explain the development of the skyscraper. Equally important was a developing popular taste for the geometrical sublime. In New York and Chicago, tall buildings were designed "to make a splash rather than to give the maximum commercial return. They were the headquarters of insurance companies, newspapers, and telegraph companies, which were often in competition with each other, and knew the value of height, splendor and a memorable silhouette in establishing their image or increasing their sales."[36] In short, a fundamental factor was the public enthusiasm for tall buildings. Because many found skyscrapers aesthetically pleasing and impressive, corporate America not only invented them but built more and built higher. Just as an earlier generation had surveyed the surface of the United States into a grid, the skyscraper projected that linearity into the air, rising first in Chicago (where the new form was born in the rebuilding boom after the Great Fire of 1871) and almost immediately thereafter in New York. Indeed, the air rights above lower buildings eventually became a commodity to be bought and sold as builders were forced to comply with city codes restricting the number of tall buildings in a given area.

Skyscrapers had symbolic uses as landmarks and icons of progress. It is hardly accidental that they emerged in part as an outgrowth of competition between newspapers. In 1875 the *New York Tribune* completed the city's tallest structure (260 feet). The *Observer* and the *Herald* almost immediately built tall office buildings of their own. In 1889 Joseph Pulitzer's *World* erected a building of 16 stories, then six more than any other. In the 1890s life insurance companies entered the competition and quickly surpassed the newspapers. These businesses needed to maintain an impressive public image, and they self-consciously erected their tall buildings in order to use them as symbols. But, as Mona Domosh has demonstrated, the builders wanted symbols that could achieve two contradictory goals: to impress the cultural elite with their good taste and to proclaim their importance to the general public. Most skyscrapers succeeded better at impressing the multitude, and "the success of these early tall buildings as attention-catching devices seems to parallel their failure as signs of legitimacy"[37] Banks often built ostentatiously low structures on corner lots, which emphasized that their wealth permitted them conservative horizontality. Other arbiters of cultural taste, such as museums, continued to embrace the Beaux Arts style. After the initial phase of skyscraper construction, when most buildings were functional, this contradiction between vernacular and highbrow taste encouraged a good deal of eclecticism as architects tried to harmonize traditional styles and the new building form.

The mayor of New York was self-conscious about the contradiction between popular and elite taste when speaking to foreign guests in 1909. He began by candidly admitting that "there are comparatively few buildings in New York City which, when taken by themselves, are not architecturally incorrect; there are only a few buildings that even by a stretch of the imagination can of themselves be called beautiful." Yet, despite the individual inadequacies, he declared the whole to be much more than the sum of its parts: "Take the city altogether, the general effect of the city as a whole, the contrast of its blotches of vivid color, with the bright blue of the sky in the background, and of the waters of the harbor in the foreground, the huge masses of its office buildings, towering peak on peak and pinnacle above pinnacle to the sky, making of lower Manhattan, to the eye at least, a city that is set on a hill, and New York does have a beauty of her own, a beauty that is indescribable, that seizes one's sense of imagination, and holds one in its grip."[38] This statement was made when the Singer Building was the tallest in New York. The Woolworth Building would not be completed for three more years, and the heights reached by the Chrysler and Empire State buildings were scarcely imagined; yet already the mayor had articulated what was to become a standard refrain of praise for the skyline, which seemed to elevate and magnify the metropolis into the biblical "city on a hill." With this reference, the mayor claimed for New York what the early Puritans had wished for their colonies: to be a model for the world. The "huge masses of office buildings" were thus metamorphosed from undistinguished commercial structures into representations of destiny. The skyline became an expression of the national will. Montgomery Schuyler had already expressed a similar sentiment in 1903: "We can imagine quarters and avenues in New York in which a uniform row of sky-scrapers might be not merely inoffensive but sublime."[39]

An unintended outcome of skyscraper building was a new view of Manhattan from the harbor, from the East River, or from the Hudson. By the 1890s, the skyline had emerged as a new visual category. The term 'skyline' was used prominently in an 1896 article in the *New York Journal*, but it can be found in scattered sources more than two decades earlier.[40] By the late 1890s 'skyline' was in general use, and in 1897 Schuyler was among the first critics to admit that the silhouette of Manhattan visible from the surrounding shores had a new symbolic quality that was not a conscious intent of architects but a concatenation of the whole: ". . . it is in the aggregation that the immense impressiveness lies."[41] Schuyler's generation was responding to the creation of an artificial horizon, a completely man-made substitute for the geology of mountains, cliffs, and canyons. New York, Chicago, and later other cities had begun to make their distinctive mark on the sky. Arnold Bennett voiced the typical response to this new landscape: ". . . a great deal of the poetry of New York is due to the sky-scraper. At dusk the effect of the massed sky-scrapers from within, as seen from any high building uptown, is prodigiously beautiful, and it is unique in the cities of this world. The early night effect of the whole town, topped by the aforesaid Metropolitan tower, seen from the New Jersey shore, is stupendous, and resembles some enchanted city of the next world rather than of this."[42] Bennett also clearly understood that "in the skyscraper there is a deeper romanticism than that which disengaged itself from them externally."

This romanticism often began with the architect or engineer. John A. Roebling, a reader of Swedenborg and Hegel, made a series of notes for a never-completed book on "the harmonies of creation" in which he argued that mankind was passing out of the metaphysical period and into the era of a pragmatic search for truth. Men would become "the lords of creation" as they multiplied production to create a new technological Eden. Roebling and other prominent engineers also wrote articles linking engineering with religion. For example, Robert H. Thurston, the first president of the American Society of Mechanical Engineers, wrote in the *North American Review* of the "scientific basis of belief."[43] Thurston and Roebling valued ingenuity, efficiency, and elegance in solving practical problems, but they also inscribed their pragmatic values in a spiritual order, and they often yoked machines to millennial hopes.

As Merritt concluded, such "engineers believed in the interdependence of spiritual values and material development."[44] Likewise, Louis Sullivan, an avid reader of Emerson, sought a way to unite modern steel construction with organic form. During the ascendancy of modernism, critics usually described early skyscraper architects as hard-headed engineers or proto-modernists devoted to Sullivan's dictum that form follows function. However, the recent shift toward eclecticism and post-modernism makes it possible to appreciate these early architects more fully. Sullivan, because of his beautifully written *Autobiography of an Idea*, was always understood as a romantic imbued with transcendentalism. Modernists tended to dismiss his philosophy along with his rich ornamentation as a regrettable lapse or as an anachronistic survival of another era, but Sullivan was not the only early architect to seek an organic form for the tall building or to embrace romantic ideas. Both John Root and the prolific author and architect Claude Bragdon were Swedenborgians, and Hugh Ferriss, whose drawings in *The Metropolis of Tomorrow* influenced a generation of skyscraper architects, had mystical leanings.[45] Rather than view these men as incomplete modernists, Thomas van Leeuwen has shown that they were visionaries seeking an adequate form. Seen in this light, the frequent use of gothic ornamentation, notably in the Woolworth Building and in the Chicago Tribune Building, is not a deviation in a line of development from Sullivan to the glass boxes of high modernism; it is an expression of indigenous transcendentalism and belief in organic form.[46] While Sullivan brilliantly created his own ornamentation, many other architects resorted to gothic detail. The result could be pleasing. As Goldberger writes, the Woolworth Building manages, "thanks to Gilbert's delicate detailing, to give an impression of considerable lightness." Indeed, "the Gothic style's insistent, almost urgent shouts upward seemed particularly appropriate for the skyscraper."[47]

Aside from the romanticism of architects, who tended to see their buildings in harmony with both nature and classical tradition, there was the egotistical romanticism of wealthy individuals, such as F. W. Woolworth and Walter P. Chrysler, who used buildings as personal and corporate symbols.[48] In 1913 Woolworth paid $13.5 million dollars in cash for his 792-foot building, the tallest in the world. Yet, John Nichols notes, "[the] contractor who built the Woolworth Building was deeply upset on his client's behalf because he knew that as an investment it could never make a proper return.

'Then Mr. Woolworth let me into his secret—that there would be an enormous hidden profit outweighing any loss. He confessed that the Woolworth Building was going to be like a giant sign-board to advertise around the world his spreading chain of five-and-ten-cent stores.'"[49] This "giant sign-board" was also lavishly decorated within. The dome of its three-story main foyer was covered with marble and glass mosaics. The building's four Corliss engines provided enough electric power for a city of 50,000. Woolworth's personal office was a copy of the Empire Room of Napoleon's palace in Compiégne, and on the wall hung a portrait of Napoleon copied from the original at Versailles. William Taylor points out that Woolworth was a shrewd publicist who focused on "what the public would perceive at street level" and who "wanted his building to be at once imposing, instructive, and self-aggrandizing."[50]

Woolworth considered himself a captain of commerce and made every effort to turn the building into an advertisement for himself and his stores. He did this most notably at the inauguration of the building. President Woodrow Wilson gave the signal from the

New York viewed from the Woolworth Building. Courtesy of Collections of Library of Congress.

White House for the official opening. When his telegraphic signal was received at 7:21 in the evening, 80,000 lights went on throughout the 55-story structure, and a sizable crowd outside cheered as it flashed out of the twilight. Inside, an orchestra struck up the national anthem for the 900 guests, including 100 U.S. senators and representatives, assembled on the 27th floor for a banquet in honor of the architect.[51] Critics were nearly universal in their praise of the new structure. Schuyler declared the Woolworth Building every bit as good as a gothic cathedral. Indeed, the structure soon came to be popularly known as the "Cathedral of Commerce"—a name Mr. Woolworth gladly adopted. Just as churches had raised the highest towers of the traditional city, each corporation now raised its skyscraper.

A decade later, with this in mind, Walter Chrysler and his architect, William Van Alen, intentionally set out to build the highest building in Manhattan, which by the late 1920s seemed to require 925 feet. But when Van Alen's former partner announced a slightly higher structure, Chrysler decided to add to his building a steel tower consisting of six stainless steel arches that tapered to a slender spire. The redesigned structure, 1048 feet high, was an Art Deco monument embellished with gargoyles based on radiator ornaments from Chrysler's cars. Even this building was New York's tallest for less than a year; the soon-completed Empire State Building was 150 feet higher.

Such buildings became perpetual advertisements for their owners, but they had other less strictly economic functions. They catered to the romanticism of the average viewer, who often had never heard of the owner or the architect but who nevertheless responded to a skyscraper's immediate presence. In the first decade of the twentieth century the Flatiron Building, designed by Daniel Burnham and erected at the intersection of Broadway and 23rd Street in 1902, was for a few years was the tallest building in the city. Its triangular shape, dictated by the narrow sliver of land available, made it appear like a ship plowing through the city, or a sheer wall. From the ground such a skyscraper seemed to rise vertically far more sharply than most mountains and to reach much further into the air than it actually did. The *New York Tribune* noted: "Since the removal last week of the scaffolding . . . there is scarcely an hour when a staring wayfarer doesn't by his example collect a crowd of other staring people. . . . No wonder people stare! A building 307 feet high presenting an edge almost as sharp as the bow of a ship . . .

The Flatiron Building. Courtesy of Collections of Library of Congress.

is well worth looking at."[52] The building was fortunate in its situation. Not only did it stand at the intersection of two wide streets, Broadway and Fifth Avenue, but there was a public park opposite its narrow end, and the view of the building from this point quickly made it a landmark of the city. The Flatiron was frequently photographed from this vantage point, most famously by Edward Steichen and Alfred Stieglitz. The latter wrote that, on a snowy day, "it appeared to be moving toward me like the bow of a monster ocean steamer—a picture of new America still in the making."[53]

Early skyscrapers were designed to be seen from the ground, looking up, and this tradition continued into the 1920s. Gradually, however, architects began to realize the appeal of looking at the city from atop its tallest structures, and observation areas at or near the top became part of the design, most notably in the Empire State Building. Like the Natural Bridge, a tall building could be appreciated from two radically different perspectives. Dean MacCannell has noted that a guidebook description of the view from the Eiffel Tower emphasized "the wonderful quality of seeing actual objects as if they are pictures, maps or panoramas of themselves."[54] Such sites permit the tourist to defamiliarize the world and then play a game of recognition, searching for the known in a new miniaturized version. A variety of aids were devised to assist this game, ranging from maps and diagrams to labels affixed to the windows.

From this new vantage point, one seemed to be master of all one surveyed. The cityscape from a skyscraper became a central part of popular iconography. Millions of postcards were sold of views from the most famous tall buildings in each American city. Artistic photographers such as Alfred Stieglitz explored the new upper world of glass and steel. The penetration of the skyscraper view into popular consciousness was certainly complete by the late 1920s, when a view from a skyscraper window became a recurring motif in magazine advertising. These images often silhouetted an executive against a window, as he looked out over the city; the reader was expected to identify with this captain of commerce. Roland Marchand notes: "The panorama view through the window was always expansive and usually from a considerable height. It was never obstructed by another skyscraper across the street." As these "fantasies of domain" suggested, the skyscraper "had become the artist's shorthand description for the concept 'modern'."[55] And this modernity emphasized how the businessman's gaze dominated the new man-made landscape. In European cities such high vantage points had been restricted to church and king, and in most cities the cathedral had long been the tallest building permitted.

It was not a new technology that created the skyscraper, nor was it impersonal economic forces. Rather, it was the social ascendancy of the businessmen, who employed architects to transform the urban landscape. As they ascended higher into the city, two new visions emerged. The view of the skyline, already noted, was an unintended outcome of the concatenation of cultural forces that called the skyscraper into being. But the new vista of the city glimpsed from the upper floors of these buildings was intentional, and it quickly became an important perquisite for executives. By the 1920s the olympian perspective from their offices was immediately recognized as a visualization of their power. The logic hidden within this representation resembled that of Bentham's panopticon, transforming the city into a site controlled from above and dividing its populace into the majority who scurry along the ground and the few who survey them from above. If the skyline suggested the creation of an artificial nature, the olympian gaze from atop a pinnacle of commerce suggested subjugation as the obverse side of mastery.

Even as the public acquired a taste for the geometrical sublime and celebrated the new skyline as an ornament to the city, some began to ask if there could be too many skyscrapers. They seemed sublime to tourists, architects, and owners, but many city residents ceased to appreciate them as architecture. Skyscrapers were in many ways an anti-social form, if one considered their effects on the amenities of daily life. Most obviously, they concentrated a dense mass of people in certain locations and weakened the sense of neighborhood. The proliferation of tall buildings destroyed the human scale of the city and cut off much of the sunlight, turning the street into a dark canyon subject to winds and downdrafts. An urban transportation system already taxed to the limit could not cope well with the sudden increase in traffic that a skyscraper brought to an area. A city full of skyscrapers seemed massive, cold, crowded, and impersonal, as was suggested by the work of some photographers—notably Berenice Abbott.[56]

As these problems became clear, citizen groups pressured city governments. A conflict developed between proponents of the skyscraper and those who wished to preserve the horizontal city, in which the tallest buildings were little more than five stories high and all spaces were accessible by foot. This vision of the central city remained dominant in much of Europe—Paris, Copenhagen, Amsterdam, London—at least until World War II. Henry James was one spokesman for this genteel vision of the city. *The American Scene* records his return to New York after an absence of many years, during which the skyscrapers had come into being. Though James tried to find redeeming features in the new buildings, he could not. He recognized that foggy weather "worked wonders for the upper reaches of the buildings, round which it drifted and hung very much as about the flanks and summits of emergent mountain-masses," but on the whole he found the new towers "all new and crude and commercial and over-windowed." Collectively, the towers struck him as "some colossal hair-comb turned upward [and] deprived of half its teeth." In Boston he complained "of the detestable 'tall building' again, and of its instant destruction of quality in everything it overtowers" and of "the horrific glazed perpendiculars of the future."[57] Likewise, William Dean Howells condemned the emerging city skyline as "vast bulks and clumsy towers, the masses of those ten-storied edifices which are the necessity of commerce and the despair of art." They were an "agglomeration of warring forms, feebly typifying the ugliness of the warring interests within them."[58] Such criticisms were not confined to a refined minority. Eventually, New York and other cities gave up on the horizontal city, but not without a fight. The horizontal city had been, after all, the ideal of the Chicago World's Fair, which had presented the city of the future as a series of Beaux Arts structures of uniform height, surmounted by a few neoclassical domes.[59] Sullivan and other architects of the new tall buildings discovered to their dismay that the "White City" was popular with the public.

This conflict between genteel and vernacular architects became embroiled in a whole series of economic issues. Girouard notes: "The first skyscraper boom in Chicago lasted only two years. In 1893 buildings over ten stories high were prohibited there by law."[60] The law passed easily enough, because several pressure groups favored it. Property owners in the business area wanted to stop vertical expansion in order to encourage lateral growth, and owners of undeveloped sites in the business area objected to being assessed for taxes on the basis of their 16-story potentiality. Existing skyscrapers wanted to keep a monopoly. The height limit was raised to 260 feet in 1902, but in 1911 it was reduced to 200 feet because of an oversupply of office space. In New York, Schuyler noted similar patterns of opposition and support based on self-interest, and lamented the failure of "well-meant efforts to fix a limit by legislation to the altitudes which are converting the slits of street between them into Cimmerian and wind-swept ravines."[61] Even architects who had designed skyscrapers often felt that they should not exceed a certain limit. For example, Ernest Flagg, architect of the Singer Building, New York's tallest structure in 1908, testified in that year before a committee charged with drawing up new zoning laws. He declared that, in order to secure adequate light, skyscrapers either ought to be restricted to occupying only three-fourths of their building sites or they should not be allowed to rise more than 100 feet above the street. These and other details of his plan would have created a rather uniform skyline of 100-foot structures. Real estate developers, in contrast, favored a completely laissez-faire building code, permitting buildings of any height and size so long as they were safe. At the other extreme, a Committee on Congestion of the Population of New York predicted that because of skyscraper development the city would soon need three layers of roadways and sidewalks, and would remain congested nevertheless.[62]

A similar conflict erupted at hearings on the New York zoning law, passed in 1916, that resulted in the famous regulations regarding set-backs from the street. In part the law was a response to the Equitable Building. Completed the year before, it was a particularly overbearing structure, covering a city block and rising 40 stories straight up from the street. The new rules outlawed such structures and dictated that "no building shall be erected in excess of twice the width of the street, but for each one foot that the building or a portion of it sets back from the street line, four feet shall be added to the height limit."[63] This was a compromise among several parties, with none entirely satisfied by the results. Architects preferred not to have arbitrary restraints, yet recognized the need for regulations. Owners of existing skyscrapers feared that taller buildings might block their view and cut off their light. Developers felt that only the sky was the limit. Public demands for more air and light found some allies among architects and more among property owners concerned to preserve the value of their holdings. The resulting zoning laws required setbacks that guaranteed the public a view of the sky and softened the impact of new skyscrapers on existing properties. In the 1920s Hugh Ferriss argued for broad open spaces between skyscrapers, an idea that was echoed later in the utopian visions of the New York World's Fair of 1939 and by Frank Lloyd Wright. In practice, however, skyscrapers rose higher and were crowded more closely together.

The long struggle to provide zoning for skyscrapers repeated a familiar pattern. As in the case of the railroad, a technological innovation excited the popular imagination at its inception and overcame all opposition. Yet, as had been the case with the railroad, a few even associated the skyscraper with the devil, and saw the skyscrapers of New York and Chicago as cities of the damned. The novelist Henry Blake Fuller depicted the Chicago of skyscrapers as an inferno, and the journalist Henry Saltus similarly thought New York could best be understood by reading Dante.[64] Yet, as with the railroad, such attitudes were short-lived. In 1916 Carl Sandburg declared: "By day the skyscraper looms in the smoke and sun and has a soul. . . . It is the men and women, boys and girls so poured in and out all day that gave the building a soul of dreams and thoughts and memories."[65] In New York the Stieglitz circle embraced the skyscraper as a symbol of modernity. Alvin Langdon Coburn photographed the Singer Building from below and photographed the city from its tower; John Marin painted the Woolworth Building in dancing movement.[66] Within a generation, many considered the tall building to be sublime, particularly seen from the top, just as the railroad was popular when considered from the passenger's point of view. In each case, using a new technology, Americans defamiliarized a known landscape and invested it with new meanings. The geometrical sublime, like the dynamic technological sublime before it, provided an olympian sense of perspective that could be immediately translated into a sense of power over nature. In each case, the human cost of achieving that power was literally invisible to the inhabitants of the new technological structure.

The geometrical sublime came to be a dominant way of seeing and understanding the city after the First World War. During the 1920s a series of exhibitions, magazine articles, displays in department stores, and works of art manifested a public enthusiasm for the new buildings that Merrill Schleier calls "skyscraper mania." The image of the skyscraper as a sublime object, he notes, "was adopted by optimists and pessimists alike. The numerous paintings and photographs of boundless towers rendered from disorienting perspectives were manifestations of the simultaneous amazement and inability to grasp the skyscraper's monumental proportions and symbolic implications."[67] This interest reached its peak during the construction of the Empire State Building, which remained the tallest building in the world for a generation after its completion in 1931. It was constructed in less than 2 years on the site of the old Waldorf-Astoria Hotel, which had been purchased for $16 million and immediately torn down. The whole city followed the work with interest. For most New Yorkers the building was the crowning achievement in the erection of a vertical city, and many observed it closely using telescopes installed in Madison Square Park. From the same kinds of instruments that stood at the Grand Canyon or Niagara Falls, the curious saw the steel skeleton of the building go up in an astonishing 25 weeks during 1930. The walls of the upper floors were put up even faster, averaging one story a day overall, with the highest floors done faster yet: 14 stories of brick laid in only 10 days.[68] The speed was made possible by a new construction technique in which the stone on the outside of the wall did not need to be put up at the same time as the windows. Vertical metal strips "served to divide the window from the wall. To one side of any strip there could go up a column of stone, to the other side a stack of windows with aluminum panels between. No longer did the window have to be fitted to the stone" and the stone could be delivered "from the quarry ready-cut. . . . There was almost such a thing at the Empire State as a factory assembly of standard units."[69] Indeed, much of the handling of the materials was inspired by the assembly line. The builders constructed narrow gauge tracks around the perimeter of each floor. Using a hoist to bring carloads of materials up, they "deposited brick alongside the bricklayers, without having been handled from the time they came into the building until the bricklayer placed them in the wall." The building was erected in 11 months. There was a human cost to this achievement, however. Workers found the pace numbing, and the building contractor suffered, in his own words, "a rather severe nervous breakdown."[70]

The scale of the Empire State Building was so vast that newspaper reports attempted to capture its immensity through a romance of numbers. The *New York Times* recited facts calculated to overwhelm the reader: the building was sheathed in 10 million bricks and "200,000 cubic feet of Indiana limestone"; it contained "2,158,000 feet of rentable area and 37,000,000 cubic feet of space," and it weighed "about 303,000 tons." The builder, Paul Starrett, proudly noted that the building contained 67,000 tons of steel—"the largest order ever awarded for structural steel for building use."[71] These numbers stretched or exceeded the imaginations of most people, who could call to mind no object remotely as large. Such figures, often posted on the walls of observation points, served less to explain than to mystify the building.

For the grand opening, on May 1, 1931, a large crowd assembled outside the Empire State Building at 11:30 A.M. as President Herbert Hoover, in Washington, pressed a button to turn on its lights. As the president's participation in the event suggests, the completion of the building was regarded as a national event. A large radio audience heard Hoover's opponent in the 1928 election campaign, Alfred E. Smith, declare: "The Empire State Building

The Empire State Building from the ground. Photograph by Irving Underhill, 1931; courtesy of Collections of Library of Congress.

stands today as the greatest monument to ingenuity, to skill, to brain power, to muscle power, the tallest thing in the world today produced by the hand of man."[72] While there was no parade to mark the event, civic pride was everywhere apparent, not least in frequent tributes to the workers who had put up the enormous structure. Lewis Hine had made a series of photographs of construction workers in precarious conditions high above the city. His dramatic images appeared in national magazines and were then collected in a book titled *Men at Work*. These photographs, together with extensive radio, newsreel, and newspaper coverage, made the erection of the Empire State Building into a great national event, demonstrating willpower and determination in the face of great obstacles. The heroic difficulty of the task and the speed with which it was accomplished helped establish the sublimity of the building.

To many, the building seemed to prove that Americans would be able to pull themselves out of the lingering depression. An editorial in the *New York Times* declared: "All must feel also that the Empire State Building is a monumental proof of hopefulness. Those who planned and erected it and found the funds for it must have been firm in the belief that the future of New York is assured. . . ."[73] Edmund Wilson was less sanguine, noting in *The New Republic* that this "pile of stone, brick, nickel and steel, the shell of offices, shafts, windows and steps," which had managed to rent out only one-fourth of its offices, was "being advertised as a triumph in the hour when the planless competitive society, the dehumanized urban community, of which it represents the culmination, is bankrupt." Yet, despite these harsh words, Wilson found the building to be "New York's handsomest skyscraper." The building's "towering plinth" appeared graceful and light, "bisque pink" on a warm afternoon and "semi-translucent like a cake of ice" on a winter morning.[74]

On opening day a carnival atmosphere reigned in the area. Clusters of people craned their necks to study the new structure as far away as Bryant Park on 42nd Street. People jostled for a chance to inspect the marble lobby and buy a ticket to the observation platform on the 86th floor. They crammed into the elevators and shot up to the observation deck, to emerge from the relative darkness of the car into another world. The sudden break with the ordinary cityscape known at ground level could hardly be more dramatic.

One man is quoted as saying "Central Park looked like a small pasture dotted with tiny puddles of rainwater," another as remarking that "the Chrysler Building, its spire gilded by the sun, appeared small and like a toy." Official visitors on opening day had a similar reaction as they "viewed Manhattan Island and the metropolitan area from a new pinnacle": "Few failed to exclaim at the smallness of man and his handiwork as seen from this great distance. They saw men and motor cars crossing like insects through the streets; they saw elevated trains that looked like toys."[75] The *New Yorker* noted "Bryant Park is a pancake and the Statue of Liberty something to throw at a cat."[76] People standing on the observation point felt how enormous the building was, and, by extension, felt themselves suddenly enlarged in relation to the rest of the world. They were particularly impressed to see that "the shadow of the building stretched across the East River and into Brooklyn."[77] The sun projected their own shadows just as far. Even in bad weather the building attracted visitors. One reporter for the *New York Sun* found himself unable to see more than twenty stories below. "Street noises came up to him, but his sensation was that of having been wrapped in cotton batting and suspended between earth and sky."[78] The intimation that one has moved into closer contact with heaven was an important aspect of the skyscraper, expressed in its very name.

After almost 6 months of operation, visitors were paying a total of $3100 a day to see the view.[79] The sight became a central part of knowing the city, with a yearly average of approximately 1.4 million visitors.[80] An article devoted exclusively to the new view carried the headline "Other Skyscrapers Are Dwarfed":

A new panorama of the metropolitan district—a vast panorama of shimmering water, tall towers, quiet suburban homes and busy Manhattan streets—was unfolded yesterday to visitors who ascended to the observatory above the eighty-fifth floor of the Empire State Building. From this highest vantage point steamers and tugs which appeared to be little more than rowboats could be seen far up the Hudson and the East River. Down the bay, beyond the Narrows and out to sea, a ship occasionally hove into view or faded in the distance. For miles in every direction the city was spread out before the gaze of the sightseers. . . . In Manhattan the tall buildings which from the streets below appeared as monsters of steel and stone, assumed a less awe-inspiring significance when viewed from above. Fifth Avenue and Broadway were little more than slender black ribbons which had cut their way sharply through masses of vari-colored brick. Among them lilliputian vehicles jockeyed for position, halting or moving forward in groups, often like a processional.[81]

The new vantage point seemed to empower a visitor, inverting the sense of insignificance that skyscrapers could induce when seen from the ground. The observation platform offered a reconception of urban space, looking down on buildings that formally had seemed like monsters, miniaturizing the city, making it into a pattern. From the top floor of a skyscraper the congestion of the streets becomes a fascinating detail; from "a height of more than 1000 feet pedestrians were little more than ants and their movements hardly could be detected."[82]

The vision of humanity as a swarm of insects reappeared continually. In 1905 a journalist wrote of the view from the Flatiron Building that below "are things that you would take for beetles, others that seem to you ants. The beetles are cabs, the ants are [human] beings—primitive but human, hurrying grotesquely over the most expensive spot on earth."[83] The brightness at the top heightened the sense of contrast between the observer and the people below. The perpetual twilight in that nether realm became a pleasing contrast to the powerful light at the top of the city, which was literally removed from the dirt, din, and darkness of the surface; the people below became tiny figures without personal characteristics, mere insects whose humanity had disappeared. If the geometrical sublime made one appreciate human skill and ingenuity, at the same time it divorced the observer from the sense of connection with others. By making the city into a shimmering hieroglyph calling out for interpretation, attention was displaced from human beings and the apparent pettiness of their lives. Lifted up to the sky, the visitor is invited to see the city as a vast map and to call into existence a new relationship between the self and this concrete abstraction. As one reporter from Boston wrote, "You may be awed by it, you can even be a little afraid of it; you cannot deny that it is Today in Steel and Stone."[84] The same could as easily have been said of the Eads Bridge, the Brooklyn Bridge, or the earlier skyscrapers. In the experience of the geometrical sublime, the awed observers felt they were confronting the technological present in all its manufactured fullness.

Roland Barthes' reflections on the Eiffel Tower are suggestive here:

[T]he Tower overlooks not nature, but the city; and yet, by its very position of a visited outlook, the Tower makes the city into a kind of nature; it constitutes the swarming of men into a landscape, it adds to the frequently grim urban myth a romantic dimension, a harmony, a mitigation; by it, starting from it, the city joins up with the great natural themes which are offered to the curiosity of men: the ocean, the storm, the mountains, the snow, the rivers. To visit the Tower then, is to enter into contact not with a historical Sacred, as is the case for the majority of monuments, but rather with a new nature, that of human space. . . .[85]

But the human relation to this new nature, is not, as Barthes suggests, that of beauty or romance; it is that of power. Appropriate to this new nature is a new form of the sublime. The skyscraper completes the formation of the city as the double of nature, providing a spectacular perch from which to contemplate the manufactured world as a total environment, as though one were above it or outside it. The paradox that the tall building is at once within and outside the scene makes it attractive to the tourist as a literal focal point. The panoramic vision permits one not merely to view the city as a series of individual parts, but to read it as a total structure. To make sense of the panorama requires a series of comparisons and mental triangulations, a species of abstract thought that manipulates the visible landscape. The skyscraper made cities comprehensible as "intelligible objects, yet without—and this is what is new—losing anything of their materiality; a new category appears, that of concrete abstraction. . . ."[86]

The skyscraper defined the new variant of the mathematical sublime that had been glimpsed from the walkways of the Eads Bridge and the Brooklyn Bridge. As in the natural sublime, there was an element of terror in looking at the city from a high place, gazing down a sheer wall. In addition, there was the sudden shift in viewpoint and scale. A person whisked up by elevator to stare out at Minneapolis or Boston or Chicago from the top of its highest building suddenly saw a much wider urban horizon and often experienced this as a corresponding psychological expansion as he or she struggled to link the details of the scene below with knowledge of the city seen from the ground. The mental activity of triangulation does not require, as in the mathematical sublime, that the scene appear infinite in size. Rather, it requires an apparently infinite series of mental transpositions of scale and orientation, forcing the viewer to perform olympian calculations. The vast region visible from the top of a skyscraper appears intelligible, offering itself for decipherment like a huge hieroglyphic. Yet, as with all sublime landscapes, its full meaning remains unutterable. For this very reason, the sense of power does not abate; it can be constantly renewed as one looks out over the metropolis.

The view provides a sense of mastery, particularly for those who own the skyscrapers and look down on the city. Individual skyscrapers not only promoted the corporate image to the general public; they also satisfied psychological needs of the owners and tenants, who preferred upper floors and the widest possible vistas. Accordingly, rents ascended along with the elevator. The hero of James Oppenheim's 1912 novel *The Olympian* reaches his apotheosis when he first goes to the top of the skyscraper he has constructed:

He was utterly alone in the skies. Below him rose the skyscrapers, giving slanting glimpses of deep streets busy with tiny black people and darting traffic, and from their tips curled white smoke in the boundless swim of sunlight. He saw the waters that circled the city like a hugging arm of the sea, and on the level stretches harbor-craft and ocean-liners. He saw the bridges suspended between Long Island and Manhattan, Brooklyn beyond; he saw the Jersey heights. . . . All the mighty metropolis stretched like a map below him, crowded to the circling horizon with millions of human beings.

Perhaps because Oppenheim had lived through the rapid construction of the New York skyline, he was able to express directly the usually hidden contents of this vision:

It was Science tearing off the crust of the earth and releasing the powers and riches of Nature. Busily the race seized on these, a chaos of rough enterprise—mines, manufactories, laboratories, exchanges. And in the swift trade that followed three mighty gods began to roughly organize the chaos—Steam, Electricity, Steel. The railroad came, the post, the mill and farm machinery, the typewriter, the telegraph, the telephone, the automobile. And all these were like nerves and blood vessels laid out through the chaos till it began to coalesce, the parts aware of each other, the Earth gradually shaping into one body.[87]

The vision of the city from the top of a skyscraper materialized a new historical relationship between human beings and their environment. The new "body" that fuses nature and culture presents a historical vision of technological progress as a sequencing of objects. A new organism emerges, built out of materials wrested from nature. It tantalizes the viewer with a vision of the totality of civilization, expressed as Barthes' "concrete abstraction." This vision appears to be the logical development of economic and technological forces. Yet the skyscraper literally cannot be understood as the product of the marketplace. Not merely a center of commerce, it is a symbolic structure: from the outside a corporate icon, from the inside a site of the magisterial gaze. To experience either the jagged skyline in the distance, the immense vistas aloft, or the insect life of the street below validates that power. The geometrical sublime and its fantasies of domain thus altered the phenomenology of the city.

5

The Factory: From the Pastoral Mill to the Industrial Sublime

The history of American industrial development can be divided into three periods on the basis of its power sources: water, steam, electricity. Each of these forms of power generation dictated the size, the location, and the form of industry, and each bore a distinct relationship to the sublime.

American industrialization began by relying on water power, which dispersed factories into the hinterland. At first the water mills almost seemed to blend into the pastoral world, and they were regarded as sublime because they orchestrated large rooms of intricate machinery and produced so much. As late as 1838 there were only 2000 steam engines in the entire United States, and two-thirds of them were mounted on steamboats and railways. Stationary steam engines did not contribute a significant part of the nation's industrial power until after 1840. They spread rapidly after the middle of the century. At first, steam-driven factories were subsumed within the pastoral tradition, but by the end of the century they were seen as sublime precisely because they were intensely unnatural.[1] Electricity emerged in the 1880s but was not widely adopted for power until after 1900. It permitted new facilities on a hitherto inconceivable scale, and those facilities were appreciated as a union of the geometric and the dynamic technological sublime. In short, changes in production led to a new aesthetics of industry.

This was by no means a universal story. In England the steam-powered factory was already assimilated into the sublime by the 1780s, before the factory system existed in the United States. As early as 1758, only a year after Burke's work on the sublime appeared, the view of the iron works at Coalbrookdale prompted English artists to make engravings of it. These engravings treated the factory as beautiful, but within a generation other artists had fully assimilated it into the sublime. In 1788 George Robertson's six engravings of Coalbrookdale contrasted its beauties with sublime horrors. At virtually the same time, Arthur Young found the valley "a very romantic spot" enhanced by "that variety of horrors art has spread at the bottom." He found that "the noise of forges, mills, &c. with all their vast machinery, the flames bursting from the furnaces with the burning of coal and the smoak of the lime kilns, are altogether sublime. . . ."[2]

The English critique of factories, like the later critique of railroads, was framed primarily in terms of social class. In contrast, at the end of the American Revolution there were only three steam engines in the entire United States, and all of them pumped water (two in mines and one in New York City).[3] The English first came to terms with factories during the Enlightenment and the French Revolution; in contrast, American factories did not emerge until the 1820s, when they were comprehended in terms of transcendentalism and Jacksonian politics. The machine tended to be seen as an organic outgrowth of society that could flourish in a laissez-faire economic system, to the benefit of all. Such a view was difficult to sustain in England, where factories emerged earlier and where they were larger and more numerous.

Moreover, the preferred forms of power in the two countries were quite different. The Americans relied on abundant water power, the English on steam. This technical difference meant that English factories were concentrated in cities, creating an industrial landscape, while American factories dotted the countryside and became the basis for smaller communities that seemed in harmony with their surroundings. An observer of American conditions in 1808 summarized how small towns grew up wherever there were good mill sites, noting that the first building raised was often "a solitary saw-mill":

To this mill, the surrounding lumberers, or fellers of timber bring their logs, and either sell them, or procure them to be sawed into boards or into plank, paying for the work in logs. The owner of the saw-mill becomes a rich man, builds a large wooden house, opens a shop, denominated a store, erects a still, and exchanges rum, molasses, flour and port, for logs. As the country has by this time begun to be cleared, a flour-mill is erected near the saw-mill. Sheep being brought upon the farms, a carding machine and fulling-mill follow . . . the mills becoming every day more and more a point of attraction, a blacksmith, a shoemaker, a tailor, and various other artisans and artificers, successively assemble. . . . So a settlement, not only of artisans, but of farmers, is progressively formed in the vicinity; this settlement constitutes itself a society or parish; and, a church being erected, the village, larger or smaller, is complete.[4]

This description, based on developments in Maine, suggests the centrality of water power in settling the country. Likewise, in western New York State "mills and manufactures formed the first rudiments of almost countless villages and towns."[5] American industrialization would not be based on the steam engine, nor would it concentrate people in cities. Rather, it would employ clusters of workers in small towns in the foothills of the Appalachians, where rivers and streams drop rapidly toward the coastal plain. Steam engines were essential to industrialization in much of England, but water power "was to be had almost everywhere in the northeastern United States, where other conditions for industrial expansion were most favorable."[6] Thus the British and American industrial systems produced distinctive landscapes.

In 1814 Wordsworth condemned the new industrial system in his poem "The Excursion," describing child labor in factories and the spread of manufacturing towns over pastoral landscapes:

the smoke of unremitting fires
Hangs permanent, and plentiful as wreathes
Of vapour glittering in the morning sun.[7]

The English romantics decried such developments, which until half a century later scarcely had counterparts in the United States. An American was virtually unable to find a similar landscape before 1850. A mill site seemed not so much a foreign imposition as the seed from which a community grew.

Even Wordsworth appreciated the achievement that industry represented. In the same poem in which he criticized the blighted industrial landscape, he also sounded the note more commonly heard in America. He realized that some could "exult to see/An intellectual mastery, exercised/O'er the blind elements, a purpose given,/A perseverance fed; almost a soul/Imparted—to brute matter." He continued: " I rejoice,/Measuring the force of those gigantic powers/That, by the thinking mind, have been compelled/To serve the will of feeble-bodied Man."

In the United States eulogies to human reason predominated, and they were often articulated as replies to Wordsworth. Not only was the American critique of factories more muted than the

English; the American exultation of intellectual mastery over nature was stronger. As with the railroad, Americans at first believed that factories might not pose a contradiction to the natural world but might extend and complete it.

The mills of Lowell, Massachusetts—the first large factories in the United States—were based on water power. In the plans of those who built them, and in the early works of art depicting them, these factories were imagined as parts of a middle landscape that blended with the agricultural world that surrounded them.[8] By 1833 some 5000 workers were employed in nineteen mills, with more under construction. Lowell had grown spectacularly, becoming the second-largest town in Massachusetts.[9] John Kasson adopted Erving Goffman's concept of the total institution to describe these mills, which not only provided work but also established a complete social system, including boardinghouses, churches, and moral supervision of the laborers. Lowell was an administered way of life, regulated by the factory bell. An institution of "republican reform in the antebellum period," Lowell "promised to resolve the social conflict between the desire for industrial progress and the fear of a debased and disorderly proletariat."[10] The proprietors sought to prevent the formation of a permanent working class by hiring primarily young women and employing them for only a few years before marriage. They lived in communal boardinghouses, where they were compelled to eat and which were locked up for the night at 10 o'clock. They worked 12 hours a day, beginning at 5 in the morning. Anyone who missed church often or who was "notoriously dissolute, idle, dishonest, or intemperate"[11] could be fired. Such women had little opportunity to indulge in sinful pleasures. The French traveler Michel Chevalier admired this moral-industrial order on the whole but noted sadly that the "rigid spirit of Puritanism has been carried to its utmost in Lowell."[12]

To judge by the evaluations of early visitors, Lowell succeeded by its own standards for about three decades. Harriet Martineau visited along with Ralph Waldo Emerson, who lectured to the mill girls after their shifts in the mills, and was astonished to see them "all wakeful and interested, all well-dressed and lady-like."[13] Charles Dickens found Lowell a stunning contrast to English working towns; he declared that the young women were "healthy of appearance, many of them remarkably so, and had the manners and deportment of young women, not of degraded beasts of burden." Dickens went so far as to "solemnly declare, that from all the crowd I saw in the different factories that day, I cannot recall or separate one young face that gave me a painful impression," not one that "I would have removed from those works if I had had the power."[14] Many other visitors to Lowell agreed with such assessments, with hardly a dissenting voice among foreign observers.[15]

Part of the attraction of such total institutions for a visitor was their combination of complexity and order on a massive scale. Even before one entered them, the factory buildings were impressive from the outside as architecture. Burke argued that "a successive disposition of uniform parts in the same right line" could be sublime. For example, he noted that to produce a "perfect grandeur" the repetition of a form, as in a colonnade, was far more effective than a bare wall of the same length. "When we look at a naked wall, from the evenness of the object, the eye runs along its whole space, and arrives quickly at its termination. . . . "He argued that a bare wall is "only one idea, and not a repetition of similar ideas: it is therefore great not so much on the principle of infinity, as upon that of vastness."[16] These observations are particularly applicable to the Amoskeag mills, which were of nearly uniform design and which stretched for half a mile along the Merrimack River in Manchester, New Hampshire. A British visitor found Manchester

Exterior view of the Amoskeag Mills. Courtesy of Smithsonian Institution.

quite unlike its English namesake: "It has clean air, clear water, and sunny skies, every street is an avenue of noble trees. . . . Perhaps the handsomest, certainly the most impressive, buildings in Manchester are the Amoskeag and Manchester Mills. They are built of worn red brick, beautifully weathered, and form a continuous, curved facade."[17] A visitor to Lawrence, Massachusetts, first saw the town from "elevated ground" and found it "almost astounding to discover at one view large factories extending nearly a mile, alongside a broad river, and behind these nearly a square mile of ground, covered with dwelling-houses, shops, and all the other requirements for a population of 17,000 inhabitants, most of whom had been attracted thither during the ten years of the growth of the mills."[18]

Impressive as the factory towns were from the outside, the external architecture was less remarked upon than the floors of machinery inside. The early modern factory presented a system of linked machines that many ordinary observers could barely grasp, virtually compelling their admiration for the ingeniousness of its mechanical contrivances. The writer John Pendleton Kennedy was "lost in admiration" before "the vast engineering" and "infinite complication of wheels" in a factory.[19] Most visitors to early textile mills testified to the powerful impression created by many rows of spinning

Interior view of an electrical spinning mill. Courtesy of Smithsonian Institution.

machinery, each containing uniform lines of spindles that became a visual metaphor for the promised cornucopia of industrial production. (In later years photographers often exploited such repetition in depicting the mills.) In an age when one seldom saw more than one or two machines together at one time, the view of a large factory room humming with incessant activity created astonishment at the ingenuity and apparent perfection of the arrangements. Just as Jefferson's description of the Natural Bridge in Virginia resorted to measurements of its height, visitors to the early factory often attempted to comprehend the scene before them mathematically. Counting the number of yards of cloth woven or the tons of raw cotton processed was one way to grasp the immensity of the factory compared to hand production in the home. In 1833 the city of Lowell boasted that nearly a million bricks had been used to build just one mill, and that the city produced 15,300 miles of cloth a year from 7 million pounds of cotton. By the end of the nineteenth century, the Amoskeag Mills claimed to manufacture cloth at a rate of 4 million yards per week, or 50 miles of cloth per hour.[20] Such statistical claims were a corollary of the mathematical sublime.

Yet for many the most impressive fact of the mills was not the architecture, the intricate machinery, or the vast production, but the cohorts of workers who seemed to be marshalled into an ideal order. The perfect expression of this disciplined work force being presented to the world occurred when Andrew Jackson visited Lowell in 1833, the first president to make a factory tour. Four thousand "young females" dressed in white turned out to greet him, "massed four deep along the route"—"a veritable 'mile of girls.'" After this "army" marched by in review, Jackson asked innumerable questions about the number of women employed, the mills and the economics of the operation. He became so interested in the ingenuity of the arrangements and the power of the water wheels that he asked to see the mills. They had been closed down for the day, but one was hastily set in motion. The mills made a powerful impression on Jackson, as did a demonstration of firefighting equipment. The president spoke of the mechanical marvels and the well-dressed mill girls for the rest of the day.[21]

Such were the responses of outsiders visiting a factory, not of workers inured to its discipline. The casual visitor may have seen the mill as a sublime demonstration of man's ingeniousness and superior intellect, but—in contrast with the perpetual movement of Niagara

Falls—the factory functioned only if workers tended the machines. From the point of view of a woman in a textile mill, the industrial sublime was something experienced only on the first days at work. A 15-year-old girl who came to the Lowell mills in 1845 at first liked it "very well" in the spinning room, but after two years she was in ill health and felt overworked.[22] Just as the person who had always lived in the countryside did not see it as picturesque or sublime, and just as the farmer whose lands were flooded by a new mill did not see it as being in harmony with nature, laborers usually saw the factory not as a landscape but as a process. Outsiders might perceive it as a form embodying certain abstract ideas, but to the laborer the factory was a place of action.

In the first years of Lowell's existence the corporations attempted to shape perceptions of the new environment—notably through the publication of the *Lowell Offering,* a magazine written by mill girls but uniformly complimentary to the owners who financed it. In an early issue, one mill girl echoed the rhetoric of the technological sublime: "In the mill we see displays of the wonderful power of the mind. Who can closely examine all the movements of the complicated, curious machinery, and not be led to the reflection that the mind is boundless, and is destined to rise higher and still higher; and that it can accomplish almost any thing on which it fixes its attention!"[23] Yet even in the *Offering* this is an isolated example. Admiration for the machinery was more common among visitors to the new mills than among the workers.

By the end of the early 1840s, the idyllic vision of a new kind of factory work that permitted young women to become educated and save money for a dowry was fading. Many of the workers had varicose veins from standing for 13 hours a day. Many suffered from respiratory problems, having worked in badly ventilated rooms where the air was filled with wisps of cotton fiber. On Chester Creek in Pennsylvania, "none of the mills employed exhaust fans, although they were known in the Philadelphia area, and on cool or windy days the windows were kept shut. In two places, the picker room and the card room, the problem of ventilation was particularly severe. The air was filled so thick with 'flyings' that breathing was difficult and workers developed a constant cough. . . . Windows in these rooms could not be opened because the breeze would blow around even more of the fibers."[24] Noise was almost as serious a problem. The din was deafening, and the rumble and clank remained in the workers' ears long after they left work. One woman described the sound as "crickets, frogs, and Jew-harps, all mingled together in a strange discord." A physician who studied Lowell's Merrimack Company in 1849 under the auspices of the American Medical Association discovered that, despite the company's threats to blacklist those who failed to complete a one-year contract, workers lasted on average only 9 months. He found the factories unhealthy and "less cared for than our prisons."[25]

A Pittsburgh physician wrote that Pennsylvania's factories were "ill-ventilated" and their air "highly surcharged with the most offensive effluvia—arising from the persons of the inmates, and the rancid oils applied to the machinery." Worse yet, the mills were hot, "rising to eighty and even ninety degrees in summer. Their atmosphere is constantly filled with floating particles of cotton; the finer the yarns to be spun, the higher the temperature must be. . . . The cotton wool, when impregnated with the oil used to diminish friction in the machinery, and in the usual temperature of the rooms, emits a most offensive fetor. . . which none but those accustomed to it can respire without nausea."[26] People worked from 12 to 14 hours a day in this atmosphere, most of the time on their feet. As early as 1837 an employee testified to the hardships of children in these conditions, and a sympathetic observer summarized his remarks as follows: "The children are tired when they leave the factory; has known them to sleep in corners and other places, before leaving the factory from fatigue. The younger children are generally very much fatigued, particularly those under twelve years of age. . . . Has known the children to go to sleep on arriving home, before taking supper; has known difficulty in keeping children awake at their work; has known them to be struck to keep them awake."[27]

Workers organized against such conditions. By 1845 the operatives in the Boston area had their own newspaper, the *Voice of Industry,* and on the Fourth of July that year more than 2000 working men and women from Lowell, Boston, and Lynn met in Woburn. They heard an address by Miss S. G. Bagley, who attacked the conditions in the mills, the lengthening hours, and the danger of "eternal slavery" for operatives, who had no time for education. Just as important, she charged that the *Lowell Offering* had refused to print articles critical of the mills, and that it was "controlled by the manufacturing interest to give a gloss to their inhumanity."[28] Her address touched off a public debate about working conditions and did much to discredit the *Offering.*[29]

The factories did not immediately create a proletariat, however. The primary loyalties of the women in the early textile mills were to their parents and their extended families. While years in the mill inured them to rigid factory discipline, working by the clock, and cash payment, factory "girls" usually worked elsewhere before and after these experiences. They were not passive victims of industrialization but active agents in the labor market, changing jobs often in accord with their own and their families' needs. As Thomas Dublin has noted, "mill employment could be turned to individualistic purposes," including the desire to save money for a dowery or an education.[30] Mill girls regarded their work as a temporary expedient.

After a generation, however, factory work was fast becoming a permanent condition and formal class lines had become more pronounced. American cities doubled and redoubled in population in the first half of the nineteenth century. New York grew from 202,000 in 1830 to just short of a million in 1870. In the same years, eighteen cities grew to 100,000 or more.[31] The intimate walking city that blended housing, commerce, government, and small-scale industry evolved into a distended and differentiated structure, and it was no longer possible to include a whole city in a public event. Workingmen had been integral to the civic celebrations and processions of the 1820s for the Erie Canal and the Baltimore Railroad, but in the larger cities of mid-century America artisans and factory workers had disappeared from the processions. Separate holiday celebrations signaled the break between classes. While workers tried to preserve the carnivalesque qualities of their culture, professionals and the middle class preferred more solemn events, which often celebrated industrialization.[32] From the middle of the century forward, the industrial sublime often served to gloss over the difficulties of the worker.

The Fourth of July began to change. By the 1850s the solemnities of earlier years were giving way to parody and satire. Diana Appelbaum notes that the previously fixed pattern of "processions, exercises, and dinners" was "being replaced in the 1850s by more diffuse and lighthearted forms of celebration," including balloon ascents, regattas, band concerts, and steamboat excursions.[33] Smaller towns retained the trappings of the Jacksonian celebration, but the cities increasingly fractured along class lines. Instead of joining in a common civic parade, workers developed their own traditions, held their own speeches, and exploded their own fireworks. In a case study of Worcester, Massachusetts, Roy Rosenzweig found that the middle and upper classes, which at first had tolerated the militant and often rowdy July 4 celebrations, gradually became alarmed during the later years of the nineteenth century. Fearing that the day was becoming incendiary and might spark civic unrest, middle-class reformers sponsored a "Safe and Sane July Fourth" movement that "sought to repress and replace the distinctively boisterous and ethnic commemorations by the working class."[34] This attempt to impose middle-class values on the holiday often failed, but the workers' celebration nevertheless faded in importance, overtaken by popular amusements. Just as the working class turned away from the "dignified" performances of Shakespeare that were introduced by their "social betters" during the Victorian period, and just as they often spurned the high culture of a world's fair for the midway, they turned to amusement parks and professional baseball to celebrate the Fourth. Militancy was increasingly reserved for May Day, which became the time for industrial action. For example, in 1886 demands for shorter hours brought strikes to much of the nation. On May 1 more than 20,000 workers marched in the streets of Baltimore and many more in Chicago, where initial success was derailed by the Haymarket bombing and its aftermath. By then, ritual expressions of solidarity between workers and manufacturers had all but disappeared.[35]

Despite increasing worker militancy, most nineteenth-century authors depicted factories as harbingers of progress and tended to ignore the workers. A patriotic volume from 1889, *Picturesque Sketches of American Progress*, stressed the number and the size of the factories in each American city, with no suggestion that any was unsightly or unpleasant. This attitude is particularly striking because the volume devotes its final 30 pages exclusively to the Knights of Labor. While the author approvingly quotes the Knights' attack on the "alarming development and aggressiveness of great capitalists and corporations" (which, "unless checked, will inevitably lead to the pauperization and hopeless degradation of the toiling masses"), he does not criticize the factory system *per se* or the new modes of production then being introduced.[36] As in Edward Bellamy's *Looking Backward* and other utopian writings of these years, the system of manufacturing was seen as a boon to mankind, capable of providing a life of ease for all.[37] The improvement of machines in precision and reliability became a metaphor for social improvement.

As factories continued to grow in scale, they relied increasingly on steam. Before 1850 steam engines remained identified with railroads and steamboats, but after then steam power rapidly became central to industrial expansion—particularly in the Middle West, where water power was less abundant than along the eastern fall line. As operations increased in scale, the earlier emphasis on small, cheap, inefficient steam engines gave way to demands for larger engines that performed more precisely and efficiently. Low purchase price became less important than low operating cost and reliability. In response to these changing needs, American engine builders created the automatic variable cutoff engine, "the first major advance in stationary steam engineering since the general adoption of the high-pressure noncondensing engine" 40 years before.[38] While many contributed to this development, the central figure was George H. Corliss, whose shop in Providence produced 481 large steam engines between 1847 and 1862. As the premier engine builder of the era, Corliss set a standard not only for quality but also for service. At the Paris Exposition of 1867 his engine won the highest honors, beating 100 other entries.[39] By 1870 there were over 1000 of his machines operating in the United States, and his firm was a logical choice to produce the great steam engine that would drive all the machinery at the Philadelphia Centennial Exposition of 1876.

The machine Corliss created was perhaps the most famous steam engine ever made in the United States. The exposition's managers contracted with him to build, install, and maintain it for $77,000, considerably less than the $112,000 it cost him to do the job. For this sum Corliss created two beam engines joined by a crankshaft. Together the engines weighed 600 tons. The main driveshaft was more than 300 feet long, and in addition there was almost a mile of line shafting. What immediately caught the attention of any onlooker, however, was a giant geared flywheel, 30 feet in diameter, that made 36 revolutions per minute. This was just slow enough so that the eye could follow its rapid, silent, seemingly effortless and endless rotation. The combination of the huge engines and the inexorably moving flywheel proved a mesmerizing vision for most fairgoers, though it aroused some scorn from engineers, many of whom correctly discerned that the future lay with faster engines (such as the Porter-Allen, which routinely operated at 150 or more revolutions per minute). The engine that made Corliss a household name did not always impress his peers. The journal *Manufacturer and Builder* commented: "On the whole, we admire the engine more for its great size than for its plan and design, which are decidedly inferior, and not up to the present 'state of the art.'"[40] This may

The Corliss engine at the Philadelphia Exposition, 1876.

have been correct from an engineering viewpoint, but for the general public the enormous engine worked according to principles visibly embodied in its many movements. As Eugene Ferguson notes: ". . . there was a great deal more movement to be observed than in most steam engines. The rods and cranks and levers and latches that opened and closed the values that admitted steam to each end of the cylinders and exhausted it again . . . were all in plain sight, providing overtones and counterpoint to the steady rhythm of the great machine's chief members."[41] This showy monument moved just slowly enough to be watched with pleasure and studied in silence. Its enclosed condenser eliminated puffing exhaust; all one could hear was a soft murmuring from the highly polished working parts, which *Scientific American* called "the poetry of mechanical motion."[42] It became the central icon of the fair, endlessly reproduced in posters, praised in every popular guide, and eulogized by William Dean Howells in the *Atlantic Monthly* as "an athlete of steel and iron with not a superfluous ounce of metal on it."[43]

A guidebook ordered readers to begin their visit to the Centennial Exposition at Machinery Hall: "The first thing to do is to see the tremendous iron heart, whose energies are pulsating around us."[44] What made the Corliss engine especially attractive to the imagination was the fact that such a wide range of machines were set into motion from one powerful center—the "iron heart."

The Corliss engine was the focal point of the exposition's opening ceremony. On the platform beside it stood President Ulysses S. Grant, the emperor of Brazil, and George Corliss. Under the inventor's direction, each head of state turned a small crank. The great flywheel began to spin, to emotional applause from the massed spectators. *Scientific American* commented that "perhaps for the first time in the history of mankind, two of the greatest rulers in the world obeyed the order of an inventor citizen."[45] The engineer was now at the center of the pantheon of American heroes.

In Machinery Hall the hundredth Fourth of July was celebrated with traditional speeches and a reading of the Declaration of Independence. A large band played, a choir sang a hymn that Oliver Wendell Holmes had composed, and the poet Bayard Taylor read his "National Ode." Simultaneously, an unauthorized group of women led by Susan B. Anthony read their own Declaration of Independence for Women, demanding that they also be acknowledged. In all these proceedings, the word was central: the word of the orator, the poet, the founding fathers, and the would-be founding mothers.[46] Yet ultimately, as Martha Banta has noted, the Corliss engine was "the primary statement male America had to make about itself,"[47] and it made a more lasting impression on most fairgoers than any speech. The aged Walt Whitman "ordered his chair to be stopped before the great, great engine . . . and there he sat looking at this colossal and mighty piece of machinery for half an hour in silence . . . contemplating the ponderous motions of the greatest machinery man has built."[48]

Such a finely honed specimen of the technological sublime found a buyer at the close of the exposition: George Pullman, who wanted it to drive the machine shops of his sleeping-car company. Until 1910 it remained on display to passing trains in Chicago, visible through huge plate-glass windows.

One appeal of the engine was its apparent ability to work with little or no human intervention. Many commentators treated it as though it were alive. Kasson notes that the public did not need to understand the engine's operation to be impressed by its bulk, power, and near silence: ". . . fairgoers did not approach the engine as an immaculate work of engineering, to be judged by its efficiency alone. Rather, in a way characteristic of popular reactions to powerful machinery in the nineteenth century, their descriptions frequently became incipient narratives in which, like some mythological creature, the Corliss engine was endowed with life and all its movements constructed as gestures."[49] Such descriptions implied automatic machinery and in effect eliminated workers from the landscape of production. Likewise, accounts of industrial production during the decades after the American centennial often say little about workers, who tend to remain nameless servants of a larger process.[50] More and more, the creativity of production was becoming the province of the engineer and the inventor, and laborers merely machine attendants. Giving life to the machine and celebrating the inventor were not only characteristics of popular descriptions of the Corliss engine but also standard ploys of many writers. Machines sat, stood, or crouched; they devoured fuel; they spewed out smoke; they served faithfully; most dramatic, they moved. Trains galloped, rushed, and careered along. In such metaphors, Americans asserted that the machine had natural features and functions. For them, technology had become what Emerson called "the double of nature."

Machines such as the Corliss engine became synecdoches for the new form of the factory system, which was moving away from water-powered rural mills toward industrial concentrations. The new facilities expressed the demand of new corporations for mass production at a few sites, with economies of scale and greater subdivision of labor. Compared to anything Americans had known before 1880, these factories were enormous. Where once a facility of more than a few hundred workers (such as Amoskeag or Lowell) was unusual, the new factories of General Electric, Westinghouse, Ford, and United States Steel employed thousands. They were miniature cities, with their own railway lines, streets, recreation facilities, power stations, police forces, medical services, newspapers, telephone networks, and educational programs. Whereas the old factories had blended into the small towns architecturally, the new steam-driven facilities were total environments, visibly distinct and self-contained.

Two quite different literary traditions developed to deal with the new industrial landscape. Some—notably William Dean Howells, Robert Herrick, and such magazines as *The Survey* and *Harper's Weekly*—demanded social reform. As Stilgoe points out, in most novels "sensitive, educated men and women see only pain, suffering and exploitation in the industrial zone."[51] They usually saw little or no sublimity in the industrial scene, and instead took the side of

workers in their struggle for higher wages, better working conditions, shorter hours, and an end to child labor. This tradition,
which grew directly out of the antebellum critiques of cotton mills,
often "read" the new landscape as a blighted contrast to the pastoral world that had preceded it. Pittsburgh had once assumed with
pride the title of "The Smoky City," but by 1912 reformers cried out
against the ill effects of coal smoke on public health.[52] In 1910
Margaret Byington began her survey of a Pittsburgh-area factory
town in this reform spirit: "Homestead gives at the first a sense of
the stress of industry rather than of the old time household cheer
which its name suggests. The banks of the brown Monongahela are
preempted on one side by the railroad, on the other by unsightly
stretches of mill yards. Gray plumes of smoke hang heavily from the
stacks of the long, low mill buildings, and noise and effort dominate what once were quiet pasture lands."[53] This vision resembles
the older English critique of industrialization, in which the railroad
and the mill have usurped the natural world, polluting it with
smoke and noise. Furthermore, one of the bloodiest labor con-
frontations in American history had occurred at Homestead. In
1892 striking workers were attacked by a private army of Pinkerton
guards, hired by Andrew Carnegie's steel corporation. The workers
repelled the attack, but later lost the strike after the governor of
Pennsylvania sent in the state militia.

It was impossible to see these factories clustered along the railroad lines as harmonious parts of rural America. But the districts
that had sprung up outside Pittsburgh, Chicago, and the major
cities of the Northeast could be represented as a mysterious realm.
In a 1883 survey entitled *Peculiarities of American Cities*, Willard
Glazier advised: "By all means make your first approach to
Pittsburgh in the night time, and you will behold a spectacle which
has not a parallel on this continent."[54] He went on to rhapsodize
over the "thousand points of light" reflected in the rivers and the
"fiery lights" from numberless furnaces and chimneys. While admitting that "Pittsburgh is a smoky, dismal city at her best," Glazier nevertheless proclaimed that smoke was a "sure death of malaria" and
a preventative for many other diseases. Arnold Bennett was writing
in this tradition when he called the outskirts of Toledo "a misty,
brown river flanked by a jungle of dark reddish and yellow chimneys and furnaces that covered it with shifting canopies of white
steam and of smoke, varying from the delicatest grays to intense
black."[55] In such language, cavernous factories draped in smoke
became sublime. Their size, obscurity, and danger were converted

into assets. Whereas the early New England mills had been praised for their clean appearance and the way they blended into the landscapes of small towns, these industries seemed exciting because they were vast, dim worlds of iron, brick, and smoke. The new landscape was read as an objective correlative of man's new powers of transformation. It was understood in terms drawn from the tradition of the sublime—particularly as it had developed in Victorian England, where "the hugeness and massiveness and unashamed arrogance" of industrial sublimity made it "the aristocratic taste of the time."[56]

In 1905 Pittsburgh's Carnegie Institute set aside funds for the largest commission ever given to an artist for a mural and chose John White Alexander to paint "The Crowning of Labor" on the walls of its great hall. In the lower part of the mural, powerfully built, heroically depicted men, stripped to the waist and surrounded by flames and smoke, work blast furnaces. Above them, in a column of vapor and smoke, the colors are lighter. A host of female angels in diaphanous robes descend, bearing fine crafts from "the looms, the workshops, and the studios of the world."[57] Banta concludes that in this image "all the American 'Pittsburghs' (male and materialistic to the core of their smokey depths) are saved through the vision of grace conveyed by the American Girls swirling down from the skies."[58] This traditional sexual polarization is reminiscent of Kant's early writing on the sublime, in which beauty is female and the sublime male.

Such a conception of industry was hardly limited to the Carnegie Institute or to managers who read *Factory* and *Cassier's Magazine*. When Theodore Dreiser motored from New York to Indiana with a friend, at several points along the way they made unscheduled stops to admire massive works of industry. In Pennsylvania it was a great stone railway bridge at Nicholson. Driving through Buffalo, "not stopping to revisit the Falls or those immense turbine generators," they came to "a grimy section of factories on a canal or pond, so black and rancidly stale that it interested [them]." "Factory sections," Dreiser noted, " have this in common with other purely individual and utilitarian things,—they can be interesting beyond any intention of those who plan them." Dreiser was fascinated that the oily pond "constantly emitted bubbles of gas which gave the neighborhood an acrid odor." This was an anti-nature, seen without ecological reservations, where warehouse roofs "rose in such an unusual way and composed so well" that Dreiser and his companion stopped to sketch them. After this pause, they drove only a few blocks before halting again to view "an enormous grain elevator, standing up like a huge Egyptian temple in a flat plain." Dreiser continues: "Before it, as before the great bridge at Nicholson, we paused, awestruck by its size and design."[59]

Increasingly, writers and artists admired whole factory districts. By 1900, cities that once had gloried in a single bridge celebrated the possession of many bridges or bragged of the sheer complexity and scale of new factories. This new view could coexist with the more traditional merging of industry into a pastoral landscape. In 1895 *Harper's* published an article entitled "The Industrial Region of Northern Alabama, Tennessee and Georgia," which began by describing the sublime natural landscape seen from Lookout Mountain and which, after several pages, focused on the coke furnaces and steel mills near Chattanooga.[60] But more often, as in the writings of Dreiser and Bennett, the industrial world was presented as a separate realm, fascinating because it was utterly unnatural.

The industrial sublime combined the abstraction of a man-made landscape with the dynamism of moving machinery and powerful forces. The factory district, typically viewed from a high place or from a moving train, thus combined the dynamic and geometrical sublimes. The synthesis evoked fear tinged with wonder. It threatened the individual with its sheer scale, its noise, its complexity, and the superhuman power of the forces at work. Yet, as with other forms of the technological sublime, this scene ultimately reaffirmed the power of reason—but not in Kant's sense. Rather than provoke an inward meditation that arrived at a transcendental deduction applicable to humanity as a whole, these landscapes forced onlookers to respect the power of the corporation and the intelligence of its engineers.

Manufacturing districts were aesthetically pleasing once one found the proper vantage point along a canal, from the top of a skyscraper, or from a train window. According to Karen Tsujimoto, Elsie Driggs' darkly precise painting "Pittsburgh" was inspired by "a vivid childhood memory of passing through Pittsburgh by train at night."[61] In James Oppenheim's 1912 novel *The Olympian* an ambitious young man travels through Pittsburgh on the train; Oppenheim writes: "A vision shown and passed, swallowed in the night; the sublime spectacle of window-lit mills at the riverside girdling with darkness the fierce flaming of the Bessemer converter, whose several swelling tongues of fire licked at the flaring clouds and crumbled in showers of golden snow. Against that burning a lonesome one-armed telegraph-post was silhouetted, a voice in the darkness, passing."[62] The clean air and the comfort of the smoothly rolling train reduce this threatening landscape to a pleasing vignette. It is not incidental that this glimpse of the steelworks occurs at night. As Burke observed, "darkness is more productive of sublime ideas than light"[63]—particularly if there are powerful contrasts. Later in *The Olympian*, when the same young man has risen to an administrative position in the steel industry, he again examines the mills at night. This time he stands with his bride atop a bridge inside the works, and they "ha[ve] unfolded to them the Vision of Steel":

. . . over the vast acreage they saw the shadowy outlines of a dozen immense buildings nested in a network of switches and tracks. Over those tracks clanked yard-engines with rattling trains of flat cars; red and green signal lamps winked and glistened; and laborers, swinging lanterns, hurried to and fro.

Some of those looming buildings glowed at the windows as if they were eaten by fire; some—the converter sheds—were like craters with waving manes of flame and rolling clouds of luminous vapor. Everywhere they saw sheets of fire, leaping white tongues, glare and smoke and steam, while lightnings flashed at the cloudy skies. And over it all a hundred black chimney-pipes looked through the changing lights.

This passage builds up a series of contrasts between light and darkness—between flames, glare, glow, lanterns, and lightnings and the surrounding night, cloud, chimneys, and darkness. The play of these contrasts suggests the powerful effect of seeing a steel mill, the molten metal giving off a blinding light. The imagery seems drawn from the tradition of the sublime in painting.

In the above passage, the workers were reduced to picturesque details in the vast industrial landscape. Dreiser did the same in his description of a steel mill outside Buffalo. He concluded: "This matter of manufacture and enormous industries is always a fascinating thing to me, and careering along this lake shore at breakneck speed, I could not help marveling at it. It seems to point so clearly to a lordship in life, a hierarchy of powers, against which the common man is always struggling, but which he never quite overcomes, anywhere."[64] This vision was not a grim finality but an exhilaration, glimpsed "at breakneck speed."

The public's fascination with production was at its height in the early years of the twentieth century, and many corporations began to throw open their doors to visitors. The H. J. Heinz Company, one of the first to do so, had 20,000 visitors a year as early as 1900. Tourists at Niagara Falls also made excursions to nearby factories, most of which were driven by electrical power generated from the falls. By 1907, visitors to the Shredded Wheat Company—advertised as "the factory with a thousand windows," whose size and complexity, brochures declared, rivaled the falls themselves—numbered 100,000 per year.[65] General Electric, National Cash Register, and Hershey Chocolate soon established regular tours, which they found to be an effective form of public relations. A company could explain far more about itself in a tour than in an advertisement, making a deeper impression.[66]

In effect, the factory tour asked people to parade through the production facility. It was a long way from the 1828 Baltimore procession, in which workers had paraded by the assembled townspeople, to these pilgrimages of 1900. In the early stages of industrialization the skilled artisan had been a central figure, but in the new factories complex tasks were increasingly subdivided and the work process was organized by managers. Labor's loss of status can be charted by means of the photographic representation of workers. In 1890 they

The Heinz Company's buildings. Courtesy of Smithsonian Institution.

faced the camera frontally and took a variety of individualized poses. A generation later, workers more commonly were marshalled into orderly ranks, often in company uniforms that effaced their differences.[67] The arrival of tourists inside the factory was another sign that management controlled the life of the plant. Tours made workers into minor actors in the drama of production, which focused on systems of machines and the cornucopia of goods produced.

With the success of factory tours, it was a natural step to rethink the displays at world's fairs, which had been devoted almost exclusively to products arrayed side by side with those of competitors. By the 1890s corporations began to bring models of their factories to the fairs. For example, General Electric had a model of its main Schenectady factory on display at Buffalo in 1901. However, as Marchand has pointed out, full-scale industrial displays soon became standard fare at expositions. The coal companies of Pennsylvania set up an operating anthracite mine at St. Louis in 1904 as part of a 13-acre "Mining Gulch" described as "a shallow ravine, extending south from the Mines and Metallurgy building [and] filled with mining and metallurgical exhibits in actual operation." Nearby stood a functioning cement factory and "oil boring outfits in actual operation." The Palace of Machinery contained a 500-ton steam engine as large as a three-story house, and other "great engines of all descriptions, nearly every one larger than the famed Corliss type."[68] St. Louis was the apotheosis of heavy industry at world's fairs, and it consumed more than 10 times the power used at Buffalo just three years earlier. There was nothing unusual in scale or intention when U.S. Steel erected a working blast furnace and Ford set up a full-scale assembly line at the San Francisco Panama-Pacific Exposition of 1915. The assembly line operated for 3 hours each day before a capacity crowd. Thousands of people, mostly male, crowded in to see this new marvel of manufacturing technique. The Ford exhibit's manager, Frank Vivian, later recalled: "They always said at that fair, 'If you're looking for a woman on the fair grounds, go up to the Food Building, but if you want a man, you've got to go down to Ford's.'"[69]

The assembly line epitomized the productive capabilities of electrified factories. The smoky, grim landscape of the turn of the century gradually gave way to a cleaner industrial district as the steam engine was replaced by the electric motor. Inside and out, electric power made possible entirely new factory layouts. Just as elevators facilitated the rise of the skyscraper, electric cranes, conveyor belts, and lifts made it possible to move huge quantities of materials in continuous flows. The assembly line was impossible before electricity, because the overhead driveshafts dictated rather rigid arrangements of machinery. But an electric motor could be plugged in anywhere, and machines with unit drive could be arranged in any sequence to suit the flow of the work. Electric power eliminated the maze of overhead shafts and belts that once had cluttered the visual field, opening up the factory space; moreover, it was cleaner. The buildings that housed the new production lines were much larger than the factories of the steam age; at the same time, they were lighter and airier. For all these reasons, the electrified factory was far more efficient, permitting simultaneous increases in productivity and real wages.[70] In comparison with what preceded it, the electrified factory appeared as an immaculate, uncluttered, rational space.

The most popular of the new factories was Ford's Highland Park Plant in Detroit, which opened on January 1, 1910. Although not originally designed with the assembly line in mind, it was a new-style facility. Electrical drive was used throughout the plant. The floor plan was more flexible than those of older facilities. The buildings were spread over a large area, providing managers with space that encouraged innovation. Electric craneways moved material efficiently from railway cars to manufacturing areas. The spaces were well-lit and uncluttered, as overhead drive, belts, and gearing systems had all been eliminated. Henry Ford, who had been almost unknown, became a national figure after introducing the assembly line (1912–13) and the $5.00 day (1914) at this factory.

The excitement caused by the automobile itself should not be overlooked here. In 1899 an automobile trip from Cleveland to New York was such a novelty that it merited extensive daily press coverage, and crowds turned out all along the route. When the car arrived in New York, a million people turned out to see it.[71] That only a decade later ordinary people might be able to own an automobile seemed wonderful to most Americans, and they wanted to see the new means of production by which all those cars would be built.

A full-time staff of 25 people took visitors around the Ford plant. Even before 1920 a typical day saw 3000 or more visitors touring the factory. President William Howard Taft came and pronounced the assembly line "wonderful, wonderful." Many businessmen made special trips to Detroit as though to a shrine. One journalist declared that the factory had become "a national landmark and a new Niagara Falls."[72] The larger River Rouge Plant, which Ford built in the 1920s to cope with the increasing demand, was an even more popular tourist site. Just outside Detroit, it was the world's largest factory, and it did not merely assemble motorcars. At this single location Ford smelted steel, produced light and power, made virtually every part for the automobile, and assembled these parts on the assembly line. Here was a facility built on a vast scale and making a product that manifestly was transforming the United States. In his best-selling autobiography, Henry Ford brashly proclaimed: "We shall learn to be masters rather than servants of Nature."[73]

In 1923, when Ford was at the height of his popularity, William Stidger published *Henry Ford: The Man and His Motives*, which contained a meditation on the River Rouge Plant.[74] Stidger (a clergyman) celebrated the "romance" of the factory, which seemed to be a "miracle" as it turned raw materials into automobiles in just three days. To gain a perspective on the factory, Stidger stood on a high place above the grounds, obtaining a geometric view of the whole structure. A full tour of the factory almost invariably invited the visitors to share the magisterial overview of the plant so as to see the whole range of individual tasks as part of an enormous design. From his vantage point, Stidger described the "four great piles of raw products below us in the dim light of a snowy winter's day; iron ore, coal, limestone, and wood." "They looked," he says, "like a miniature mountain range in themselves, these great piles, gaunt and grimy and raw—dirty and elemental; elemental as the crater of a volcano into the mouth of which I have looked down in Java."[75] He later refers to them again as "four great mountain peaks," thus underscoring the notion that he is looking at a man-made nature. In this vision, nature is but raw material—grimly impressive, volcanic, stigmatized as raw and dirty—in comparison with the "romance of the River Rouge plant." Stidger presents the factory in terms of the natural sublime. Gazing into the boilers of the steel mill, he writes, "we looked through blue glasses and saw typhoons of flame leaping white against the brick walls, Niagaras of tumbling, turbulent, tumultuous, white waters of flame and fire, awe-inspiring, soul-subduing romance! Romance! Romance of Power!" Kant saw the sublime in a hurricane; Stidger sees the steel mill as a man-made cataract of flame and fury. Kant's natural forces reduce man to a feeling of weakness; Stidger feels intoxicated by the revelation of human mastery—"Power! Power! Power! That is the source of the romance of the River Rouge plant." The factory combined the geometric and the dynamic sublime. Seen close up, its productive processes reveal frightening yet exhilarating forces under human control. Seen from a height, it was a vast man-made nature reduced to geometric piles of materials.[76]

Stidger, like many writers of the first decades of the twentieth century, used the word 'romance' virtually as a synonym for 'sublime', referring to a startling transformation of "raw" nature into disciplined order on a massive scale. A power plant or a steel mill was a particularly apt subject for such writers. Moreover, this industrial sublime reveals, like the natural sublime, a spirit that animates the scene and operates behind it. In *The Olympian* Oppenheim calls this spirit "the industrial mother":

And it seemed to these two [observers] as if there had been bared to them, in this amazing spectacle of men and flame and machinery, the very birth-throes of the anguished Industrial Mother. Her ingots roared in the "wringers," her engines shrieked and clattered in the yards, her rolls and wheels and mighty furnaces crashed and clanked and screeched.

They beheld the birth of Steel. Guide-led over the peril of the switches, engine-wafted from one mill to another on either side the river, they stood first in this pit, then in that, flamelit midgets in Vulcanic sheds.[77]

In this description workers are absorbed into a larger whole, and "the huge and sweating laborers" are "like midwives assisting the mother." These men are "scorched and blinked in a glare of fire"; far from being alienated, however, they are "all passionately intent, desperately speeded" and assisted by "the black arms of roof-lost cranes" while "the white-hot ten-ton ingot or the splash and vapor of fluid iron or the bumping of the dinky engine writhes round them monstrously." Oppenheim continues:

Like Dante following Virgil these two invaded circle after circle of this lurid, beautiful Hell—the blast-furnace releasing fluid iron that fell with shower of white flakes and cloud of white smoke glaringly into the ladles; this sputtering iron poured into the egg-shaped Bessemer converter that blew air through it till it changed into steel, while the heavens above flamed and shuddered; the fluid steel caking into ingots in the molds and carried like a row of little men on the flat cars to the next mill; the hand of the crane seizing ingot by ingot and lowering them like lost souls into the withering-white soaking pits in the floor;

the same electric crane lifting them when they were reheated white-hot, pushing them on the rolls that swung them back and forth through the "wringer" until they were pressed into steel sheets, while the hot metal roared like hungry lions.[78]

Whereas Wordsworth had seen "almost a soul imparted—to brute matter," in the industrial sublime Oppenheim sees the worker absorbed into the landscape of production (a "lurid, beautiful Hell"). As with the Corliss engine, machines are given human qualities. The cranes have hands, the furnaces screech, the molten steel roars, and the ingots have souls. The plant is metaphorically transformed into a female body, the "Industrial Mother," that gives birth to steel. This rhetorical strategy naturalizes the steel mill even as it becomes terrifying and sublime. Feminization has the effect of making technology less threatening. The metaphor of birth suggests that man is only drawing out of nature powers that lie latent inside her.[79]

Between 1830 and 1930 the factory had passed through three distinct stages of development that corresponded to the three forms of power employed: water, steam, and electricity. All three measured man's power over nature, but distinctive architectural forms and aesthetics were suited to each. A water mill had to be placed beside a stream, and the stream's flow limited the factory's size. Even the Amoskeag and Lowell factories, which reached impressive proportions, were at first perceived to be in harmony with the natural order. In contrast, the grimier steam-driven factories, located along the railway lines, dominated their surroundings and were understood to be dynamic, unnatural environments. The electrified plants were still larger and more productive but seemed serene when compared to the smoke-billowing coal-fired factories, and many observers were struck by this paradox. The huge electric stations and dams that emerged after c. 1910 produced power for flexible use anywhere. They seemed to require virtually no workers, and they suggested a quiet, streamlined, antiseptic industrial landscape. The power station, and more particularly the hydroelectric dam, recaptured some of the qualities of the pastoral mill. Seen from the outside, both were serene and silent, and both seemed more in harmony with nature than the steam-driven factory. But the huge scale of the new power stations and their potential for transforming society made them more imposing than anything that had preceded them.

The public toured new electric power stations in much the same spirit in which they visited other industrial plants. Power plants seemed to epitomize new values that were emerging as the industrial system converted from steam to electricity. Less than 4% of all factory power was electrical in 1900; just over 50% was in 1919.[80] The new power stations became symbols of modernity and the future. On his tour of the United States, Arnold Bennett encountered in New York, along with the skyscraper, this new American tourist site:

. . . an enormous white hall, sparsely peopled by a few colossal machines that seemed to be revolving and oscillating about their business with the fatalism of conquered and resigned leviathans. Immaculately clean, inconceivably tidy, shimmering with brilliant light under its lofty and beautiful ceiling, shaking and roaring with the terrific thunder of its own vitality, this hall in which no common voice could make itself heard produced nevertheless an effect of magical stillness, silence, and solitude. . . . It was a hall enchanted and inexplicable. I understood nothing of it. But I understood that half the electricity of New York was being generated by its engines of a hundred and fifty thousand horsepower, and that if the spell were lifted the elevators of New York would be immediately paralysed, and the twenty million lights expire beneath the eyes of a startled population.[81]

The immense machines seem to function almost without human intervention. Man has tamed nature, extracting power from "useless" raw materials. The "conquered and resigned leviathans" do his bidding.

Powerhouses became tourist sites. Just as George Pullman displayed the Corliss engine outside Chicago, Henry Ford was so proud of the powerhouse at his Highland Park factory that he designed it with large glass windows so passersby could marvel at the great machinery, which, like that in New York, was kept immaculately clean. Stilgoe notes that "in the central station urban Americans first glimpsed the beginning of a new era of ultra-efficient industrial production."[82] These powerhouses, "in their almost smokeless chimneys, tiny work forces, and brick and concrete walls, objectified the futuristic electrical force created within their walls." A critic wrote enthusiastically in *Architectural Forum* that "there is a feeling of grandeur and of poetry and of beauty in the orderly assembly of this modern, efficient and economical equipment."[83] The scale of the new powerhouses also called attention to them.

Like Ford's River Rouge factory, with its mountain range of raw materials, electrical stations maintained immense coal reserves to ensure that power could be generated despite bad weather or strikes.

Hydroelectric power stations could also arouse popular awe, so long as they did not interfere with established icons of the natural sublime. One of the first great battles between industrial and conservation interests broke out over the Hetch-Hetchy dam project near Yosemite. This was followed almost immediately by a fierce controversy over the uses of Niagara Falls, whose waters were rapidly being diverted to power stations. Yet the great electric stations at Niagara were also popular tourist sites, attracting many to see how Westinghouse had harnessed the waters to the largest hydroelectric power station in the world. In 1905 the two powerhouses at Niagara produced one-tenth of all the electric power generated in the United States—much of it consumed by industrial Buffalo, 30 miles from the point of generation.[84] The Niagara-Buffalo area offered a striking contrast between the natural sublime of the falls and the industrial landscape that grew up around it. After 1895 aluminum and chemical companies, including American Cyanamid, Union Carbide, Pittsburgh Reduction Company (aluminum), International Acheson Graphite, and the Carborundum Company, began to build large factories nearby to tap the inexpensive power.[85] Rail passengers traveling from Buffalo to Niagara contrasted the new factory districts with the cataracts and meditated on the meanings of industrialization.

For H. G. Wells, as for Theodore Dreiser, the new industrial developments were more inspiring than the falls themselves. Peter Conrad notes: "For Wells, the interest of Niagara lies not in the sublime frothing of the waters but in the 'human accumulations' which mark a scientific victory over the cacophonous waste: 'The dynamos and galleries of the Niagara Falls Power Company . . . impressed me far more profoundly than the Cave of the Winds; are, indeed, to my mind, greater and more beautiful than the accidental eddying of air beside a downpour.'"[86] Wells, Conrad points out, insisted that the power station was the human "will made visible, thought translated into easy and commanding things." He especially liked the turbines that whirled in "cloistered quiet." For such observers, the hydroelectric power station was a clean and silent industrialism. Conrad concludes: "From the natural sublime of the nineteenth century, which muddies the senses and wastefully disrupts nature,

Wells has passed to the mechanical sublime, which regulates the mind and technologically supersedes nature. . . . The turbid ghost of the falls now inhabits the sleek machines. . . . Niagara, to Wells, is a laboratory in which America re-invents itself, outgrowing the grandiloquently primitive nature" that it inherited. Wells articulated the attitude of the engineer, whose journals lavished attention on the new power plant. Local capitalists made the electrical transformation of the falls the major theme of the 1901 Pan-American Exposition in Buffalo.

But to many Americans Niagara's hydroelectric projects were problematic emblems of progress. As plans for more power stations went forward, the press discussed the cataracts' imminent demise. In *Cassier's Magazine* one author lamented: "Niagara Falls are doomed. Children already born may yet walk dry-shod . . . across the bed of the Niagara River."[87] Such sentiments may appear overwrought, but in fact by 1900 the state government had granted charters for companies to use virtually all the water flowing over the falls, and some engineers frankly proposed their total conversion to economic uses. By 1908 a single company was withdrawing from the river 8000 cubic feet per second, lowering the water level by half a foot. The exercising of further options held by corporations would shrink the crest of Horseshoe Falls from 2950 feet to 1600, placing them entirely on the Canadian side. Even many of the proposed remedies displayed a preference for technological solutions. One engineer suggested a dam across the Niagara River. Letters to *Scientific American* suggested that, although in the future the falls might be bone dry much of the time, the waters could periodically be diverted from the power stations "for scenic purposes."[88]

If many engineers and businessmen viewed the conversion of Niagara to industrial use as a logical development of the marketplace for energy, the general public was not prepared to treat the falls as just another piece of real estate. As Roderick Nash notes, Americans worried that their civilization was becoming entirely devoted to "Mammon," or the mere pursuit of wealth.[89] The dams at Hetch-Hetchy and the canals siphoning water from Niagara became symbols of crass commercialism to Frederick Law Olmsted, to the Colonial Dames of America, and to the American Civic Association. One protest article argued that the public was "not necessarily insensible to the glory of having at Niagara 'the power center of the world,' or blind to the fascination of unique hydraulic problems magnificently executed." Yet the author declared that the public also "finds a glory and magnificence in the sight of what nature has done here, which, compared with the success of a few industrial enterprises, is vastly for the greater good of the greater number." He concluded that "mournful indeed would be a mechanical triumph over this international inspiration."[90] Nevertheless, as late as 1911 *Scientific American* editorialized against the "hysteria" over Niagara and found it "strange" that "in all this discussion we hear nothing whatever of the good to come to humanity from allowing this immense falls to work out its board and lodging."[91] The technological object had collided head-on with the quintessential object of the natural sublime. If the American public was forced to choose, it chose the natural sublime. The result was the same in the 1960s—Congress faced a popular revolt when it briefly considered putting a dam in the Grand Canyon.

Yet in most cases new dams did not threaten to deface national icons, and the Army Corps of Engineers, contractors, and private utilities did not meet much opposition as they constructed hundreds of dams in the first decades of the century. During the 1920s and the 1930s hydroelectric dams became symbols of both technological progress and economic prosperity, both for the private sector and for the government. For example, the Philadelphia Electric Company constructed the Conowingo Dam across the Susquehanna River in the 1920s.[92] This 4648-foot structure impounded a 14-square-mile lake and produced 1.3 billion kilowatts of electricity a year. The federal government worked on an even larger scale with the many dams on the Tennessee[93] and the Colorado. When in 1936 President Roosevelt "pressed a golden key which put into the operation the first generator in the giant power house" at Hoover Dam, he emphasized the need for such technological projects "to create a new world of abundance."[94] In fact, the construction of the dam was a powerful stimulus to the local economy. The building contract for Hoover Dam was the largest yet to have been signed by the U.S. government.

The public did not understand the dams on the Tennessee and Colorado in merely utilitarian terms. Frank Waters declared Hoover Dam "the Great Pyramid of the American Desert" and "the Ninth Symphony of our day." To him and many others it appeared "in its desert gorge like a fabulous, unearthly dream. A visual symphony written in steel and concrete—the terms of our mathematical and machine-age culture—it is inexpressibly beautiful of line and texture, magnificently original, strong, simple, and majestic as

the greatest works of art of all time and all peoples, and as eloquently expressive of our own as anything ever achieved." Yet this
work of art is "wholly utilitarian and built to endure, it is the greatest single work yet undertaken to control a natural resource."[95] This
Emersonian formulation unites utility and art.

The public embraced Hoover Dam. As soon as its construction
began, in 1931, thousands of tourists came to see it rise from the
floor of the Black Canyon and to stare down into the chasm and
watch cranes lowering tons of concrete in huge swiftly moving buck-
ets that discharged their contents more than thirty times an hour.
Many preferred to watch the night shift, not only to avoid the searing heat but also to see the emerging dam bathed in artificial light.
In 1934–35 there were 750,000 visitors, making the dam as popular
as the Grand Canyon. For the dedication, thousands more came to
appreciate the dam from vantage points above it, stand on the roadway for the first time, drive across the structure, and take a guided
tour down to the hydroelectric stations at its base. The 10-acre powerhouse exemplified the dynamic sublime, with 17 enormous tur-

Hoover Dam under construction. Courtesy of Smithsonian Institution.

The visitors' gallery at Hoover Dam. Courtesy of Special Collections, University of Nevada, Las Vegas.

bines generating 5 billion kilowatt hours a year. It became a fixed point on the regional itinerary.

Hoover Dam had many of the qualities that Burke expected of a sublime building. Its construction could immediately be seen as a problem of great difficulty. Its dimensions alone made it a candidate for the mathematical sublime. It was the tallest dam yet built anywhere in the world, standing 726.4 feet above bedrock. Its base is as thick as two football fields, and its top is wide enough for three lanes of traffic. Each of the four water-intake towers behind the dam is 33 stories tall. The reservoir it created was the largest in the world, covering 227 square miles.[96] At its inauguration, "as the machinery started, a double waterfall, thirteen feet higher than Niagara, poured from the two opposite canyon walls below the powerhouse."[97] This was understood as the victory of human reason over a periodically rampaging river. Holding back such a flood behind its immense curved wall was a triumph of the mathematical sublime.

New Deal dams were not only works of functional engineering but carefully crafted landscapes. Richard Guy Wilson has pointed out that "the approaches to the dams were carefully studied, and while engineering and hydrological considerations dictated a dam's location, the visitor was brought to the dam in a dramatic manner, always seeing the structures within the landscape, as symbols of man's control of nature."[98] Certainly this generalization holds true for Hoover Dam approached from the Nevada side. It is first glimpsed from the road well above the site, and subsequently disclosed in more detail as one negotiates a series of switchbacks. As Joseph Stevens has noted: "The experience, especially the first glimpse of the dam from one of the hairpin turns in the road zigzagging down through the red and black cliffs, seldom fails to elicit a visceral response. The sight is unearthly, particularly at night, when recessed lamps illuminate the expanse of concrete and the tailrace below in a blaze of dazzling golden light. Confronting this spectacle in the midst of emptiness and desolation first provokes fear, then wonderment, and finally a sense of awe and pride in man's skill in bending the forces of nature to his purpose." It would be hard to find a more succinct restatement of the technological sublime, whose larger implications are contained in the following words from Stevens's book: "In the shadow of Hoover dam one feels that the future is limitless, that no obstacle is insurmountable, that we have in our grasp the power to achieve anything if we can but summon the will."[99]

Like many artists of his generation, Charles Sheeler explored the apparent omnipotence of industrialization. In six paintings commissioned by *Fortune* he depicted the central objects of the technological sublime: the water wheel, the railroad, electrification, and flight. Martin Friedman has observed of the series: "Sheeler always depicted power at absolute stasis. In his hermetic visualizations, power is not treated in terms of crashing strength but as an intellectualized concept with its mechanisms always in mint condition."[100] The immobility of these paintings creates a tension between the static forms and the reader's knowledge that all these objects move at great speed. Published together in a single issue of *Fortune*, Sheeler's images were presented like a photographic essay. The accompanying text asserted: "The heavenly serenity of Sheeler's style brings out the significance of the instruments of power he here portrays. . . . He shows them for what they truly are: not strange, inhuman masses of material, but exquisite manifestations of human reason." This instrumental "reason" should not be confused with Kant's transcendental reason. The text continued: "As the artists of the Renaissance reflected life by picturing the human body, so the modern American artist reflects life through forms such as these; forms that are more deeply human than the muscles of a torso because they trace the firm pattern of the human mind as it seeks to use co-operatively the limitless power of nature."[101] *Fortune* echoes Emerson: the machine stencils the pattern of the mind onto the world, allowing human beings to make nature an extension of their will. Sheeler's "Suspended Power" depicts one such exquisite machine: an enormous water turbine about to be installed in a TVA hydroelectric plant. It hangs above the precisely calibrated hole where it will lie perfectly level, receiving the rushing waters that will turn its blades. The surfaces are so clean and uncluttered that the turbine becomes almost an abstraction, which in fact it is when compared to the photographic studies Sheller made in preparation for the painting.

The next painting in Sheeler's series, "Conversation—Sky and Earth," presents transmission lines at Hoover Dam as mediators in the hydraulic cycle between earth and sky. The powers of this cycle are contained for human use by the dam, whose concrete sides and rigid steel towers thrust sharply up into heaven and dominate the earth. Both the upward angle of vision and the sharp contrast between the intensely blue sky and the steel gridwork emphasize technological dominance. Man seems to master nature, taking control of elemental forces and siphoning them off for use. *Fortune* emphasized that the painting would "give the spectator so close a sense of reality that he is thrown into a reverie such as the grandest natural scenes can sometimes induce." Instructing readers in the technological sublime, the text explained that the mood induced "is not a vague, subjective reverie of the sort an autumn sunset may inspire in an adolescent. It is rather an active and inquiring wonderment, subtly directed by the artist. . . ." For *Fortune* the painting suggested rhetorical questions about man's ability to grasp and direct the power of nature:

What is this incredible, elusive power that man has taken so magnificently from the waters and the hills? What unguessed secrets of the universe are hinted by its transmission unchanged through unchanging strands of copper or aluminum? In what way have men's minds, grappling with the raw phenomena of lightning and magnetism, managed to contrive so swift and carefully guarded a channel for a force that no one can fully apprehend?. . . It is not surprising that the modern scientist, confronted with such questions and with their partial answers, which open up still further questions, should often be a man of deep religious feeling. And it is not surprising either that the modern artist, depicting such a scientist's handiwork, should put a devout intensity into painting. This is as truly a religious work of art as any altarpiece, or stained-glass window, or vaulted choir.

Religion remained a part of the American sublime, as it had been for the pilgrims to Niagara. *Fortune*'s editors cited Henry Adams: " . . . electricity has become as universal an element of modern life as fire, water, earth, and air were in the ancient world. Sensing this universality, Henry Adams saw a relation between the powerhouse and the Cathedral of Chartres, between the Virgin and the dynamo."[102] By citing Adams *Fortune* sought the luster of profundity, but the core of his argument was missing. Adams had feared the new "kingdom of force" the dynamo represented, and he had preferred the medieval symbol of the Virgin. Adams had doubted the power of reason to comprehend and control the world, which he feared was accelerating toward entropy and chaos rather than toward progress and order. Yet the majority followed not Adams but Ford and *Fortune*. The sense that the machine might rage out of control was restricted to a minority, while the dynamo and the electrified factory became icons of the technological sublime.

6

The Electrical Sublime: The Double of Technology

Hoover Dam was part of an electrical landscape, built after 1880, that had spread from expositions and special events into all areas of public life. Electricity was invoked as the panacea for every social ill and the key to a whole range of social and personal transformations that promised to lighten the toil of workers and housewives, to provide faster and cleaner forms of transport, and to revolutionize the farm. Beyond the obvious utility of electricity in virtually every sector of the economy and in domestic life, the new forms of lighting transformed the appearance of the world. Dramatic demonstrations of arc lights began in the late 1870s and seemed to offer visible proof of the coming changes. The Brooklyn Bridge or a skyscraper, when studded with electric lights, took on an entirely new appearance. As these changes filled the urban night, a shimmering new world came into being. The electrified urban landscape emerged as another avatar of the sublime.

This landscape was first visible at the international expositions, and even a cursory examination of these fairs reveals at least two forms of sublimity. The Philadelphia Centennial of 1876 marked a high point of the technological sublime at world's fairs, though it was still a strong element of the fairs in Chicago, Buffalo, and St. Louis. At these later expositions, electrical technologies, particularly spectacular lighting displays, embellished and then eclipsed the great machines.

The anthropologist Burton Benedict found that expositions "promulgated a whole view of life. They created a world in which everything was man-made. Nature was excluded or allowed in only under the most rigorously controlled conditions. . . . At world's fairs man is totally in control and synthetic nature is preferred to the real thing."[1] Individual machines were less impressive than large systems, such as the mechanical power train of shafts and belts driven by the Corliss engine at the Philadelphia Exposition, a large railway shipping yard, a spinning mill, or an orderly cityscape seen from a skyscraper. However, mechanical systems had limits beyond which they could not expand without the aid of electrical technologies, and those technologies proved an ideal medium for the expression of human control over nature. At expositions they facilitated the integration of displays. Uniform electric lighting could provide a sense of coherence to the whole, and electric motors required little attention and could be easily located almost anywhere, thus permitting the realization of display ideas that were impossible with mechanical drive. Much as a railroad or any other large enterprise virtually required an extensive electric communications system (first the telegraph and later the telephone), modern factories and expositions were impossible without electric motors, ventilation systems, elevators, and artificial lighting, which together permitted expansion beyond the limits of older buildings.[2] While electricity itself was invisible, electric lighting became a visual representation of the new force.

Spectacular illumination had long been a feature of public ceremonies, but the earlier displays had obvious technical limitations. Torches, oil lamps, and bonfires, used in ceremonies since ancient times, required continual refueling and created disagreeable smoke. Fireworks, introduced into Europe from China, became a common feature of state occasions and an effective way of demonstrating social power through conspicuous display. Though the early forms of spectacular lighting lasted for only short periods, they temporarily made it possible to triumph over darkness, which Burke argued was imbued with fear and terror: "We have considered darkness as a cause of the sublime. . . . In utter darkness, it is impossible to know in what degree of safety we stand; we are ignorant of the objects that surround us; we may every moment strike against some dangerous obstruction; we may fall down a precipice the first step we take; and if an enemy approach, we know not in what quarter to defend ourselves." In this situation, "strength is no sure protection; wisdom can only act by guess; the boldest are staggered, and he who would pray for nothing else towards his defense, is forced to pray for light." For most of human history, people confronted darkness as an inevitable part of the daily round. It was a kind of natural force, and early forms of lighting were but temporary stays against its overwhelming power. At another point in his argument, Burke noted that while darkness was sublime, "a quick transition from light to darkness, or from darkness to light" produced an effect on the mind that was still more powerful.[3] Sudden and extreme contrasts were the essential ingredients to luminous sublime effects, which could be further enhanced by movement— particularly very rapid movement, as with lightning. Fireworks possess these qualities, and this suggests why they became an essential part of state occasions.

In the nineteenth century fireworks formed the closing act of any important celebration, such as the Fourth of July. Early fireworks displays should not be overlooked because they were small by comparison with later efforts. The distance measured by the sublime is that between ordinary sensory experience and the irruption of a novelty. The first sight of rockets and Roman candles produced astonishment and awe. During the second evening of the Boston Railroad Jubilee of 1851 a special exhibition of fireworks was given in front of the Revere House, where the American president and the governor general of Canada were staying. In addition to the "usual display of rockets, bombs, Roman candles, and the like, there were several set pieces; —one of which was a representation . . . of a steamship, decorated with flags flying." A larger fireworks display was staged in East Boston for the thousands of onlookers who lined both sides of the harbor, where the "reflection of the fires from the intervening waters gave additional splendor to the view."[4] As was noted in an earlier chapter, fireworks were prominent at the dedications of the Eads Bridge and the Brooklyn Bridge. The organizers of public events looked for other forms of spectacular illumination, including gas lighting and, in the theater, limelight. In the 1870s, powerful electric arc lights were invented and proved ideal for new kinds of displays. Electric lighting, unlike fireworks, was not confined to a few moments, and it quickly became a constant and dominant element of spectacles and processions in the late nineteenth century.

Dramatic lighting made possible the revisualization of landscapes, filling them with new meanings and possibilities. It took the technological sublime in a new direction, displacing attention from particular machines or man-made structures to a set of visual effects. Between 1880 and 1915 electrical engineers found ways to re-present virtually any object with light, so that cultural meanings could be altered as the object was written upon, edited, highlighted, or blacked out. The engineers who developed these effects became showmen. Two important early figures in this development,

Luther Stieringer and William Hammer, exemplify the small class of men who would later be known as "illuminating engineers." Stieringer had worked with Thomas Edison in designing the distribution system for electric lighting. From 1883 until his death in 1903 he specialized in exhibition lighting, working as either chief or consulting engineer at virtually every important American exhibition or world's fair. Hammer was another Edison alumnus. After exhibition work in the 1880s, he pioneered the field of electrical signs, making many of the technical innovations that led to the spectacle of the "Great White Way" in New York. These two men were among the first to realize incandescent lighting's combination of safety and theatricality. In comparison with gas, there was considerably less risk of fire or explosion, no flickering, and no consumption of air, so that interiors could be safely lighted. Electric lights also offered ease of control and a variety of colors for special effects.[5]

By 1907 the profession had its own association and its own journal (*The Illuminating Engineer*). By the 1890s every major electrical corporation had two or three such experts who created electrically illuminated displays for expositions. These technological artists manipulated light, shadow, and color to emphasize a fair's central buildings and symbols. Between 1885 and 1915 they created night landscapes that awed a generation of Americans, and they did so in full awareness of the pre-electric uses of lighting to create spectacular effects. An early issue of *The Illuminating Engineer* carried an extensive article on "Spectacular Lighting in the Past" that sketched developments from ancient Rome to eighteenth-century France.[6]

The profession developed as each exhibition attempted to outdo its predecessors in producing new man-made wonders. This competition was further fueled by the new national corporations contesting for public attention. By the 1880s electricity was drawing vast crowds to expositions that previously would have closed at dusk. At the Louisville Southern States Exposition of 1883 the Edison Electric Lighting Company installed 4600 light bulbs, creating a sensation and drawing huge crowds. For a few months, Louisville was the brightest spot in the United States.[7] This was the last exposition that considered using gas for illumination. Electrical effects were quickly perfected at a series of smaller events in the ensuing decade. The following year, the International Electrical Exhibition in Philadelphia featured a fountain, 30 feet in diameter, whose fifteen jets were periodically made the center of attention as all other illuminations were extinguished and special electric lights, both colored and filtered, were trained on the streams of water. Equally striking was the Edison exhibit, a 30-foot column covered with more than a thousand bulbs. It too was lighted only at specified times, and here the special effect was that the lights appeared to wind around the pole.[8] The *Journal of the Franklin Institute* noted an advantage of decorative lighting that was soon grasped by every exposition planner: it had a remarkable ability to produce "artistic effects." The article continued: "By its aid, the crude buildings hurriedly erected without any attempt at finish for a temporary purpose, were transformed into a temple of light, which at the first glimpse evoked expressions of delight from every beholder."[9]

This observation was not lost on later organizers. In 1888, at the Centennial Exposition of the Ohio Valley and Central States, 100 arc lamps were used outside the exhibition buildings. Inside, a local newspaper reported, William Hammer "made the incandescent light as facile in his skilled hands as a brush in the fingers of a painter." Indeed, he was rightly understood as an artist: "He produces color effects with the skill of an aquarellist. He produces water effects with his light that will startle the piscatorial inhabitants of the fountains and lakes at the Exposition. . . . Mr. Hammer will astound the people with the marvelous exhibition he has perfected."[10] Some of these "marvels" seem quaint in retrospect, such as putting light bulbs in the eyes of carved animals, hanging illuminated Japanese fish high in the air, or setting up an illuminated Christmas tree. Hammer also staffed the entrances to the hall with attendants wearing "helmets surmounted by a powerful electric lamp" which were connected by wire to the heels of the attendants' shoes so that they blinked on whenever the attendant stepped on a steel plate. Hammer's most impressive effect was an electric fountain beneath the dome of the main building that shot water 60 feet into the air, illuminating it with colored lights. The room was periodically darkened to dramatize the effect for the spectators, whose applause suggests the validity of Burke's observations on contrast and celerity in lighting. Although these special effects may seem tame today, in 1888, when less than 1 percent of American homes had electric light, they were startling novelties.

By 1894 the centrality of electric lighting at fairs was firmly established. The Columbian Exposition in Chicago had more lighting than any city in the country. Some visitors saw more electric light in a single night at the fair than they had previously seen in their entire lives. Special effects had been developed further—notably the electric fountain. At Chicago, fountains shot water high into the air and wove complex patterns against the night sky. To further dramatize the scene, the spotlights were fitted with filter systems that permitted the operators to create symphonies of color. Two giant electric fountains, one at each end of the Court of Honor, spewed 44,000 gallons of water a minute into kaleidoscopic variations. Perhaps because the photographic technology of the day could not capture these displays, they have been overlooked by later historians. Yet by all accounts they were among the most popular events. One observer noted that long before 6 P.M. crowds began to gather around the Court of Honor:

Slowly yet continuously came the troops of people . . . eager to secure a good position from which to behold the illuminations, when the mantle of night should have fallen upon the White City. On these great occasions there is so much to see, the attractions cover so wide a range of territory, that it is no easy matter to obtain a position where all can be surveyed. The electric fountains and Administration Building in a blaze of glory are at the west end; the magnificent pyrotechnic display is eastward of the lake; the surface of the grand basin is covered with floats from which shoot up numberless fiery serpents; all along the roofs of the Agricultural and Liberal Arts Buildings are lines of flickering flambeaux. Long before the display begins the Grand Plaza and the margin of the basin are crowded with the expectant throng.[11]

The most spectacular view was that from atop another technological marvel invented for the Chicago Fair: the Ferris Wheel, itself studded with light bulbs. It afforded a panoramic view of the exhibit buildings and the amusement zone. Even Henry Adams, who later would meditate on the electrical dynamo as the symbol of a new force let loose on the world, was not immune to the fair's charm. He wrote to a friend: "We haunt the lowest fakes of the Midway, day and night. We have passed our evenings on the water in the administration launch, looking at fireworks and electric fountains; we have turned somersets in the Ferris Wheel, and have been robbed of our surviving dollar."[12] Theodore Dreiser preferred the view from an electric launch, where "a feeling of the true dreamlike beauty of it all came to me, at first only as a sense of intense elevation—not wonder, but elevation at being permitted to dwell in so Elysian a realm. . . .Then followed an abiding wonder."[13] The scene also entranced the American impressionist painter Childe Hassam, who produced several works based on the nightly illuminations.[14] As these responses suggest, electrical illumination was a powerful tool that distracted and charmed even the most critical observers.

These displays in Chicago were so successful that the organizers of Buffalo's Pan-American Exposition of 1901 selected electricity as their central theme, in order to celebrate the first hydroelectric power station at Niagara Falls. At the center of the fairgrounds they built a 400-foot electric tower, covered with 40,000 small bulbs, with a 60-foot model of Niagara Falls gushing from its side. Rather than blind visitors with powerful lighting, Buffalo's engineers decided to use 200,000 small incandescent lights in the fair's Grand Court, bathing it in an even diffusion of light so that it seemed a huge impressionist painting. The exclusive use of incandescents instead of arc lights permitted precise control over highlighting, contrast, and color, making the exhibition grounds into a subtle work of art. The Electric Tower and the Grand Court became the scene of a ritual at dusk, when all lights were extinguished and the waiting crowd fell silent, anticipating the illumination. This carefully orchestrated event began with a pink glow like the flush of dawn on the Electric Tower, which then deepened into red and shifted slowly to yellow as other lights flashed on across the grounds.[15]

A night view of the Pan-American Exposition at Buffalo, 1901. Photograph by C. D. Arnold; courtesy of Collections of Library of Congress.

At the Louisiana Purchase Exposition of 1904, lighting remained a central part of the fair experience. As an official history put it, it was "best to see the illumination at first from a considerable distance":

Riding around a curve on a trolley car, or topping the brow of a hill, one suddenly became aware of something wonderful in the distance, a mighty bouquet of light blossoming out of the darkness. For half a mile the flowers of light sparkle in the murk, clear, clean-cut, golden. The distance not only lends enchantment to the view, but mellows the scene to a soft glow, soothing to the eyes. One beholds glowing though the darkness, long lines of little lights, broken here and there into fantastic designs. Now a huge star breaks out, made of many lights. Yonder is circle after circle of gleaming brilliances, far up in the sky. Still higher up is outlined a skeleton framework of lights, and you know that it is the illumination of a tower, though you see nothing whatever of the tower itself.[16]

Electricity dematerialized the built environment of the fair, transforming its buildings into enchanting visions, here naturalized as "flowers" and "blossoms." On closer approach, "the darkness gradually melts from the vicinity of the little lamps, and you perceive the ivory-tinted exteriors of the huge buildings, glowing in the light of thousands of lamps. Stepping into the edges of the main picture, you are entranced by the scene. Lagoons and plazas and broad

The St. Louis Exposition at night, 1905. Stereograph by H. C. White Co.; courtesy of Collections of Library of Congress.

thoroughfares for promenade are made as bright as day." Electric fountains use "all the colors of the prism," and the "greensward . . . becomes a plain of fire." Further, "the flower beds take on fantastic hues," and "the lights change and change in bewildering variety."[17]

Extensive special effects required the organizers of expositions to work closely with local utilities and manufacturers. After 1892 the electrical manufacturing industry had been reduced to the duopoly of General Electric and Westinghouse, and each of these companies accepted large blocks of utility stock as payment for new generating equipment. The industry was held together not only by these formal ties of ownership but also by the National Electric Light Association, which took an active part in organizing fairs. Almost every fair included a vast electricity building, the size of a football stadium, devoted to the latest technologies. No one could visit a world's fair between 1883 and 1915 without seeing spectacular lighting displays and ingenious electrical devices—electric elevators, boats, trains, amusement rides, and even moving sidewalks.

The spectacular electrical effects at world's fairs announced the emergence of a new form of the sublime that may be understood as an extension of Burke's aesthetic. Although Burke died several generations before the advent of spectacular electrical illuminations, he observed that "lightning is certainly productive of grandeur, which it owes chiefly to the extreme velocity of its motion."[18] The dramatic lighting in Chicago and Buffalo relied in good part on such sudden transitions, and in 1904 the St. Louis World's Fair introduced a new technique by which lighting was made to flash out from a central point across the fairgrounds, moving like lightning on the man-made horizon. Such demonstrations no longer sought to use technology to transform nature into Emerson's "double of man"; rather, they used electricity to create a double of technology, transforming a scene that was already spectacular into a more powerful vision—one that was colored, edited, simplified, and expanded.[19] The illuminations stunned crowds in Buffalo and St. Louis, where applause was followed by silent admiration. Spectacular illuminations combined the mathematical and the dynamic sublime as the spectator encountered both extreme magnitude and irresistible power.

The use of electric lighting to dramatize both man-made and natural landscapes became widespread, often being employed at tourist attractions. From 1895 on, amusement parks used huge amounts of electricity to create dazzling environments and to drive their rides. They copied from Chicago the Ferris Wheel, the electric fountains, and artistic lighting.[20] Such parks manufactured an environment of illusion and special effects which the individual could seek at stated times. In effect, the electrical sublime made a single site the location for two quite different experiences. F. Scott Fitzgerald noted this in his short story "Basil: A Night at the Fair," in which a young man visits a state fair at two different times: "Once again the fair—but differing from the fair of the afternoon as a girl in daytime differs from her radiant presentation of herself at night. . . . The substance of the cardboard booths and plaster palaces was gone, the forms remained. Outlined in lights, these forms suggested things more mysterious and entrancing than themselves, and the people strolling along the network of little Broadways shared this quality, as their pale faces singly and in clusters broke the half-darkness."[21] At night the fair's lighting transforms the tawdry and elevates ordinary people into temporary residents of a mysterious realm.

Kant's sublime made the individual humble in the face of nature, the technological sublime exalted the conquest of nature. The electrical sublime represented a third kind of experience, as it dissolved the distinction between natural and artificial sites. In blurring or even erasing this line, it created a synthetic environment infused with mystery. For example, in the 1920s the Natural Bridge of Virginia was opened for impressive nighttime ceremonies in which electrical technologies were combined with Old Testament theology. Visitors were escorted to seats beneath the arch in total darkness:

Then very slowly a dim light increased in intensity until the arch was clearly to be seen. Soft music swelled in volume as the light grew brighter. Then a man's voice, amplified to fill all the space, began the verses in the first chapter of the book of Genesis, reading with impressiveness, and pausing as the music, becoming of greater volume, echoed in rich, sonorous tones from cliff to cliff beneath the mighty arch, classical selections from the old masters, Verdi, Liszt, Wagner, Chopin. . . . Except for the music, and except for the murmur of the stream no sound might be heard; the reverent silence was a higher tribute than would have been the loudest applause.[22]

Awed silence is, of course, the sublime response. The reading from Genesis and the classical music contextualized the event; technology created a new sensory experience of the Natural Bridge.

Scenery, orchestral music, dramatic lighting, and biblical oratory were fashioned into an overwhelming display that contained its own interpretation. Whereas Burke had imagined light as a minor element of architectural design, as a contrast to darkness, the electric sublime was a manufactured experience that reduced the night to a background. Whereas Kant had expected the individual to draw the correct transcendental conclusion from a sublime encounter with nature, the electrical sublime produced awe on demand and ensured it would be understood within the interpretive framework of the impresario. Such overdetermined displays were hardly limited to natural monuments—one could be developed for any site, even a site as large as New York City.

The Hudson-Fulton Celebration held in New York in 1909 was one of the most spectacular public events of the first decades of the twentieth century. Authorized by the state legislature in 1906, its organizing committee included generals, prominent politicians, members of distinguished families, and several of the wealthiest men in the United States, including Andrew Carnegie and J. Pierpont Morgan.[23] The legislature called upon all the cities in the Hudson Valley from New York to Albany to commemorate both Henry Hudson's voyage of discovery in 1609 and Robert Fulton's steamboat, first launched on the Hudson River in 1807. Together, Hudson and Fulton represented the discovery of America and its industrial development. The celebration staged to recall their achievements included "the greatest fleet of war vessels ever gathered in American waters . . . forming a line nine miles long"—52 American warships plus flotillas from England, Germany, France, Holland, Mexico, Argentina, and other countries.[24] The French magazine *L'illustration* interpreted this event in terms of the United States' emergence as a great naval power,[25] but the Hudson-Fulton Celebration can also be understood as part of a domestic tradition of public festivals. Indeed, the celebration marks the full expression of a new form of spectacle that emerged out of the tradition of republicanism.

Whereas the natural sublime was oriented toward the eternal now and the technological sublime toward the future, the Hudson-Fulton Celebration was intended to have a dual focus, looking toward the past in its parades but embracing the future in its electrical displays. Like the Erie Canal celebration of 1825, the Hudson-Fulton Celebration was one of the largest public events yet staged in the United States. More than a million people journeyed to New York to see it, driving up prices for hotel tickets and for good seats at the parades, banquets, illuminations, and other special events that went on for an entire week. Nor was this merely an event for the wealthy or for out-of-towners. A reporter noted extensive decorations in shops in the poorer sections of the city, including not only flags and "the familiar Hudson-Fulton monogram" but also hand-made objects. On Mulberry Street a storekeeper displayed a miniature warship made of copper. The owner of a drugstore on Ninth Avenue tried to sell hair tonic by displaying "a number of pictures of [the balding Robert Fulton and Henry Hudson] of the 'before and after' style" with the inscription "It would have done them some good. It will help you." Wandering further, the reporter found that on the East Side "most of the decorations [were] mingled with Hebrew characters explaining the reason for the celebration." One delicatessen had "a large papier-mâché bologna surrounded by dozens of small electrical lights."[26]

The celebration began with a naval parade on Saturday. As the replicas of Hudson's *Half-Moon* and Fulton's *Clermont* sailed along the shores of Manhattan, they were greeted with a continuous cannonade from the assembled warships. On Monday an official reception was held at the Metropolitan Opera House for 2000 guests, who gave a huge ovation to Julia Ward Howe, author of "The Battle Hymn of the Republic." By 1909 she was in her nineties, but she had written a new work celebrating the steamboat especially for this event. This 46-line poem, "Fulton," was not memorable, but Howe's presence was an impressive survival from another era. When she finished reading, a large glee club sang the "Battle Hymn," joined by most of the audience on the choruses.

On Tuesday a huge didactic pageant was staged, focusing not only on Hudson and Fulton but also on other figures in American history. A series of historical floats proceeded from 110th Street to Washington Square. Despite rainy weather, more than 2 million people turned out. Wednesday was education day, with special exercises in all the schools and colleges in the New York area. On Thursday, 25,000 troops from the participating nations marched in a military parade, and the fine weather drew more than 2.2 million spectators. On Saturday the celebration ended with a series of smaller parades, one in each school district, in which more than 300,000 children participated; these were followed by carnivals and a final, extravagant illumination.[27]

This event was quite unlike the Jacksonian observances of Independence Day, in which the local community had been more actively involved. Nor had the earlier yearly celebrations carried the burden of creating good citizens; the rural life itself was thought to do that. The earlier observances had permitted the many trades and professions to display themselves and rearticulate the body politic. The Hudson-Fulton Celebration, in contrast, was organized from the top down, mandated by the legislature, and orchestrated by an elite. Nor was it an annual event; it was the first celebration of any consequence for Hudson or Fulton. Frankly educational, it sought to inculcate civic virtue. The contrast between the *Half-Moon* or the *Clermont* and the ironclad warships of the modern navies underscored the advances made in three centuries, as many observers noted. Moreover, unlike a civic ceremony such as the Fourth of July, the Hudson-Fulton Celebration required the extensive cooperation of public and private institutions—most obviously the utilities, General Electric, and the corporations that owned the illuminated skyscrapers and the many electric signs used in the event. The centrality of private interests in the celebration marks the shift from a Jeffersonian polity anchored in face-to-face democratic debate to a society in which corporate public relations influence both politics and culture.

Historical reconstruction represents quite another impulse from commemorative ceremonies. Independence Day was an annual event, a memorial to the written and spoken word, embodied in the Declaration of Independence and the speeches that marked the day. A few revolutionary songs might be played or sung, but the event could achieve its meaning without extensive dramatic props. In contrast, the Hudson-Fulton Celebration occurred but once, and meanings had to be invented for it and inscribed within the event. No universally acclaimed document such as the Declaration granted it a special status, and the artificiality of the event was evident in its name alone. Hudson and Fulton belonged to different centuries. Their achievements were quite distinct. Even the anniversary was inexact—Fulton's steamboat should have been celebrated 2 years earlier.

In view of these shaky underpinnings, creating a great public event required extensive stage props. These fall rather easily into two groups. The editor of the magazine *World's Work*, Walter Hines Page, listed four great attractions of the celebration: "The rebuilding of Hudson's *Half-Moon* and of Fulton's *Clermont*, and the appearance on them of persons in the dress of the period of each; the dazzling effects on land and on water of artistic designs and devices in electric light; the amazing achievements in decorative illumination by electricity; the aptness that we are showing in the representation of pageantry of historical events."[28] Page's list easily resolves itself into two features: historical reconstruction and technological display. The technological elements were the great parade of ships, the first airplane flights over New York, and the spectacular illuminations of the city. The historical elements were the parades with themes taken from New York's history and from European high culture.

Such parades were not a novelty. Starting in the 1870s, tableaux vivants had become a common element in local historical pageantry. As David Glassberg notes, the tableau vivant began as a popular form of parlor entertainment involving the recreation of historical paintings, such as Benjamin West's *Penn's Treaty with the Indians*, John Chapman's *Baptism of Pocahontas*, or Archibald Willard's *Spirit of '76*.[29] The centennial of the Declaration of Independence in 1876 gave an impetus to this practice, which moved from the parlor out to the village green. It only remained to put the tableaux into motion as parade floats.

Organizing such a parade was not easy. While the Manhattan parade went off without a hitch, a restaging in Brooklyn a few days later was a complete disaster, underscoring all that could go wrong. Only a few of the floats could be unloaded from their barges, which were at the wrong height for the docks. While most of the parade remained marooned in the East River, some of the floats that reached the shore were unable to proceed because overhead trolley cables blocked their way. Hundreds of historically costumed Brooklynites, "in despair," climbed aboard smaller floats, creating ludicrous tableaux that jumbled together figures from different centuries. The crowd of half a million "went away disappointed, many of them angry."[30] This Brooklyn disaster is a reminder that when such an event did come off it reflected remarkable cooperation among disparate groups, including labor unions, longshoremen, politicians, electrical utilities, traction companies, civic organizations, the city council, and the police.

The 54 floats displayed in 1909 were a virtual catalogue of popular history. They included such old reliables as George Washington taking his oath of office, the trial of Peter Zenger, the legends of Sleepy Hollow and Hiawatha, and the Statue of Liberty. Special floats represented the *Clermont* and the *Half-Moon*. New York consciously modeled this display not on the Quebec tercentenary of a few years before but on New Orleans' Mardi Gras parades. In Quebec the tableaux were arranged in fixed positions, and admission was charged, with the result that a disappointing number of people saw them. New York decided on a parade with free admission. To ensure its success, A. H. Stoddard, a Mardi Gras expert, and a number of float builders from New Orleans were hired to instruct 160 workers in papier-maché techniques and costume design.[31] The subjects chosen revealed an attempt to engage the sympathies of as many citizens as possible. Not only were there tableaux representing memorable moments in the history of the city and state; the organizers were conscious of the need to include "the representation of as many as possible of the nationalities composing the cosmopolitan population of the State, so as to make them feel that the heritage of the State's history belongs to them as well."[32] For example, one float represented the Staten Island house in which Giuseppe Garibaldi had briefly lived in 1849 after his unsuccessful revolutionary efforts, and another depicted the Marquis de Lafayette during his popular tour of America in 1824–25.

This attention to ethnic diversity should not be understood as an expression of what is now called multiculturalism, since many groups—Jews, Eastern European immigrants, African-Americans— were left out. The ethnic heritage that remained was thought to merit a separate Carnival Parade. Displaying Northern European Protestant ethnicity, it was organized by the 60,000 members of the German societies of New York. The official historian of the event noted that the Carnival Pageant "illustrated that great body of Old World Folklore. . . . Although the legends and allegories represented were not indigenous to America, yet they form a real part of our culture, inherited, like the cumulative facts which constitute our progressive civilization, from the past. . . . The Carnival Parade, therefore, was something more than a jollification and a merrymaking. It was designed to recall the poetry of myth, legend, allegory, and in a few cases of historic fact, which, while foreign in local origin, is an heritage of universal possession and belongs to all nations."[33] A majority of these "universal myths" were from the German-speaking nations. There were floats representing Lohengrin, Lorelei, the Death of Fafnir, Götterdämmerung, the Valkyries, and Tannhäuser. Sprinkled among them, for contrast, were floats drawn from Greek myths (such as the Corinthian musician Arion and the Æolian Harp) and from biblical stories (such as Solomon's meeting with the Queen of Sheba). On the whole, however, the Carnival Parade expressed Germanic high culture. It did not contain American folk figures (Paul Bunyan, Mike Fink, Johnny Appleseed) or regional figures (the Connecticut Yankee, the cowboy, the pioneer). The parade was part of the larger effort to import European genteel culture into the United States, also expressed in the preference for Beaux Arts architecture, symphonic music, and the assembly of large collections of European art in neoclassical museum buildings.

The more central Historical Pageant dealt exclusively with America. It was divided into four historical periods, each represented by a section of the parade. One float anticipated the dimensions of that story. On it sat a Native American woman holding an open book that represented history. Behind her stood a wigwam overshadowed by a skyscraper; in front of her "issued the prows of a canoe and a battleship, symbolizing the progress in shipbuilding and navigation."[34] Above her head was an owl, symbolizing learning. The symbolism was easily "read" by the million people who viewed the parade: the white race had replaced the Indians, who were inevitably doomed because they were technologically backward. Knowledge meant progress and control of one's destiny. The rest of the 54 floats dealt with four periods: Indian (floats 2–10), Dutch (11–22), Colonial (23–43), and Modern (44–54). This organization suggested that the core values of the United States were firmly rooted in a distant time. There were three floats devoted to the founding of the Iroquois Confederacy, conventionally understood as the model for the American federal union. These were followed by four floats representing the Indians living in a timeless world shaped by the four seasons. The Iroquois appeared both as wise legislators and as happy children of nature. The final float presented them in a war dance.

The Dutch section of the parade began with four floats representing ships: one standing for Holland, then the *Half-Moon,* next Hudson adrift in a longboat amid icebergs, and finally the first vessel built on Manhattan Island. Together these suggested immigration and its hardships. The remaining Dutch floats showed the purchase of Manhattan, the purchase of the Bronx, Peter Stuyvesant's arrival in New Amsterdam as Governor-General, merry-making on Bowling Green, a typical Dutch house, the surrender of New

Amsterdam to the British, and Saint Nicholas with six reindeer. This section appears to have been modeled on the fiction of Washington Irving, whose famous stories "Rip Van Winkle" and "The Legend of Sleepy Hollow" provided themes for floats that appeared elsewhere in the parade. These Dutch, wearing broad-brimmed hats and surrounded by wooden shoes, windmills, and milking pails, embodied a pastoral idyl lost in the mists of time.

The twenty Colonial floats dealt almost exclusively with the revolution and Washington's presidency. Nothing hinted that many in New York City had favored the British cause, particularly in the early years of the conflict. The floats focused on individuals who had defended republican values such as freedom of the press and "no taxation without representation" and celebrated military leaders and acts of heroism such as the protest against the Stamp Act, Alexander Hamilton's attempt to quell an angry mob, and the death of Nathan Hale. By emphasizing individual acts of heroism, the floats articulated the central values of republicanism. The revolution was depicted as an act of political virtue that culminated in the figure of George Washington. In this story, entrepreneurs had no place.

The national period was allotted only half as many floats as the revolution, suggesting that the character-forming events of American history took place before 1800. The Civil War and the more recent Spanish-American War were omitted entirely, and technological achievements received only moderate attention. Aside from Fulton's *Clermont* and his ferry which once carried passengers to New Jersey, only two floats represented technological advances: the Erie Canal and the introduction of water from the Croton Reservoir to New York. Remarkably, the railroad and other emblems of the technological sublime were omitted. Rather, the genteel, the patriotic, and the sentimental prevailed in the parade, though not elsewhere in the celebration. The procession concluded with the Statue of Liberty and Father Knickerbocker. In this vision, the United States had been forged in the revolution, and what had followed was inevitable progress—often suggested by quaint artifacts from the past, such as a hand-pumped fire engine and the "Old Broadway Sleigh." The latter was described by the event's official historian as from "a period long gone by," a time "before river tunnels, subways, elevated roads, electric trolleys or horse cars, when winters were more severe than now, when the snow was not removed from the streets."[35] As these lines suggest, the parade as a whole projected an affectionate relationship to a past untroubled by political conflicts, racial tensions, or class divisions.

The Hudson-Fulton procession did not represent human labor, industrial production, or the banks and corporations that dominated the financial life of New York. It bore only an oblique relationship to the economic or political lives of its audience, and it presented a severely edited version of both past and present. It emphasized the white middle class and Northern European ethnic groups, both in its themes and in its participants. For example, the first four floats in the fourth division of the parade were accompanied by a Commission Band of 100 pieces, 500 men from the Patriotic Order of the Sons of America, 500 men from the Irish-American Athletic Club, a Scottish Color Guard, the United Scotch Societies, a French Band, and 200 men from the United French Societies. The rest of this section of the parade contained no fewer than 2000 marchers, including five more bands.[36]

In addition to being represented in papier-maché, Hudson's and Fulton's vessels were replicated at full scale for the naval parade, not only out of a desire to recover and pay tribute to a fading reality but also in order to emphasize the achievements of modern technology. The scale of modern war vessels made these tiny craft seem minuscule. Compared to a modern 40,000-horsepower warship, the *Clermont* was driven by less than 20 horsepower, and just in case the wind was favorable or the steam engine broke down it had an auxiliary mast and sail. The *Half-Moon* was just as diminutive, and it seemed almost incredible to spectators that such a vessel had been sent out on an important expedition. Such contrasts were needed to gain a purchase on the public imagination. Without the powerful and familiar text of the Declaration of Independence to anchor this October ceremony, the organizers combined dramatic reenactments of early voyages with a new pantheon. The modern heroes assembled in 1909 included the arctic explorers Frederick Albert Cook and Robert Edwin Peary[37] and the aviators Glenn Curtiss and Orville Wright.[38]

As Walter Page reported to his readers, despite the enormous efforts to impress the public with historical pageantry, spectacular technology proved even more popular in the New York celebration. The parades were part of a larger effort to promote the City Beautiful and Beaux Arts architecture by importing both into an actual urban environment. Classical architectural and decorative styles were being offered to the public as the appropriate aesthetic forms to articulate a grand vision of the city. At the same time, the event demonstrated how electrical technologies could reshape the city's appearance. These two kinds of spectacle could coexist in the event without apparent discord, no doubt because each was allotted ample time and different groups worked on each. The Hudson-Fulton ceremony was a vast undertaking. To create a unified effect required extensive coordination of government, private citizens, theaters, banks, corporations, and advertising firms. Their cooperation intimated a glorious future for the city, modeled on expositions. In addition to the usual lighting of the city, 1.2 million additional incandescents were installed, plus 7000 arc lights and batteries of searchlights. The 6-mile parade route was lined with lights placed 2 feet apart on both sides of the street. On Fifth Avenue between 40th and 42nd Streets was the Court of Honor: 36 white stucco Corinthian columns surmounted by gilded balls and festooned with lights. The parade route intentionally avoided Broadway, with its garish commercial advertising, because the organizers preferred Beaux Arts architecture. The exposition style expressed dignity through the almost exclusive use of white light and the decision to illuminate primarily classically inspired buildings. Powerful searchlights were focused on Grant's Tomb and on neoclassical structures such as the Brooklyn Institute of Arts and Sciences. The Washington Square Arch, the Great Waterside Electric Station of the Edison Company, and the Soldiers and Sailors' Monument had bulbs placed along their edges, cornices, and columns. Thirteen thousand bulbs illuminated the contours of the Brooklyn Bridge

The Court of Honor at the Hudson-Fulton Celebration. Stereograph by Stereo-Travel Co., 1909; courtesy of Collections of Library of Congress.

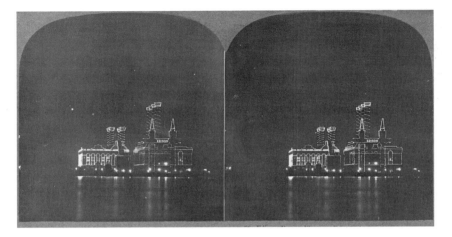

The Edison Power Plant on East 35th Street in New York. Stereograph by Stereo-Travel Co., 1909; courtesy of Collections of Library of Congress.

alone, and the other Manhattan bridges received corresponding treatment, so that from a distance each was an elegant outline. On Staten Island a thousand bulbs spelled out WELCOME in 20-foot-high letters; the Colgate Clock sign on the New Jersey coast was rewired to create a giant picture of the *Half-Moon* with electric waves dashing against it. The Plaza Hotel was festooned with lights from top to bottom, and the Waldorf Astoria had "long chains of lights arranged in the form of a pyramid."[39] Large displays and signal fires were mounted on the New Jersey side of the Hudson River and in all the towns along its banks 150 miles upstream to Albany and Troy.[40]

Private organizations—theaters, banks, corporations, office buildings, advertising firms—paid more than 80 percent of the cost of the lights. The list of sponsors immediately suggests the powerful backing of both old social and new corporate elites.[41] As a result of their efforts,

The Hudson River from the Statue of Liberty on Bedloe's Island to Spuyten Duyvil, was ablaze with light. The entire Jersey coast, from a point opposite the Battery to where the last of the line of warships is stationed, was illuminated. Lamps had been strung along the hulls and rigging of the big liners at their docks, and the fronts of manufacturing concerns abutting the shore were a mass of incandescent lamps. As the light of day faded and the millions of incandescent lamps flashed into view, the outlines of towering skyscrapers and bridges were merged into the night and became magic structures painted into the sky or slender ribbons of fire.[42]

New York's lighting effort was unique only in terms of its scale. The year before, Philadelphia had celebrated a national Elks convention with city-wide illuminations, including the outlining of prominent buildings with bulbs, new signs, and other special effects. Likewise, the annual Mardi Gras in New Orleans spent a great deal for spectacular illumination. What distinguished the Hudson-Fulton Celebration from earlier displays was the sheer extent of its installations, plus the appearance of an innovation: fireless fireworks. This idea came from W. D'Arcy Ryan, a General Electric engineer who had earlier illuminated Niagara Falls. His tools consisted of an enormous bank of searchlights and projectors on Riverside Drive, the modification of a design originally created for the Charleston (S.C.) Exposition a few years before but not used then because of a lack of funds.[43] Ryan installed special steam engines to create a medium for his shafts of light to work in, making his effects seem bright and substantial.[44] This "Electric Steam Scintillator" had many nozzles and valves, permitting the operator to release steam for coloration through the aptly named Niagara nozzle, fan nozzle, snake nozzle, plume nozzle, column nozzle, pinwheel nozzle, and sunburst nozzle.[45] Combining the steam scintillator with projectors, prismatic reflectors, flashers, and filters, Ryan could produce a tremendous range of special effects. He had transferred the techniques of the electric fountain to a new medium. A colored bulb painted in bands of different colors could be placed in a reflector to project a flower-like shape. As one expert noted, "by massing reflectors of different sizes, having lamps colored in this way, a veritable luminous bouquet can be produced."[46] Nor were these effects static. The pinwheel nozzle and the snake nozzle made swirling movements that mimicked fireworks. The elaborate machinery required 50 operators, whose coordinated performances simulated conventional fireworks—plus an enormous peacock's tail, a false sunrise, and other special effects—in a kaleidoscopic series that left New Yorkers breathless.[47]

Virtually the entire citizenry of the greater New York and the Hudson Valley saw parts of the illuminations, and in addition many ferries and the steamers of the Fall River Line brought sightseers each night. The success of these preparations cannot be doubted from the extensive newspaper attention, including front-page coverage every day and lavish photographic spreads in the Sunday editions. The illuminations were extensively praised on every occasion, and the last performance was an encore, restaged by popular demand. The *Times* reported:

New York burned the incandescent phantasmagoria of its week of celebration into the memory of its thousands of visitors last night in one last general illumination of fleet and river, city and highway of pageantry. After all, this is the electric age, and electricity was called upon to furnish a fitting climax to what has gone before. . . . From the northernmost extremity of Manhattan Island to the far-away Jersey coast, the city was one long, wide banner of light that sparkled and scintillated in the crisp night air, paling to insignificance a moon of harvest splendor. . . . The illuminations which stood out above all others . . . were of course the array of battleships in the Hudson and the jeweled cobwebs of light which outlined the big bridges spanning the East River. The battleships held the crowds on Riverside Drive, from Seventy-second Street to Washington Park, and held them there until after midnight.

The press in the subways before and after the illuminations was terrific. On the New Jersey side crowds lined the Palisades, gazing at a city which had been transformed into a cubist skyline, a dazzling vision that literally left them speechless with admiration. These throngs of onlookers heard no orations. They were not enjoined to be virtuous citizens, nor were they presented with historical parables. This was a ceremony without words. It invited them not to active participation but to silent wonder, not to emulation but to recognition of the superior technical skill that made the display possible.

The new skyscrapers, including the *New York Times'* own building, provided some of the most favored vantage points:

The Hudson River lay like a necklace of vari-colored jewels when night came on. . . . From the Times Tower the line stretched up river further than the eye could see, the distant lights dim and blurred in the haze of distant illumination. From the north, like an aurora borealis, a fan of light shot into the sky from the battery of search lights, alternately crimson and blue and gold. Below, Broadway stretched north and south glittering like a river of fire, in which the crowd and traffic were floating straws. Eastward, past the tall illuminated skyscrapers, came the bridges' golden lacework against the black background, and in the midst of all, vying with the moon, the huge light above the Metropolitan Building's tower shone like the eye of Polyphemus.[48]

The spectacle displaced nature, as the harvest moon paled to insignificance. Indeed, the electrical effects got so much attention from the press that the dedication the large Palisades Park along the Hudson River was scarcely noted, although it was an official part of the proceedings.[49] This act of preservation, like that at Niagara, had required pressure on the state legislature and the cooperation of many private citizens. Yet far more impressive to most of the citizenry was the man-made aurora borealis that proclaimed the arrival of the city as a supreme artificial construction. The electrical sublime had turned New York into a unified work of art that dazzled the public imagination, making explorers, inventors, and electricians the ideal citizens of American history. Moreover, the display announced engineering's supremacy over the Beaux Arts tradition. The *London Times* noted: "Skyscrapers are not by day remarkable for grace or beauty of line, but by night under the electrician's skill they were shown to be capable of transformations which suggested rather palaces and dreams than the sober realities of the modern Land of the Dollar."[50] As the geometrical sublime was intensified, the city temporarily dissolved into a shimmering artificial pattern. The Hudson-Fulton Celebration created a text without words, rewriting the city as a sublime landscape.

The Hudson-Fulton Celebration marks the penetration of the electrical sublime into the pageant movement and the deployment of a powerful technology in a civic ceremony. This development alone does not account for the decline of historical pageantry in the following two decades, however. Rather, the pageant form was radically changed after America's entry into World War I. As David Glassberg summarizes: "Shaped by the pressure to be of use in 'organizing the soul of America' for war . . . historical pageantry changed greatly from 1917 to 1919. Productions featuring a full-length, dramatic reenactment of episodes from local history, which had flourished through 1916, virtually came to a halt with America's entry in the war."[51] This shift can be measured by comparing the Hudson-Fulton Celebration of 1909 with the way Washington, D.C., celebrated July Fourth as an International Festival of Peace in 1919. Many of the same elements of spectacle used at the Hudson-Fulton Celebration reappeared in this festival, but these elements had all been reshaped. Again the organizers combined their representations with spectacular lighting effects. Again many foreign nations took part, principally allies in the war. Instead of sending battleships, they contributed floats to a parade. The procession had completely severed its connection to the local community and become international. Simultaneously, it had abandoned its historical character and become future-oriented. These radical changes in emphasis were self-conscious. The American Pageant Association had decided during the war to abandon local historical events and to support mobilization "through the publication and widespread distribution of short allegorical masques."[52] The more successful of these were performed several hundred times each, often for the benefit of the Red Cross. The masques employed unambiguous allegory and large abstractions to express what the promoters called "the spiritual aims of the nation."[53]

These aims were military. Whereas the floats of the Hudson-Fulton Celebration had omitted any reference to the Civil War or the Spanish-American War, in 1918 Joseph Lee designed a model July Fourth pageant with only three scenes: one focusing on the Revolution, one on the Civil War, and one on World War I. In this and other such events, less dialogue was employed than in the older festivals of local history, and music took on a larger role. A symbolic form was evolving in which text was increasingly subordinated to strong visual and musical appeal. At the same time, these new pageants no longer celebrated pluralism. The diversity of immigrant groups that had been partially acknowledged at the Hudson-Fulton Celebration was completely submerged in calls for 100-percent Americanism. The masque stressed "what America has given the immigrant" rather than what the immigrants had contributed. July 4, 1918, was designated "Loyalty Day" and focused on organized parades of immigrant groups who declared their allegiance to the United States.[54] In short, the pageant movement had been transformed into what Glassberg calls "coercive mobilization" as Americans were pressed to merge into a new form of "mass demonstration uniting performer and audience."[55] This unification was quite unlike that evident in Baltimore 90 years earlier, when workers, politicians, capitalists, and merchants all displayed themselves as part of one community, with each controlling the terms of its self-representation. In Washington in 1918, each group was reduced to an abstract symbol, and the actors in the allegorical drama had been shorn almost completely of the specificities of race, occupation, and national origin.

This development harmonized well with electrical effects, whose dramatic possibilities were ideal for mass spectacles. To celebrate the end of World War I, several cities erected brilliantly illuminated portals. Chicago's "Altar of Jewels" consisted of two 90-foot candelabra studded with 30,000 cut-glass crystals, which gave off a rainbow of colors when struck by the beams of five searchlights. From the tops of these "candles" issued steam, which was lighted in red and orange tones to create the illusion of flame. Between the two "candles" hung a sunburst of arrows surmounted by a shield, and behind them eight searchlight projectors formed a fan of light that played across the heavens.[56] This "altar of victory," which used lighting techniques pioneered at the Hudson-Fulton Celebration and the San Francisco Fair, was a powerful yet mute proclamation.

"Fireless fireworks" reappeared in Washington's 1919 International Festival of Peace. The official organizer of the event was the War Camp Community Service, which had worked throughout the war with the American Pageant Association to stage events for soldiers.[57] It was locally supported by the same groups that had mounted these wartime displays, notably the Red Cross and the Daughters of the American Revolution. The aesthetic form of their voluntary efforts had been inspired in part by the bond drives, posters, and propaganda activities of the Wilson administration.

The Altar of Jewels, Victory Liberty Way, Chicago, 1919. Courtesy of General Electric.

Most of the cabinet and many members of Congress attended the procession of floats, heard the concert, and saw the allegorical spectacle that followed. The administration's goal of a League of Nations was harmonious in almost every way with this vision of a new democratic world, and the Festival of Peace is properly understood as part of the effort to extend wartime mobilization into the postwar era in order to promote Wilsonian foreign policy.

To help pay for and publicize the event, a small army of schoolboys sold programs throughout Washington on the days immediately beforehand. On the day itself, a good many people went to the Washington Monument in the morning either to see or to help form an enormous "human flag." Afterward, medals were awarded to local war heroes.[58] The rest of the festivities began at 5 P.M., when the lengthening shadows made Washington's scorching summer heat more bearable. A series of static tableaux could be seen in prominent locations, such as "Bugle Call to World Service" at the Red Cross Building and "Call of Liberty" at the D.A.R. Building. Other tableaux included "Call of Art," "Call of Children," "Call of the Land," and "Call of Commerce, Business, and Professions." As these themes suggest, this was an expression of Wilsonian liberalism. An "Offering of Peace" said to be "symbolic of the contribution of the Negro race to reconstruction" was displayed in front of the National Museum.[59] This emphasis was consistent with pageant developments during the war: African-Americans were given small roles, and Negro spirituals were used at times. The public response to the tableaux was surprisingly strong, and as a result they were restaged the following evening, one after another, at the base of the Washington Monument.

Like the parade at the Hudson-Fulton Celebration, the Washington parade presented a series of allegorical moments. However, rather than emphasize American history and ethnic pluralism, it sought a common vocabulary for all nations. The organization was neither geographical nor alphabetical but apparently random. The first float, devoted to peace, represented mustered-out soldiers returning to their civilian jobs. It was followed by allegorical floats whose messages, to the modern eye, were at times less symbolic of peace than of conquest. For example, the Danish contribution consisted of a Viking ship, which might suggest a raiding party, but it contained in the stern a "figure of Peace in ancient costume." Likewise, France recalled the reconquest of Alsace and Lorraine, represented by two women weaving garlands for the returning soldiers. The prow of Henry Hudson's *Half-Moon* formed the front of the Netherlands' float, which depicted the negotiations between the first settlers and the Indians. Many floats emphasized distinctive national symbols: Japan modeled Mount Fuji, surrounded by cherry blossoms and surmounted by a flag; England presented a figure of Britannia standing above a maypole surrounded by dancing children.[60] During the parade a squadron of twenty warplanes flew by, exciting the admiration of the crowd. There seems little doubt that the public was satisfied. The *Washington Evening Star* of July 5 reported: "The waves of applause and swelling murmurs of appreciation showed that the beauty of the floats struck to the city's heart. . . . 'Never saw anything like it,' greeted the procession time and time again from the side lines."

After the parade came a pageant on the eastern side of the Capitol Building representing the return of peace, the victory of love over hatred, and the triumph of justice over jealousy. The central roles in these allegories were played by women dressed in flowing robes that loosely quoted the drapery of classical sculpture. Similarly, a women's suffrage pageant held in Washington in 1913 had featured women dressed as "Justice," "Peace," "Hope," "Liberty," and other abstract values.[61] Yet, while such ceremonies were increasingly feminized, electrical displays and fireworks remained under male control, creating the potential for inconsistencies in dramatizations, as would to be the case in Washington.

As darkness fell, a marine band played and a chorus of 1000 voices performed half an hour of patriotic music. Such combined choruses, made up of many local choirs, had been a favorite device of the fund drives and pageantry of the war years. After this performance came the allegorical spectacles. The steps of the Capitol Building were brilliantly lighted, focusing all eyes on the performances, which began at 9 o'clock. In the first act a young woman represented "Peace" descending upon the world. Preceded by heralds blowing trumpets, she released "a dove, the emblem of peace" and was "followed by dancing girls symbolizing the joy which peace brings."[62] This sketch suggests the obviousness of the allegories, which were designed to be easily legible. In the second act, a woman representing "America" came down the capitol's steps "to lend a helping hand to the characters representing the oppressed nations," leading them up the steps to another woman dressed to represent "Liberty." The third spectacle featured figures who represented "Capital" and "Power" breaking away from "Greed" and

"Selfishness" to be guided by "Intelligence" and "Unselfishness" to join hands with "Labor," who in turn cast aside "Hatred" and "Jealousy" and literally met "Capital" and "Power" halfway, in the middle of the steps. The last spectacle visualized the League of Nations as women dancing together on the capitol steps, representing a "great chain of the nations."

The pageant was followed immediately by fireworks, which combined traditional rockets with the special effects that W. D'Arcy Ryan had developed for the Hudson-Fulton event and refined at San Francisco in 1915. In contrast to the women's performance, they provided a militant patriotic display that undercut the message of peace in the rest of the ceremony. "Electric bombshells" burst in "a riot of 700 colored serpents" that hissed high in the air over an area equal to three football fields, accompanied by "repeated flashes of lightning, resembling the darting tongues of serpents." Manufactured thunderclaps, aerial electric fountains, cascades of dazzling fire, and golden rain lead up to the climax, in which three eagles rose slowly into the air and then broke into a "a canopy of golden colors" darted by "electric flashes and red, white, and blue streamers of fire."[63] The next day's *Washington Star* declared that the fireworks had "easily surpassed any pyrotechnic display ever seen in Washington."[64] These pyrotechnics reasserted a masculine vision of war, underscoring a major fault line in the progressive movement: the division that separated male progressives like Teddy Roosevelt, who were intent on stiffening the national resolve, and female reformers, who envisioned the transformation of America once women gained the vote. The performance not only suggested this contradiction in progressivism; it also rested on a weak organizational foundation. The International Festival of Peace had been made possible by a fragile coalition, convenient for all parties during the war, between women's groups, the military, national politicians, and nonprofit organizations such as the Red Cross. Beyond the abstractions of the masque, these groups lacked long-term common interests.

Moreover, although the crowd enjoyed it, the Washington show had excised local history and had little to do with community organizations. Relatively few people participated in the parade, compared to the numbers who had marched in Baltimore in 1828, Boston in 1851, or St. Louis in 1874. Where once virtually every spectator saw friends and neighbors in the procession, in 1919 such immediate personal identification was rare. And the visual complexity of the floats in the railroad jubilees had been eliminated in favor of a stark simplicity which demanded that the viewer abandon his sense of individuality and merge into the mass. To put it another way, this parade and pageant aspired to be a spectacular form of the sublime. Rather than celebrate a technological object as an example of the sublime, the International Festival of Peace attempted to be sublime in the terms of its own performance. It did this by eliminating the specificity of the landscape, by brilliantly lighting buildings and reducing them to abstractions and patriotic symbols. It reduced the past to a few great events that could be represented in symbolic shorthand, it wiped out ethnic differences in favor of a generic American, and it standardized the representation of all positive values as young women. Finally, it virtually eliminated speeches, and instead addressed vast crowds through a simplified language of visual allegories and striking effects. The pageant movement thus worked toward a totalization that eliminated both historical specificity and verbal explication. While to the modern eye it may seem that it did so with heavy-handed symbols, the enthusiastic contemporary accounts betray no sense of ironic distance.

Nevertheless, in 1919 the teaching of civics through symbolic drama was giving way to livelier electrical displays. One of the most popular tourist attractions of the 1920s proved to be Niagara Falls, now illuminated at night by powerful searchlights. General Electric had already demonstrated the possibility of such a spectacle in 1907, but now it became a permanent display, merging electricity and one of the most powerful natural symbols in a new version of the electrical sublime. Both the Hudson-Fulton Celebration and the International Festival of Peace registered the conflict between historical pageantry as defined by a social elite and the productions of illuminating engineers. At both events, as at the world's fairs, the sheer brilliance of the lighting overwhelmed the spectator, exacting an awed response that was only tenuously linked to the genteel themes of the event.[65] The Beaux Arts tradition and the City Beautiful movement proved less potent than flamboyant electrical displays. Women dominated historical pageantry, only to find themselves upstaged by the electrical sublime.

Spectacular lighting made possible the awe-inspiring manipulation of both nature and the man-made. As electrical lighting transformed the appearance of streets, bridges, skyscrapers, public monuments, the Natural Bridge, and Niagara Falls, it became not only the double of technology but also a powerful medium of cultural expression that could highlight both natural and technological objects and heighten their sublimity. For example, today at Mount Rushmore lighting is used each night to make the faces of the four presidents leap out of the darkness. Many of the 2 million annual visitors attend these ceremonies, which include a short talk by one of the park rangers, a 20-minute film (narrated by Burgess Meredith) on the lives of Washington, Jefferson, Lincoln, and Roosevelt, and a rousing rendition of the national anthem. After this there is a hushed moment in near-darkness, as though the program were over; then the great stone faces are suddenly bathed in intense light, looming in grandeur high above the audience. "The tourists," one reporter has noted, " sit stunned, momentarily unable to contend with their feelings."[66] This ceremony creates unity through participation even as the park ranger emphasizes that Mount Rushmore's meaning varies from one person to another. The dramatic lighting of world's fairs, the Natural Bridge, civic ceremonies, Niagara Falls, and Mount Rushmore demonstrates how technological spectacle can produce bonds of solidarity.

The Electric Cityscape: The Unintended Sublime

The planned lighting effects of expositions and events such as the Hudson-Fulton Celebration gave the public a taste of what an electrified landscape might be like. But the actual deployment of lighting in American cities was considerably less tidy and organized, expressing not a coordinated vision but competing interests: public versus private, gas versus electric, one advertisement against another. In the marketplace, the electric sign, the spotlight, and even the streetlight became economic weapons. Competition created not the majestic serenity of Beaux Arts architecture bathed in white light but a jumble of styles and a riot of color. This vibrant landscape was the product of uncoordinated individual decisions, yet it had a collective effect—a kinetic impact—that no one had anticipated. Taken together, the myriad lights produced a lively landscape with strong popular appeal. Like the accident of the city skyline, the electrified city was something fundamentally new, an unintended sublimity.

This innovation was not the product of advertising alone, though the idea of an electric landscape undeniably suggests the large signs of New York and other major cities. In fact, illuminated advertising originated in the gaslight era. In the 1840s P. T. Barnum installed outside his American Museum a large gaslit sign that illuminated Broadway for blocks. Soon, Frank Presbury explains, "the gaslit sign became evidence of enterprise. Drug stores, tobacco shops and barrooms especially made use of this form of street advertising. In size the theaters led, and the theatrical district of New York even before the advent of electricity had a degree of illumination that drew people and extended the hours during which the shops found it profitable to remain open."[1]

Broadway had a reputation as the nation's most brightly lighted street even before electrical advertising. In 1891 Richard Harding

Davis wrote a long article for *Scribner's Magazine* tracing Broadway from one end to the other. Though he did not use the term "Great White Way" even once, he did note that the theater district "is even more gayly alive at night, when all the shop fronts are lighted, and the entrances to the theaters blaze out on the sidewalk like open fireplaces."[2] Gaslight was still the dominant form of illumination. Davis noted the existence of electric street lights in passing, but they had not yet become part of a glamorous image of the city. Indeed, for Davis the electric lights were part of a tableaux of loneliness—he imagined a poor young man preparing to commit suicide late at night, "when it gets darker and the electric lights throw long, black shadows on the empty sidewalks."[3]

In 1891 Broadway's glamor was not yet identified with spectacular electric lighting, and the street had not acquired the air of a perpetual carnival. Rather, "Broadway, for so great a thoroughfare, gets people to bed at night at a very proper season. It allows them a scant hour in which to eat their late suppers after the theater, and then it grows rapidly and decorously quiet. The night watchmen turn out the lights. . . ."[4] Within a decade, however, Broadway had acquired an entirely new character, becoming the prototype of a new night landscape that the genteel organizers of the Hudson-Fulton Celebration were anxious to avoid.

Street lights alone were not enough to fashion this landscape, which required large private investments in electric signs and display lighting. A chronological history of the electric sign would begin with the simple arrangement of lights that spelled "EDISON" at the Paris Exposition of 1881. It would recount the technical developments of longer-lasting bulbs, blinking lights, and signs that simulated movement. It would describe the erection of enormous advertising displays in major American cities at the turn of the century, which were quickly imitated across the country. The illustrations would include the first large sign in New York, which proclaimed "Manhattan Beach Swept By Ocean Breezes" on a wall that later was to hold a famous advertisement for Heinz pickles. Such a history would also include the giant Colgate Clock sign across the Hudson in New Jersey. It would explain the technical developments that made larger and more elaborate signs possible,[5] and the important role of the Society of Illuminating Engineers. It would trace the spread of signs from lower Broadway and Madison Square to Times Square. No such history would be complete without a section on the enormous sign representing a Roman chariot race atop the

Hotel Normandie, one of the great tourist attractions in New York before 1914. Such a history would recall the temporary dimming of the Great White Way during World War I, which lasted until advertisers hit upon the idea of using their displays to sell war bonds.[6] It would describe the rich array of signs erected in the 1920s, and the introduction of neon tubes in that decade.[7] Such a history would be richly illustrated and would be embellished with biographical information on early sign designers and quotations from ads. Since most studies of advertising focus almost exclusively on broadcasting and the print media, scarcely mentioning electric signs, this history would be a useful addition to the literature.[8] Yet such a history would focus too much on New York, and it would concentrate too much on inventions and techniques while overlooking five other important areas of investigation that will be taken up in the following pages: the strategies of the utilities in promoting electric advertising, the relation of advertising signs to other forms of spectacular lighting, the construction of the signs by artisans, upper-class resistance to intensive electric advertising, and the way in which several

New York at night, looking uptown, 1939. Courtesy of Collections of Library of Congress.

kinds of spectacular lighting fused together in a new form of the sublime.

The first electric lights advertised themselves. When Charles Brush set up a single arc light on a street corner in Boston in 1878, curious crowds gathered nightly to see it. They were fascinated by the new light, not only because it was so much brighter than existing gas lights but also because it seemed to violate the natural order. For the first time in history, light was separated from fire. It needed no oxygen. It was not affected by the wind. It could be turned on in many places simultaneously, at the turn of a switch. The new technology proved just as great a popular success in Philadelphia, where in 1878 John Wanamaker installed 28 arc lights in his new department store.[9] Brush soon convinced his home city of Cleveland to permit him to illuminate the downtown streets, and a huge crowd turned out to see a central square lighted up in April 1879.[10] On such occasions out-of-town newspapers sent reporters, bands played, and city fathers made speeches. Wherever lights appeared—Chicago,[11] San Francisco, or a small town—the public gathered to wonder and admire.

When Thomas Edison first displayed his incandescent electric light, in December 1879, the public's responsiveness to spectacular lighting was well established. Edison's demonstration, held at night outside his laboratory in Menlo Park, was carefully orchestrated for the press. Later he would to parade men wearing electrically lighted helmets through New York at night. It is not easy for us to imagine how dramatic such a demonstration was at a time when light and fire were still inseparable in the imagination. During all of human experience until 1879, light had immediately suggested combustion, great heat, and danger. A mass of men marching along imperturbably, each with a light burning atop his head, was an astonishing spectacle.

The realization that a steadily burning bright light drew a crowd inspired promoters of all kinds—particularly the owners of department stores, theaters, and amusement parks and the organizers of world's fairs. The Paris Exposition of 1878 had closed at dusk, but in 1881 an exposition in the same city drew crowds at night with 1300 arc lights. After 1881 all fairs emphasized dramatic lighting, and many made illuminated towers their central symbols—obvious examples are Buffalo's Electric Tower (1901), San Francisco's Tower of Jewels (1915), and New York's Trylon and Perisphere

(1939). Most of the innovations in electric technology, including the electric sign, the flashing sign, the electric fountain, the searchlight, the spotlight, and the floodlight, were first displayed at world's fairs.

Promoters soon saw that, in addition to being theatrical, incandescent lighting was superior to gas in other respects: less risk of fire, no danger of explosion, no flickering, and no consumption of oxygen. Electric lighting also offered ease of control, a variety of colors, and immunity to wind or rain.

The spectacular effects first seen at world's fairs were soon incorporated into the geography of daily life. Theater managers outlined their marquees in light and used elaborate lighting effects to make their productions more spectacular. In 1882 the Bijou in Boston became the first theater in the United States to be lighted throughout by electricity. On that occasion, a newspaper reporter noted that "the Edison incandescent burners were a marked success, affording a clear and steady light, easily manageable and of a purity immaculate."[12] The New York producer David Belasco hired a full-time lighting expert, and his was probably the first theater to hold lighting rehearsals. Lighting quickly spread to the facades of theaters as well. In 1904 Adolph Zukor mounted 1000 bulbs above his theater to spell "Crystal Palace." In the lobby he covered the stairs to the second floor with a glass shell, inside of which water "cascaded over lights of different colors."[13] In other cities, however, dramatic lighting often began not with theaters or large signs but with street lights, which proved to be a powerful wedge for utilities attempting to sell other forms of electric advertisement.[14] Quite often, arc lights were installed by private associations of businessmen. For example, in 1895 the merchants of Clark Street in Chicago sought better street lighting from their utility, which agreed to cut its rate by half if 100 arc lights were installed.[15] Such an incident, taken by itself, may seem surprising. Why didn't the merchants get the city government to pay for the installation? The proceedings of the National Electric Light Association record how street lighting was promoted:

A few weeks ago after going to Wichita I heard a merchant talking to another merchant, and he said, "I have just come over from Douglas Avenue and the street is filled with people, but here we are on Main Street, and you can't see a person for blocks." I thought there was my opportunity. . . . I spoke of the idea of illuminating the street. I said, "We will make you a rate, and the company will not take any of the

honor itself. You can have all the honor." I proposed the idea of placing an arc light about eleven-and-a-half feet above the curb every twenty-five feet, which would give them a great white way second to none.
These gentlemen took to the idea and lighted up the first block on
Main Street, and within a week's time they had signed up every merchant along the street. I took pains to put a shade on each lamp so that
it lighted the street and sidewalk, not the store window; this does not
kill the opportunity of getting about $25 from signs for every twenty-
five feet. . . . Those in the next two blocks came in, too, in self-
defense.[16]

The above passage reveals a good deal about the rapid spread of
street lighting in the two decades before World War I. First, electric
lighting could easily be sold as a commercial investment to increase
the competitiveness of a business. Shopkeepers understood lighting
as a weapon in the struggle to define the business center of the city,
dramatizing one sector at the expense of others. Second, the electrification of one street quickly forced other commercial areas to
follow suit or else lose most of their evening customers. In Chicago,
the Clark Street merchants were responding to the prior electrification of other streets, and their installation in turn spurred new
lighting on North Avenue, Wells Street, Division Street, and elsewhere.[17] Third, electric companies had no interest in taking conspicuous credit for the idea of a "white way"; they preferred to work
behind the scenes through local business associations. Commercial
advantage was masked by the public-spirited desire for civic
improvement.[18] Fourth, the transformation of the street's appearance proceeded in carefully calculated stages, beginning with arc
lights that were intentionally shaded to leave store windows and
facades dark so as to prompt additional sales of display lighting for
windows and electric signs for each store front. In this strategy, electric lighting was used primarily not to sell particular products but to
transform the space of the city itself. The lighting of Wichita closely
resembled that of Philadelphia, Minneapolis, and many other cities.
In each case, public illumination was installed as a general form of
advertising that would establish the dominant business district.[19]

Overall, intensified street lighting led to demands for more private light. For example, after a "an "intensive white way" was
installed in San Francisco as part of the Panama-Pacific Exposition,
"the demand for more light resulted in increasing the intensities of
street lighting in business districts fifteen times, with corresponding
increases in window and sign lighting."[20]

For managers, there was an advantage to keeping utilities in the background when new electrical facilities were being planned: it made the achievement more a community event than a business transaction. In many cities the inauguration of a new system became a festival that emphasized the trappings of the republican tradition, with political speeches and a parade through the newly lighted district. Often these events were planned for the dark winter months—preferably before Christmas, when crowds of seasonal shoppers would swell attendance. Spurred by the local utility, Chicago's merchants paid for a new "intensive white way" that opened with three days of festivities in October 1926. When President Calvin Coolidge pressed a button at the White House, "artificial daylight" suffused the streets. Searchlights bathed buildings in various colors. Jazz bands and troops marched in a parade that featured, among other things, floats illustrating the history of artificial illumination.[21] Whereas the Hudson-Fulton event had celebrated colonial American history and Germanic culture, this parade marshalled all the energies of a July Fourth to celebrate a new technological installation.

While the public was encouraged to see a "white way" as a civic improvement, utilities presented it to businessmen as a rational expense that would bring in more customers. In addition, they argued that intensive lighting of each storefront paid for itself. In Chicago such arguments were clearly successful: the number of electric billboards and signs jumped from 100 in 1902 to roughly 2000 in 1905. Likewise, Chicago's 1926 festival immediately increased sales.

A few businesses literally covered their buildings with lighted signs. In 1903 a clothing store on a corner lot in Boston not only had illuminated windows and a sign projecting over the street; on the upper floors it had a 6-foot-high green shamrock, a 10-foot electric American flag, four lines of illuminated text running the width of the building beneath the windows, and spotlights on ten other signs mounted between upstairs windows.[22] By 1910 even that most conservative of businesses, the bank, had adopted gilded transparencies and discreet forms of electric lighting.[23] One reason such signs became so widespread was their low operating cost. In the 1920s, Chicago Commonwealth Edison's service trucks carried advertising that declared "A 100 Light Sign . . . 5 Hours per Night . . . 2 $\frac{1}{2}$ cents per Hour."[24] Starting in 1923, Boston Edison employed two men full time to facilitate contacts between advertisers and signmakers and found that "advertisers report a definite increase of from five to 20% in sales volume as a result of electrical advertising. After a trial, advertisers frequently increase, but almost never decrease, the size and number of their displays."[25] It was not only the brilliance of the boulevard that forced each business to intensify its own image. The same competitiveness that pushed businessmen as a group to intensify ornamental street lighting led them individually to install brighter and larger signs in their stores.

Even as "white ways" transformed the appearance of the city at ground level, corporations began to exploit the potential for vertical displays that the new skyscrapers offered. After 1900 some illuminating engineers specialized in "building displays," which rapidly evolved toward projecting sheets of brilliant white light on structures of all kinds. Around the turn of the century some commercial buildings were studded with lights that emphasized architectural details. This technique, pioneered at world's fairs and still prominent at the Hudson-Fulton Celebration in 1909, flourished for about 15 years.

The floodlighting of buildings began in 1907 with the Singer Tower, then New York's tallest building. Visible from all parts of Manhattan, from Brooklyn, and from New Jersey, it became a celebrated landmark. This form of publicity became widespread after spectacular effects were achieved with the illuminations of the Woolworth Building in New York (1913) and the Panama-Pacific Exposition in San Francisco (1915).[26] The Woolworth Building's highly reflective 792-foot terra cotta gothic tower was intentionally designed for floodlighting. Woolworth hired 40 engineers to plan its illumination, which emphasized architectural details. The National X-Ray Reflector Company of Chicago provided 600 special high-intensity bulbs with corrugated reflectors that diffused the light evenly over the surface, without glare. When these were first illuminated, on New Year's Day 1913, they set a new standard that other skyscrapers would imitate for 20 years.[27] From a technical point of view, such displays were especially suitable when a building was monumental, when its walls were light in color, and when it stood somewhat apart from other structures. All these conditions were met by the Wrigley Building, which became a illuminated landmark of Chicago's skyline immediately after its completion in 1921.

From an architectural point of view, the exterior lighting of skyscrapers did not require modernism and may, in fact, have retarded its development. The eclectic style, which borrowed from classical, gothic, and renaissance architecture, provided individual profiles and interesting details to highlight or to outline with individual bulbs. Major cities developed skylines that appeared to be outsized versions of an Italian city, with commercial towers taking the place of the spires of palaces and churches. The 700-foot tower of New York's Metropolitan Life Building (1909) was copied from the Campanile in Venice's St. Mark's Square. At the time of its erection, no structure nearby even remotely approached its height or its three-story illuminated clock. The brilliantly lighted tower of Chicago's Wrigley Building, copied from the Giralda Tower in Seville, was another immediately identifiable urban landmark. Its clock too was easily legible throughout the downtown area. In contrast, Raymond Hood used a quite different technique in his American Radiator Building (1924) in New York. He designed an extremely dark, almost black structure, with gold gothic trim on the set-backs above the sixteenth floor.[28] At night the building faded into the night while the lighted windows and the gold ornamentation leaped into prominence. As more skyscrapers were built which realized the possibilities of dramatic lighting, the idea of a city's night skyline became a commonplace, depicted in millions of postcards. For the most part commercial interests decided which objects were to be visible in this new landscape. By highlighting some portions of the city and leaving other areas as unimportant blanks, illumination literally directed the consumer's eyes away from poor areas toward commercial zones.

Government applied the same principles to public monuments, lighting up the Statue of Liberty, the Capitol Building, the White House, and other national and local symbols. Eventually this was extended to natural landmarks, including Niagara Falls, the Natural Bridge, Old Faithful, and Mount Rushmore.

The electrical sign was only the most eye-catching display in a new landscape that we now regard as "natural." Large signs of the sort found on Broadway were but the most obvious part—indeed in most cities the final stage—of a much larger transformation that included street lighting, display lighting in windows, floodlighting of buildings, and illuminated signs above individual shops. Each of these forms had its own economic rationale, and collectively they reshaped the landscape of the night.

By 1915 electric signs came in many forms. There were exposed-lamp signs (best suited to locations where viewers were more than

250 feet away), enclosed-lamp signs (more useful at close range and usually consisting of little more than a name), silhouette signs (in which lettering or a design was suspended in front of a lighted background), and, of course, illuminated billboards. Salesmen divided electric signs into two distinct categories: detached and attached. Detached advertising was primarily employed at central locations by large corporations in order to make a favorable impression on passersby. These displays, in places far from companies' offices and factories, were called "impression builders." Either billboards or arrays of bulbs, they were often legible at a distance of 2 miles and identifiable as markers for up to 10 miles.[29] An "attached" sign usually was smaller and visible for only a few blocks. Its main function, as a "location marker," was to indicate where a business was (although it was also intended to make an impression).

Whatever its technical form, and whether it was an impression builder or a location marker, an electric sign was not designed like a printed advertisement. A General Electric bulletin for salesmen explained this point as follows: "Circulating advertising, because it can go to the easy chair by the reading lamp, may be leisurely, argumentative, and thorough in the lesson it teaches. Display advertising, because it cannot move and because it must do its work on moving people, must be very simple, striking, and impressionistic. The one may employ logic and demand reasoning; the other must confine its efforts largely to an appeal to the senses."[30] In practice, electric signs turned out to be precursors of trends in printed media, where text would only decline in importance after 1915. In the 1890s, when "reason why" copy that included at least a paragraph was the dominant form of magazine advertising,[31] electric signs rapidly evolved away from text. They had to rely on sensory appeal, using only a few words and virtually no argument. Nevertheless, at the start electric advertising was text-oriented. The first electric sign was a word ('Edison'), and the first large sign on Broadway (it appeared in the mid 1890s) consisted entirely of words:

SWEPT BY OCEAN BREEZES
THREE GREAT HOTELS
PAIN'S FIREWORKS
SOUSA'S BAND
SEIDL'S GREAT ORCHESTRA
THE RACES
NOW—MANHATTAN BEACH—NOW

The colors of these lines, respectively, were green, white, red, yellow, blue, red, and white.[32] But what most caught the eye was the way that the sign was illuminated. First, each line appeared separately, disappearing as the next line came on; then at the end all the lines were lighted simultaneously. The series of messages stretched out the time needed to view the whole performance, giving the words time to register.

Most of the early electric signs appear to have been modeled after billboards. Of course, many of them were simply illuminated versions of earlier signs. In Boston the electric company offered this comparison: "How many of us will stop during business hours to read a bill-board advertising somebody's soap? Not one in fifty. How many of us would fail to see and read the same board illuminated at night, cut out in golden radiance and set in a frame of black, the only relief to the eye from a monotony of darkness?"[33] This contrast between the effect of the same sign by day and by night may be overdrawn, but it is essentially correct. A billboard that was lost in the clutter of the urban landscape became much more prominent when its surroundings were lost in the dark. But while billboard lighting remained common, especially in outlaying areas, in the central city it was merely a preliminary to the more striking effects that could be achieved by using light as the medium of expression.

In the first two decades of the century electric signs came into their own. More and better colors and improved flashing devices encouraged designers to create vivid moving images. The area covered by text shrank. In part this change occurred because it became possible to write several different messages in the same space, one after another. But undoubtedly advertisers also realized that the new medium did not lend itself to logical argument, but rather to striking visual effects. In 1910 one illuminating engineer summarized this succinctly: "Advertising psychologists lay down the following action of the human mind as the basis of their work: First, attention; second, interest; third, desire."[34] In line with such thinking, General Electric instructed its salesmen to keep the message as short as possible while emphasizing visual appeal. In the psychological language in vogue during the 1920s, its manual declared:

The attracting power of brightness is instinctive—its appeal to the senses is of the most elementary sort. . . .

Automatic motion is an exclusive characteristic of electrical advertising and its appeal is also dependent upon instinctive feelings.

The use of color is one of the most powerful means of creating an atmosphere—an inner feeling—a pleasant association, and these are among the chief functions of display advertising, particularly in its appeal to women. . . .

Psychologists tell us that the subconscious mind rejects the untrue and unbeautiful. It is certain that most people want to forget a displeasing impression. An effort to make the display pleasing will be rewarded many times over.

Advertising experiments have proved that newspaper space filled with a picture is seven times more effective in selling goods than is the same space in reading matter. A picture is instantly understandable to all and conveys far more than could be stated in the few words that may be included in a sign.[35]

Advertisers adopted such suggestions, and saw the electric sign as part of a coordinated strategy of reaching the public through different media. A pocket handbook for salesmen emphasized: ". . . it is not the object of the electric sign to supplant either newspaper or magazine advertising. All these branches of advertising are closely related, and in order to get the best results, each should be used in conjunction with the others. The story is told in detail in magazines and newspapers, and the spectacular electric sign follows with its flashing presentation of the trademark."[36] Since the object was not to educate but only to excite the passersby, designers tried to make electric signs curious, mesmerizing, and funny. By 1910 the most popular signs almost invariably exploited motion. At Euclid Beach, outside Cleveland, a large sign depicted a girl who appeared on a diving board for a moment and then dove into the water, which splashed and rippled as she disappeared. The caption that followed was almost superfluous: "Come on in, the water's fine." One close observer commented: ". . . the most elaborate signs . . . are almost equivalent to an entire vaudeville act. In fact, many of the schemes used in the theater to produce motion effects are utilized in electric signs; thus to show a moving vehicle, wheels are made to appear to revolve and the nearby objects to pass by."[37] In a sign for the Reo automobile, an open roadster seemed to speed along, dust "flying from the revolving wheels, the smoke from men's cigars floats away and the ladies' veils flutter in the breeze. To put the finishing touch on the realism, even the warning 'Honk, honk' is given by means of an electric horn."[38] Most signs were silent, however, merging into a kaleidoscopic array. On Broadway, "a bewildering number of pantomimes . . . kept all eyes upturned. A cork popping from a bottle,

followed by foam and the running champagne, beer or ginger ales.
. . . Fountains threw silvery columns into the air, selling the purity
of mineral water. Animals chasing each other across the sky drew
attention. . . . Every few moments a maid holding up her skirts
would be seen crossing a street in the sky and getting caught in a
shower of realistic rain. . . . A cat tangled in the ravel from a spool
of Carticelli Silk Thread. . . ."[39]

Electric signs were seldom used to introduce a new product; generally they kept a brand name before the public. In the 1930s, when
the Wrigley Chewing Gum Company erected a ten-story sign that
stretched from 44th to 45th Street, the word 'gum' did not appear
anywhere on the sign; the company confidently assumed that the
product was already known. The design suggested a vast aquarium
filled with brilliantly colored tropical fish. The Wrigley "spearman"
sat in the center and pointed to slogans that alternated in the
upper left corner, declaring that spearmint gum "Aids Digestion,"
"Steadies the Nerves," and "Keeps the Taste in Tune." The overall
effect, one critic noted, "soothes rather than startles the observer.
The fish, magnified hundreds of times, are artistic adaptations in
both form and color of authentic South Sea fauna. As they rhythmi-
cally burp large white bubbles, they seemingly swim through space,

Wrigley Chewing Gum Sign, 1930s. Courtesy of Hall of History Foundation,
Schenectady, N.Y.

the illusion being created by vertical flickerings in their scales and a running motion in the waves."[40] Because Times Square had become a visual cacophony, this static, soothing composition was more arresting than the frenetic movements of surrounding signs.

The electrified landscape was literally the product of consumer capitalism, but its manufacture should not be taken for granted. Who made and erected the Wrigley sign and other electrical displays around the United States? Utilities sold a line of standard enclosed-lamp signs, but they seldom entered the specialized sign market. After all, they profited from sales of current no matter who made the sign. Instead, firms such as Chicago's Federal Sign Company (in business as early as 1908) specialized in this area, buying current from utilities at a discount and selling both signs and service to advertisers. Similarly, one-of-a-kind signs such as those on Broadway were seldom produced by General Electric or Westinghouse; those corporations were content to supply smaller companies with the special materials needed for their construction. For example, General Electric took an interest in the Wrigley sign, creating bulbs in unusual colors especially for it. The new colors, including vermilion, orange, yellow, and metallic green, were in part designed to compliment neon lighting, then still relatively new.

Sign construction was a handcraft, and small display companies specialized in it. The oldest still operating in New York is the Artkraft Strauss Company, started in 1897. A Master Sign Makers Association united the many craftsmen. As early as 1910 some individual signs had miles of wire and thousands of bulbs. The Wrigley sign, built in the 1930s, had 1084 feet of neon tubing, bent and fastened by hand, and an elaborate wiring system for 29,508 bulbs, mounted on a 110-ton structure. Even today, only highly skilled workers can produce such complex machines; the great signs are constructed by hand and take months to build. Every neon tube must be individually heated, bent, filled with gas, sealed, and fastened in place. Every circuit must be individually wired and connected to the timing mechanism that regulates the sequential flashing of the tubes and bulbs. This artisanal work demands painstaking accuracy. Eric Sandeen, who observed signmakers at Artkraft Strauss in the late 1980s, noted: "They view what they do as a craft, not an art. Creativity is not appreciated here; precision is the criterion. Skill is measured by the ability to stick to the pattern, to manipulate the glass through all sorts of angles and make splices without causing a crack to form, and to place the electrodes exactly in the spot indicated on the pattern."[41]

Although the craft aspect of sign making encouraged small businesses, sign production became closely connected to near-monopolies of prime advertising sites and client services. Here millions were to be made. In 1926 one company alone paid $600,000 a year to rent space on Broadway.[42] By the 1920s the outdoor advertising business was dominated by the General Outdoor Advertising Company (GOAC), which controlled 80 percent of all displays in the major American cities. Independent display companies found themselves at GOAC's mercy in many aspects of their business—particularly because GOAC had a virtually exclusive contract with a sister organization, the National Outdoor Advertising Bureau, that acted as the agent for most of the sites around the country. The independents protested against this cozy arrangement, and the federal government successfully brought suit for restraint of trade. Nevertheless, GOAC remained the dominant company in the industry, handling the most spectacular signs, including the Wrigley installation.[43] The night view of the American city—depicted on postcards, celebrated in guidebooks, and thus increasingly thought of as the "real" city—was a creation of capitalism.

Some protested the powerful aesthetic impact of electric signs. In England, in 1924, George Moore complained of their "savagery": "There is one monstrosity that flaunts somebody's gin, another that pours out port, a third that insists on putting before us inescapably the name of a popular newspaper which is always advocating the beauty of London. . . . Ruby lights and electric port! It is fantastic!"[44] Others with Moore's distaste for electric signs had long before organized themselves into associations, both in New York and London, with the object of having the most objectionable signs removed and setting standards for the others. In England such groups were somewhat successful; for example, the Bovril Company was convinced not to erect an illuminated sign in Edinburgh.

In the United States, William Dean Howells was one of the first to sound the alarm. In 1896, when the Great White Way was only getting started, he complained: "If one thing in the business streets makes New York more hideous than another it is the signs, with their discordant colors, their infinite variety of tasteless shapes. If by chance there is any architectural beauty in a business edifice, it is spoiled, insulted, outraged by these huckstering appeals."[45]

Strangely, almost no one complained. Howells concluded: "It seems as if the signs might eventually hide the city. That would not be so bad if something could then be done to hide the signs." But during the next generation the signs would hide the material city in a very real sense, replacing it with selected electrical outlines.

Howells's protests were echoed by the Municipal Art Societies that appeared in the 1890s in New York, Cincinnati, Cleveland, Chicago, Baltimore, and other large cities. Their membership was primarily drawn from wealthier families of established standing. A national organization, the American Park and Outdoor Art Association, also was formed. The secretary of this association, Charles Mulford Robinson, published a spate of articles and books calling for artistic refurbishment of the city, including a color scheme for each borough in order to create a harmonious effect. Robinson demanded strict regulation of electrical displays. Writing from an upper-class perspective, he admitted that "the idea that aesthetic charm can be given to the glaring, shrieking letters of the common sign . . . seems at first fantastical," but he asserted that "something else than size and hideousness can pay."[46] Such associations favored the "white city" illumination of the 1892 Columbian Exposition or the pointilistic subtlties perfected at the 1901 Buffalo Pan-American Exposition. In their view, public lighting was to be primarily white, austerely beautiful, and refined. The Dewey Arch, in New York, was illuminated in this fashion in New York; however, much to the disgust of the Municipal Art Association, the intended stately effect was marred by the obtrusive presence of a stupendous green Heinz pickle flashing nearby.[47] As already noted, the organizers of the Hudson-Fulton Celebration similarly disparaged what they regarded as the vulgar and garish signs of Broadway, holding their parades on Fifth Avenue instead.

Because outdoor art associations had considerable social clout in the largest cities, they lobbied successfully for restrictive legislation. For example, New York City restricted signs from projecting more than 6 feet from any building. Salesmen began to discuss aesthetic standards at annual meetings of the Society of Illuminating Engineers and the National Electric Light Association. One speaker warned that all commercial lighting now had to be done as tastefully as possible, owing to "the agitation for removing signs from the streets." He continued: "It seems to me that the development of electric signs and electric display lighting should be done with the artistic effect in mind; otherwise the municipal authorities will remove all the signs, as they have done in many cities. Some of us have been figuring on signs to reach as far across the street as possible. I think that is a great mistake."[48] Salesmen were caught between customers who wanted to erect bigger and more spectacular signs and genteel society's protests over the commercialization of the night.

Yet within a few years the average American embraced electric advertising with few qualms. In Detroit, a saleswoman noted with satisfaction that, after initial resistance, "undertakers, churches, and clubs blazon their identity with the same kind of electric sign as the saloon and the corner grocery."[49] By World War I the electrified skyline was a defining characteristic of the large city and usually a source of civic pride.

Yet, although a city skyline seen from the distance was an impressive sight, at closer range the electric signs had become an element in the tug of war between signs and buildings that had characterized the main streets of American towns since the 1840s. Even before the Civil War the refinements of architecture were often obliterated by commercial signs that covered windows and exposed walls and jutted up from roofs. As vehicles moved faster, signs became bigger so that they could still be read. As horsecars were replaced by streetcars and then by automobiles, signs became enormous and their messages got even more aggressive. The automobile promoted the merging of the building and the sign on the outskirts of towns, where lax zoning laws gave free rein to fantasy and invention. Though before the 1920s a few buildings had already mimicked the service or product to be sold, in that decade motorists began to encounter large numbers of electric windmills, giant animals, huge cheeses, and other examples of what Chester Liebs calls "representational giganticism."[50] These fusions of buildings and signs invariably were lighted in striking ways to attract passersby. Gas stations erected tall, brightly lighted logos to announce their presence to drivers a mile away, giving them time to consider stopping. And a host of new electrically lighted structures sprang up along the roads in the 1920s, including miniature golf courses, drive-in theaters and restaurants, and motels. In most of the United States, commercial strips flourished with little opposition.

But many were appalled by such developments, and electric signs came under renewed attack in the 1920s. George Moore's criticisms were reported in the *New York Times,*[51] and New Yorkers may have been particularly interested in his remarks because of a local conflict. In 1922 the Fifth Avenue Association brought a petition to the Board of Alderman for the restriction of illuminated signs. A year later the 42nd Street Property Owners and Merchants Association sought an ordinance that would eliminate all signs projecting from buildings on 42nd Street between Third and Sixth Avenue. These petitions were fought by the Master Sign Makers Association and the Broadway theaters.[52] Nevertheless, signs that projected over the street were prohibited on Fifth Avenue from Washington Square to 110th Street and along certain parts of Madison Avenue. Throughout the rest of the decade this conflict simmered, culminating in May 1929 in strikes and demonstrations by signmakers and small shop owners provoked by proposals for further restrictions on electric signs throughout midtown Manhattan.[53]

Electric signs were thus inscribed in the larger cultural conflict between high and popular culture—between advocates of the reserved beauty of the "white city" ideal, who supported such events as the Hudson-Fulton Celebration, and the entrepreneurs who employed electrical displays. The Fifth Avenue Association included the owners of New York's most fashionable shops and upper-class churches. Its three-year study predictably led to the conclusion that it would be best to "encourage merchants . . . to erect artistic flat wall signs either in wood or in bronze."[54] Like the Municipal Art Associations at the turn of the century, the Fifth Avenue Association rejected the vernacular of Broadway and the roadside strip and longed for the control over electrical displays exercised by European governments.

While Americans ultimately did little to curtail the elaborate lighting of their "white ways" and the fantastic shapes along their roadsides, the city of Paris explicitly rejected elaborate American-style lighting in new ordinances passed in 1926. When combined with later restrictions, these laws eliminated all "detached advertising," permitting only location markers for goods sold on the premises.[55] In the same years, most nations on the Continent passed strict zoning laws to prevent roadside strips from developing.

In the United States, the nationally syndicated comic strip "Bringing Up Father" took note of the clash between popular American attitudes toward the electrified landscape and the disdain it provoked among the upper class and many Europeans. Jiggs takes an Englishmen to a high place that commands an excellent view of Times Square, and proudly declares: "Well, here is Broadway, the Great White Way, the Rialto of America. What do you think of it?"

The Englishman replies: "It's all right, I suppose, if one likes electric light bulbs."[56]

In 1931 a reporter visited Frank C. Reilly, the electrician who had designed many of New York's most famous electric signs, including the 400-foot ribbon of lights on the Times Building that delivered news bulletins to the crowds below. With 14,800 bulbs capable of more than 260 million flashes per hour, the Times' "zipper" had been seen by an estimated million people a day since first lighting up on election night, November 6, 1928. On the wall of Reilly's office the reporter saw a three-panel cartoon about the "zipper." "The first square," he reported, "showed three men standing in the street before the sign, reading the news. In the second, they had been hit by a taxi-cab and tossed high in the air. In the third, they were picking themselves up, staring at the letters that now marched across the building. They read: 'Three hit by taxi in Times Square.'"[57] Reilly told the reporter that the cartoon was not so farfetched. His bulletins were constantly changed to keep up with fastbreaking events, and quite conceivably an event in Times Square could be reported shortly after occurring. But the cartoon's humor hardly resides in its possible factuality. Rather, the three panels suggest one larger meaning of the electric advertisement. As the first frame suggests, electric signs distract the attention of pedestrians, drawing them into the contemplation of a display that seems more intriguing than the unelectrified, physical world. The second frame, in which the three men are hit by a taxi, suggests the penalty for such absorption in the electric landscape. The third frame returns us to the original situation as the men once again are drawn into the representational world of the sign, which now informs them of their own accident. The sign is not only fascinating and immediate; it is also responsive to the world. One might almost say it is an instantaneous commentary. Not only does it serve as a tourist attraction and a central reference point in the city; it has become a part of consciousness, translating events into display, fusing news into the edited landscape.

The public space in which such a display occurs is not merely decorated with electricity; it offers a heightened sense of reality, suggesting that the individual can leave behind the accidents and problems of daily life and merge with the flashing lights. In James Oppenheim's 1912 novel *The Olympian*, the young protagonist, just arrived in New York, is irresistibly drawn to Broadway: ". . . looking

to the east, [he] saw Broadway flaring across the mouth of the
street; and, like a fragile insect driven mysteriously, inevitably, without thought, without hesitancy, he hurried toward the lights."[58]

An important secondary effect of electric advertising is that it
directs the consumer's eyes away from poor areas and toward commercial zones. Even more important, the vibrant display of
Broadway encourages fantasy. The lights promise release from toil,
and express "something mysterious, hinting at romance."[59]
Oppenheim's unemployed protagonist passes down "a canyon of
fire: orange and gold and blue beat upon the pavements, and
through the radiance, the laughter smitten crowd was flowing up
and down." The electrical sign was an eye-catching part of a new
landscape that imparted dynamism and rhythm to a night scene
that, like Times Square, was far less impressive during the day.[60]

Nor were such responses to the luminous environment limited to
the poor. The illuminated city evoked an epiphany for the young
Lewis Mumford one evening as he walked across the Brooklyn
Bridge: "Three quarters of the way across the Bridge I saw the skyscrapers in the deepening darkness become slowly honeycombed
with lights until, before I reached the Manhattan end, these buildings piled up in a dazzling mass against the indigo sky. . . . Here was
my city, immense, overpowering, flooded with energy and light. . . .
The world at that moment opened before me, challenging me,
beckoning me. . . . In that sudden revelation of power and beauty
all the confusions of adolescence dropped from me, and I trod the
narrow, resilient boards of the footway with a new confidence."[61]
Ezra Pound, back from Europe for a visit in 1910, had a similar
response to the night skyline, finding New York "the most beautiful
city in the world" in the evening: "It is then that the great buildings
lose reality and take on their magical powers. They are immaterial;
that is to say one sees but the lighted windows. Squares after
squares of flame, set and cut into the aether. Here is our poetry, for
we have pulled down the stars to our will."[62] Mumford and Pound,
like the night crowds in Times Square, found that lights gave the
landscape "magical powers." Spectacular lighting dematerialized
the city and promised personal transformation.

The night had become a kind of negative background for artificial lighting, which glamorized the city, giving it a semi-abstract
cubist skyline that contrasted with its often drab daily appearance.
The electrified city could be grasped as a vast abstract pattern when
seen from atop a skyscraper, from the cliffs of New Jersey, or from

an airplane. *Harper's* invited its readers to assume an olympian perspective and to rejoice at technology's conquest of both earth and air: "The birdman, skimming the low skies in the mechanical marvel that the twentieth century gave him, sees, from his perch aloft, the city outlined as clearly by night as by day. The arcs pick plainly the paths of the streets, they outline the water-fronts, the great bridges that leap the rivers and span the deep valleys."[63]

An extraordinary environment had come into being, offering new visual experiences.[64] Just how novel it was seemed less obvious to Americans who had witnessed its gradual development than to foreign visitors whose cities remained comparatively dark. When the English writer Arnold Bennett visited the United States for the first time, in 1911, he marveled at the "enormous moving images of things in electricity—a mastodon kitten playing with a ball of thread, an umbrella in a shower of rain, siphons of soda-water being emptied and filled, gigantic horses galloping at full-speed and an incredible heraldry of chewing gum. . . . Sky signs!" "In Europe," he continued, "I had always inveighed manfully against sky-signs. But now I bowed the head, vanquished. These sky-signs annihilated argument." Bennet was "overpowered by Broadway."[65]

The poet Vladimir Mayakovsky visited New York in 1925 and was entranced by the extent to which electricity had penetrated into every aspect of life: "A man gets dressed by electric light, in the streets electric lights, houses with electric lights, evenly cut by windows like the pattern of a publicity poster."[66] The culmination of this new existence was Broadway.

Lights go on all along the entire twenty-five mile long Broadway. . . . This is, the Americans say, the Great White Way. It is really white and one really has the impression that it is brighter there at night than in the daytime. . . .The street-lamps, the dazzling lights of advertisements, the glow of shop-windows and windows of the never-closing stores, the lights illuminating huge posters, lights from the open doors of cinemas and theatres, the speeding lights of automobiles and trolley cars, the lights of the subway trains glittering under one's feet through the glass pavements, the lights of inscriptions in the sky. Brightness, brightness, brightness. . . .[67]

Mayakovsky exaggerated the length of Broadway, but he realized that lighting had given New York a remarkable dynamism. Yet the necklace of lights draped over the city and the glamor of Broadway evaporated with the dawn, when harsher economic and social realities

A view from Empire State Building showing the RCA Building, with its "luminous-tube" sign. Courtesy of Hall of History Foundation, Schenectady, N.Y.

reappeared. Even in the glitter of the neon night some observers, such as the poet Claude McKay, testified to a haunting loneliness:

About me young and careless feet
Linger along the garish street;
Above, a hundred shouting signs
Shed down their bright fantastic glow
Upon the merry crowd, and lines
Of moving carriages below.
Oh wonderful is Broadway—only
My heart, my heart is lonely.[68]

The urban electric landscape also made a powerful and disturbing impression on the German filmmaker Fritz Lang: "I first came to America briefly in 1924 and it made a great impression on me. The first evening, when we arrived, we were still enemy aliens so we couldn't leave the ship. It was docked somewhere on the West Side of New York. I looked into the streets—the glaring lights and the tall buildings—and there I conceived *Metropolis*."[69] Lang's expressionistic film, completed two years later, included shots of ominous, intensely lighted buildings that loomed threateningly above pedestrians. It was a completely man-made landscape of overpasses, viaducts, and skyscrapers, the space of the city having become a series of nightmarish interiors.

Yet most of Lang's and McKay's contemporaries valued the new urban landscape as a new kind of tourist site. Implicit in the responses of Pound, Bennett, Mayakovsky, and Mumford was the realization that the electric landscape was much more than the sum of its parts. As the critic Kenneth Burke noted, "Broadway is qualitatively rich; not a single light on it is worth a damn, but the aggregate of so many million lights demands attention."[70] One might analyze an individual sign as an example of arbitrary associations made by advertisers among products, trademarks, colors, and flickering patterns of lights. Yet if each sign had its special message and could be understood as a commercial venture, it was embedded in a constellation of sky signs, which in turn lay within other lighting arrays that called attention to skyscrapers, bridges, radio towers, and public monuments. The landscape was at once dynamic, chaotic, and exciting. The "White Way" must be understood not only as a series of individual messages but also as one overriding message. H. G. Wells caught a sense of this when he reported having watched New York's lighting displays emerging early one evening in 1906: "New York is lavish of light, it is lavish of everything, it is full of the sense of spending from an inexhaustible supply. For a time one is drawn irresistibly into the universal belief in that inexhaustible supply."[71] The sense of the transformation of the self was wedded to that feeling of inexhaustibility. The urban night landscape promised a perpetual cornucopia.

In the 1970s critics such as Jean Baudrillard wrote of the slippage between signifier and signified in the universe of advertising,[72] but that slippage was already manifest in 1925 on the Great White Way. Here was a literal universe of signs. Each insistently proclaimed a particular man-made product. Each was an overdetermined signifier for a product that was obviously part of the capitalist system of production and distribution. Yet no sign was ever seen alone; each was a part of an overwhelming impression produced by the constellation of city lights. Just as the many individual skyscrapers together became a skyline, advertising signs collectively became a great signifier, an important cultural marker. Collectively these signifiers lost their individual meanings and became a tourist site, flattening the city into ethereal abstraction. For those who came to stare, they were only incidentally representations of an array of products. More centrally, the White Way was a strange hybrid, combining word and image, signified and signifier, motion and stasis, illusion and reality. Although private capital had undoubtedly created this landscape, its meaning could not be derived simply from the codes of commerce. Indeed, advertisers themselves found that the effectiveness of their signs as messages was exceedingly short-lived. Much as the early sign builders had learned that their medium was poorly suited to text, those of a later generation realized that visual messages soon decayed in significance. As one reporter noted in 1938, "when a sign becomes too familiar to the crowd, it blurs into the general scene and its message is lost."[73]

The electrified landscape's meaning lay precisely in the fact that it seemed to go beyond any known codification, becoming unutterable and ungraspable in its extent and complexity. The press continually tried to express its meaning by counting the miles of wire, the millions of bulbs and kilowatts, the huge expense, and the vast daily audience. Such enumerations are characteristic of Kantian encounters with extreme magnitude or vastness, as would be the case for those seeing New York or any other large city from atop a skyscraper. But the electrical sublime was not a mere extension of the geometrical sublime, with its olympian assurances to the observer that he could turn the city into a concrete abstraction. The electrical sublime eliminated familiar spatial relationships. In the night city there were no shadows, no depth, no laws of perspective, and no orderly relations between objects. At night the urban landscape no longer seemed physically solid. An immense sign bulked larger on the skyline than a far more substantial building, and gargantuan electrified objects distorted the sense of scale. The city as a whole seemed a jumble of layers, angles, and impossible proportions; it had become a vibrating, indeterminate text that tantalized the eyes and yielded to no definitive reading.

At the same time, experiencing the White Way from within Times Square exemplified the dynamic sublime, an encounter with a violent power impossible to resist. Close up, the White Way was no longer an enormous vibrating cubist landscape. Rather, it assaulted the senses. A subject in the middle of Times Square simultaneously saw millions of flashing lights, heard the roar of traffic, and was engulfed in a restless crowd that included every conceivable human type. The brain could not process all the flashing lights that contested for attention, creating a sensory overload, a rush of impressions that rendered individual messages virtually meaningless.

Whether experienced in terms of its vast expanse or as an overwhelming immediacy, the electric landscape, like any sublime landscape, can leave the subject with a feeling of weakness and insignificance before the power of an immense and powerful object. Alternately, as with Mumford and Pound, it can provide a sense of superior self-worth. The vast city honeycombed with light prompted them to conceive something more powerful and immense than their senses could grasp. This seemed a liberating landscape that promised abundance and personal fulfillment. But to Fritz Lang and Claude McKay it threatened to become a landscape of loneliness and alienation. In either case, the specific meaning of any individual electric light or sign was far less important than the meaning of the whole. This sublime significance, unlike that which Kant had described, was man-made, creating awe and respect not for nature but for technology and for the engineers and businessmen who erected the displays. The electric landscape thus emerges as an important part of the phenomenology of industrialized society. It translated what Max Weber once termed capitalism's "romance of numbers" into a dynamic experience that redefined the historical subject. The consumer was not merely cajoled with ingenious messages; he was overwhelmed by a text without words in a display of power that taught him his role in the new social order. The civic processions of the Jacksonian era had focused on producers and craftsmen who demonstrated their trades on floats as they rolled through the streets, but in the new urban order of the Great White Way the procession had become a perpetual spectacle in which the image of the product had replaced the producer and the night city had become a tantalizing etherealization.

The night city was a wonder that urged the viewer to merge the scintillating landscape into the self. This response was engendered not by nature but by a potentially chaotic urban scene, which the engineer orchestrated into a glittering sea of images. But the exaltation of self occasioned by the electrical landscape had no metaphysical reference point. It was not only ephemeral; it was based on a game of substitution in which the physical city was etherealized and replaced by a consciously imposed mirage of lights and shadows. This landscape had emerged unintentionally from the free play of electrical signifiers, which had grown directly from, and which expressed, the marketplace, where businessmen competed by intensifying the lighting of streets, by illuminating their stores and skyscrapers, by erecting ever-more-spectacular signs, and by spotlighting public buildings as symbols of the new order. The resulting landscape can quite literally be called the landscape of corporate America. It embodied the dominant values of individualism, competition, advertising, and commodification, and at the same time it transformed these values into a disembodied spectacle with an alluring promise of personal transformation.

It remained to fuse this unintended sublime into a self-conscious construction. The first electric landscape had emerged as a coherent vision at nineteenth-century world's fairs; the fairs of the 1930s would integrate the electrical landscape into a new synthesis of sublimities.

Synthesis: The New York World's Fair of 1939

World's fairs exploited every form of the man-made sublime. Whatever was being presented—the Corliss Engine (1876), spectacular lighting (1893, 1901, 1904, . . .), a tall building (1901), an operating coal mine (1904), or an assembly line (1915)—the goal was to awe the visitor. Yet this became more difficult as innovations became more familiar.

By 1939 it was pointless to try to outdo Times Square in lighting or the Empire State Building in height. Nor did New York's "World of Tomorrow" emphasize industrial production or giant machines. Despite the spectacular lighting, the world's largest locomotive, the towering Trylon, and numerous inventions, what had once been sublime was now relegated to a supporting role. Instead, this fair emphasized miniaturized landscapes of the future and the creativity of corporate research laboratories.

The fairs of the late nineteenth century had emphasized the products of industrialization. In the first decades of the twentieth century, massive working exhibits had brought the factory to the fairgrounds. But during the Great Depression, companies emphasized their research laboratories and used vast miniature landscapes to show how their latest inventions could transform the world.[1] The miniaturized landscape took the geometrical sublime one step higher into the clouds, suggesting the view from an airplane on a day with perfect visibility; the research laboratory replaced earlier displays of hardware and production techniques with nearly magical displays of scientific prowess. These displays suggested that private corporations could solve the economic crisis and create a better world. The movement from product to production to invention suggests how quickly each form of the man-made sublime passed into the ordinary, forcing each exposition to find new ways to astonish visitors. At the same time, it traces a transition in what was considered sublime: from massive individual objects to industrial processes to simulated landscapes. These changes also measure the intentions of the exhibitors. As Roland Marchand notes,[2] corporations first tried only to sell products, then tried to educate the public about their business, and finally turned to marketing visions of the future.

The role of spectacular lighting changed accordingly. A novelty at Cincinnati, it was central to the Chicago, Buffalo, St. Louis, and San Francisco fairs. Lighting was reconfigured on the fairgrounds and in the simulated landscapes of 1939. Floodlighting of buildings, the central element in the San Francisco Exposition of 1915, was banned at the New York fair except in the case of the theme center, the Perisphere. Instead, exterior lighting was built into each building as part of its architectural design. Whereas floodlighting made buildings look much the same at night as in the daytime, the organizers of the New York fair decided that each building would have "a night appearance quite different from its daytime appearance"; they wanted "by night a structure in light, a luminous design of a static character, that by no possibility could be had by natural day light."[3] Exterior lighting emphasized the murals, sculptures, architectural details, banners, and plaques, and often intentionally left some walls in darkness, creating dramatic contrasts. Thus, not only did the fair look far different by night than by day; it also became a textualized landscape that emphasized painting, writing, flags, statues, and corporate logos, each set off by contrasting areas of obscurity.

The world's fairs before World War I and the great civic festivals such as the Hudson-Fulton Celebration had emphasized the "white city" as the logical corollary of the Beaux Arts style in architecture. In contrast, the lighting design of the World of Tomorrow represented a marriage of modernism and the vernacular of Broadway. Because the fair was paradoxically both competitive and under a single management, its design was colorful yet static. Instead of the discordant, flashing lights of the urban commercial zone, with their illusions of movement and their reiterated messages that actively competed for attention, the fair at night was a still pattern. Instead of the tremendous variety of colors of Times Square, a color scheme was imposed so that each avenue had a dominant tone. The resulting landscape aestheticized the Great White Way, preserving its form but not its surging energy. Thus, it articulated a shift from the raw energy of Times Square to a synthesis of tamed and harmonized sublime elements.

For inspiration, the designers of exhibits for the New York fair turned away from dramatic lighting and massive technological displays and toward the most prestigious technology of their day: the airplane. Flight represented the acme of human achievement. To raise a heavier-than-air vehicle into the sky was a technological marvel, the fulfillment of a centuries-old dream. Yet when the Wright Brothers flew for the first time, in 1903, almost no one saw their achievement, and only local newspapers reported it at any length. The Wrights remained secretive about their plane's design during subsequent development, seldom allowing the press to see what they were doing. Although invited to explain their invention to the American Association for the Advancement of Science and to demonstrate it at the St. Louis World's Fair in 1904, they refused. They had their eyes on commercial applications, and they were unwilling to disclose the details of their machine. Not until the

The Wright Brothers' first flight, December 17, 1903. Courtesy of National Aeronautics and Space Administration.

summer of 1908, when they held a demonstration for the army, did the public begin to understand what the Wrights had achieved. In the next few years huge throngs turned out to see them—particularly at the Hudson-Fulton Celebration.[4] Until the time of World War I, many people ran out of their houses to stare at any airplane that flew into view. In 1910, when a plane first flew over Chicago, an estimated million people went out into the streets to see it. One observer (a minister) wrote: "Never have I seen such a look of wonder in the faces of a multitude."[5] The airplane violated the natural order, defying gravity and hurling a man so high he became little more than a speck against the sky. Flight was sublime.

Human flight long remained the most exciting form of the dynamic sublime. Aviators had been the most heroic figures of the First World War. They rose above the mud and trenches, where millions of troops were trapped behind barbed wire and annihilated by the new technologies of mass extermination: the machine gun, poison gas, and the tank. The flying machine retained an element of romance, offering the spectacle of man- to-man combat in the heav-

Claude Grahame-White lands in Washington, October 14, 1910. Courtesy of National Archives.

ens and contrasting starkly with the nameless slaughter on the ground. Washington's 1919 Independence Day parade was highlighted by an overflight of twenty planes. During the 1920s, pilots who barnstormed across the United States found eager crowds awaiting their arrival. In 1924 a crowd of 200,000 turned out at Santa Monica to greet a group of military fliers returning from the first aerial circumnavigation of the world. The excitement over aviation climaxed in the tumultuous receptions held in Paris and New York to honor Charles Lindbergh's solo hop from New York to Paris in 1927 (only a quarter-century after the Wright Brothers coaxed their machine off the ground for 12 seconds, covering a mere 120 feet). In 1939 the dedication of New York's Municipal Airport drew 325,000 people.[6]

Virtually all Americans imagined that the airplane would transform everyday life. Children were encouraged to be "air-minded." African-Americans shared the enthusiasm. In 1939 the U.S. Congress banned racial discrimination in aeronautical training programs,[7] and the *Chicago Defender* celebrated this victory by sponsoring two African-American fliers on a 3000-mile tour of ten cities.[8] Yet few people had ever been up in an airplane to see the earth spread below them, and here the designers of the New York fair recognized an opportunity. To show visitors vast landscapes of the future, they simulated the view from an airplane. In doing so, they synthesized the three major forms of the technological sublime: the dynamic, the geometric, and the electric.

Forty-five million people attended the New York World's Fair in its two seasons (1939 and 1940). Even allowing for repeat visits, this was a large percentage of the American population. The fair was the best-attended event in the United States during the first half of the twentieth century.[9] But why did a country mired in economic depression conceive and plan this $155 million project? The commercial standards of profit and loss are largely irrelevant in evaluating any such public event. Burton Benedict has suggested that a world's fair resembles an international potlatch ceremony, bringing together many national exhibitors to vie with one another in the extravagance of their conspicuous expenditures. At any world's fair, groups exchange entertainments, courtesies, and rituals. "Rivals must participate," Benedict notes. "Failure to do so is acknowledgement of defeat, of the superiority of one's rival."[10] Or it is an admission of enmity—neither Germany nor Spain came to the New York fair. In contrast, Japan, Italy, and the Soviet Union entered the competition, seeking to use the fair as a showcase for their societies.[11] Mussolini sent a large exhibit, including a 200-foot waterfall, and announced that he would build a full-scale copy of Coney Island for an Italian world's fair.[12]

Yet the World of Tomorrow turned out to be dominated more by corporations than by nations. In the United States businesses had always used fairs to compete with one another, but in the past they had occupied a distinctly lower rank than American states or foreign nations. In 1939, however, the corporations did more than display their products; they took on the role of interpreting the future to the American public, telling them that the long depression and the danger of war could be overcome and that a utopian future for their children was achievable by 1960.

The corporations recognized, of course, that the public went to a fair to be entertained. After the Columbian Exposition of 1893, a fair almost invariably had an amusement park located beside it. Kasson argues that such parks served as a permanent feast of fools for the urban masses. Drawing on Victor Turner's concept of liminality, he notes that the visitor to Coney Island temporarily left behind the values of thrift, sobriety, and self-restraint. The amusement park encouraged wasteful self-indulgence. Its fun houses, exhilarating rides, and festivity permitted an escape from ordinary consciousness into a "liminal" state. The individual merged into an egalitarian crowd, restricted by few rules. Similarly the New York World's Fair offered an escape from the imperfections of present life into an ideal future. Unlike traditional societies whose rituals return to an earlier redemptive moment of creation, a rapidly changing industrial society looks not to the past but the future. At the "World of Tomorrow" the future was not merely represented by new commodities; it was a place that could be visualized and visited. The fair's forms of display had appropriated much that was on offer in the amusement zone, which did less well in competing with corporate exhibits than at previous fairs. The corporations offered not a costly feast of fools but journeys into a future world where the crisis of the Depression had been solved.[13]

World's fairs can also be understood in relation to that central regulator of cultural space and time, the museum, which they tend to imitate. The museum organized the past into collections of artifacts representative of historical epochs. In it the Victorians implicitly celebrated the forward movement of history.[14] Whereas the museum provided a space for the preservation and re-presentation of the past, the world's fair served as a site within the transitory present from which the visitor could glimpse the future. Such a goal was quite in keeping with the technological sublime. Medieval fairs had provided opportunities for the exchange of goods, and regular commercial fairs continue that function today. But world's fairs, like museums, integrated artifacts into coherent ensembles and interpreted them according to overriding themes. They marketed the idea of progress itself, providing an overall impression of coherent historical development.

The first great world's fair—the Great Exhibition of the Works of Industry of All Nations, held in London in 1851, offered the world more than a selection of the latest products and inventions. It presented these goods in a harmonious display, housed within the novel form of a giant glass conservatory—the Crystal Palace. Here, as at almost all later world's fairs, a new architectural form announced the triumph of new technologies. At the New York fair of 1939, the style adopted was a fusion of streamlining, art deco, and modernism.

The 1939 fair represented a detailed response to the Great Depression. Americans were uneasy about the present and profoundly worried about the future. Since 1929 nothing had seemed capable of restoring prosperity, except perhaps another unwanted war. To catch the public imagination, the fair had to address this uneasiness. It could not do so by mere appeals to patriotism, by displays of goods that many people had no money to buy, or by the nostalgic evocation of golden yesterdays. It had to offer temporary transcendence.

Of the fair's two major themes, only one could serve this end. The historical theme, the 150th anniversary of George Washington's inauguration as president, did not inspire any of the most popular exhibits (though it was present throughout the fair, articulating the older discourse of republicanism). The imposing 68-foot statue of Washington in colonial garb that stood in the middle of the main concourse (Constitution Mall) was accompanied by statues representing freedom of speech, of assembly, of the press, and of religion. Washington's surveying instruments, shaving set, and walking stick were displayed in Washington Hall. Venezuela displayed a lock of his hair once owned by Símon Bolivar, Great Britain an elaborate Washington family tree.[15] Actors recreated Washington's journey by horse and carriage from Mount Vernon to

Wall Street for his swearing-in ceremony. But these references to eighteenth-century republicanism could not hold a visitor's attention for long in the midst of an environment that was overwhelmingly modernist in tone. Admittedly, in the second year of the fair the historical theme became more prominent—a shift in emphasis due to the war, as most European visitors and several nations' exhibits disappeared. But during both seasons Washington and his times were almost completely eclipsed by the second theme: how science and technology would shape the future. The two themes counterpointed one another: while Washington's past was irreproachable, the immediate future threatened disaster. The organizers did not shy away from this difficulty. In practical terms, the fair's overall theme—"Building the World of Tomorrow with the Tools of Today"—required powerful exhibits that would lift the visitor out of the daily routine and the apparent hopelessness of the Depression.

As had been the case with the Hudson-Fulton Celebration, planning for the fair began years before the event and at the local rather than the national level.[16] As in 1909, a coalition of state, city, private, and foreign interests sponsored the event, but the balance between these groups had changed. And although the federal government would be involved (for reasons of prestige and to court foreign participation), control of the corporation that built and operated the fair remained in the hands of private enterprise. In 1936 the World's Fair Corporation issued $42 million in bonds. The buyers, who would eventually get back only 40 cents on each dollar invested, included Myron C. Taylor, former financial chairman of U.S. Steel; Walter P. Chrysler, founder of the corporation of the same name; Alfred P. Sloan, Jr., chairman of General Motors; Walter S. Gifford, president of American Telephone and Telegraph; and Owen D. Young, chairman of the board of General Electric.[17] Not incidentally, these same corporations secured excellent locations for their pavilions. Other subscribers included Vanderbilts, Roosevelts, and Rockefellers, publishers of major newspapers, directors of Chase Manhattan and other important banks, and many financial luminaries. Aside from the sale of bonds, New York City invested $26 million in new roads, lighting, sewers, and other services; New York State contributed $6 million. Foreign governments spent $30 million on their exhibits, but private exhibitors invested more than $50 million.[18]

Citizens might have been invited to vote on the fair's theme, to suggest its slogan, or even to buy small shares in the fair, but in fact they played little part in the preparations. Nor did the New Deal agencies (which actively promoted painting, published regional guidebooks, supported theaters, and promoted the arts) have a large role, although the Works Progress Administration did have an exhibit and there was the obligatory U.S. building. Instead, the fair's themes, architecture, and organization were in the hands of an economic and cultural elite, its membership defined by the bondholders, by New York City officials, and by large corporate exhibitors.

A month after the World's Fair Corporation was formed, a hundred writers, designers, and other "progressives in the arts," gathered for a banquet at New York's Civic Club to hear the architect Lewis Mumford and others call for a new kind of fair. Mumford wanted nothing less than to concretize "the future of the whole civilization." The fair, he argued, could tell a story through architectural form, showing the public an orchestrated future and stressing "this planned environment, this planned industry, this planned civilization." He continued: "If we can point toward the future, toward something that is progressing and growing in every department of life and throughout civilization, if we can allow ourselves . . . as members of a great metropolis, to think for the world at large, we may lay the foundation for a pattern of life which would have an enormous impact in times to come."[19]

Offered the opportunity to sell the public on their plans and their future products, virtually all the major American corporations agreed to participate, despite the high cost. Early fairs had given away exhibition space, grouping displays by industry or by type of commodity. But the New York fair's organizers adopted the system that had been profitably used for the 1933 Chicago Century of Progress Exposition, requiring participants to rent space on the grounds and allowing them to construct their own buildings rather than be housed with comparable exhibitors in theme buildings. This system made economic sense from the organizers' point of view, since it reduced costs; however, in a period of worldwide depression it limited the ability of private inventors to mount exhibits. As a result, the fair was dominated by the displays of a few large nations and many large corporations.

The exhibits were organized along an axis running from the international Court of Peace, across the Lagoon of Nations, and up

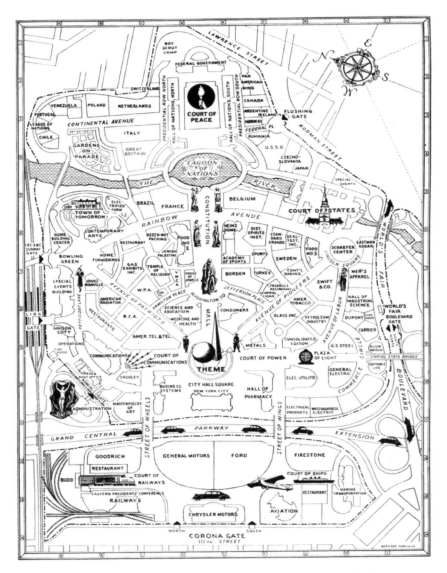

Map of the New York World's Fair. Courtesy of Smithsonian Institution.

Constitution Mall to the fair's central buildings, the Trylon and the Perisphere.[20] The Perisphere was a white ball 200 feet in diameter; the Trylon was a white tower whose three sides narrowed to a point at a height of 610 feet. The fair's major concourses all radiated from them, thus drawing visitors to them. At night colored lights and shadowy clouds played over their surfaces, making them an unmistakable focal point. Both structures were illuminated in such a way that they seemed to float in the air, and appropriate colors were used for special occasions (orange for Halloween, red, white, and blue for the Fourth of July, and so on). The area around these two central buildings was reserved not for the American states or foreign countries but for corporate exhibits.

Covering half of the grounds, the corporations' buildings were subdivided into groups—transportation, communication, electrification, production, food, and so on—each with its cluster of buildings. The corporate exhibits dominated in scale as well as in position. Either the Ford or the GM building alone was larger than any two national exhibits. The AT&T, Borden, General Electric, Firestone, Chrysler, and B. F. Goodrich pavilions were also large, centrally located, and eye-catching. Less noticeable were smaller buildings devoted to the traditional high culture that had once predominated at fairs. The exhibit halls devoted to education, religion, and the arts were on the periphery, literally marginal in the political economy of the World of Tomorrow.[21] This organization was hardly an accident. Robert D. Kohn, chairman of the fair's Committee on Theme and a member of its Board of Design, writing in the *North American Review* about the social ideals the fair was to represent, claimed that the fair's organization would be more comprehensible to the average citizen than those of earlier expositions. He rejected such "categories as science, art, agriculture, manufactures—the classic divisions of fairs for centuries" on the ground that they "would only perpetuate divisions convenient for technicians but not illuminating to laymen." He continued: "We chose to make our divisions more or less functional, the things with which the average man comes in contact in his everyday life—food, shelter, clothing, communications, education, transportation, etc. What is more, instead of isolating science and art, the planners would attempt to show them permeating all of these other things, as illustrations of their interpenetration into the functions of modern life."[22] This organization demonstrated how corporations had penetrated every realm of social life. They had assimilated art into advertising, public relations, and product design, just as they had appropriated science for their research and development laboratories. By the 1920s corporations were also sponsoring research at the Massachusetts Institute of Technology, at Carnegie-Mellon, and at other universities. For this reason, and not only because the fair's planners wished it, science and art would indeed permeate most of the exhibits.[23]

The desire for models and dioramas of the future presented a great opportunity to industrial designers. Their field was quite new, not having been institutionalized until the 1920s. Some of its early leaders had worked in stage design (Henry Dreyfus and Norman Bel Geddes), advertising (Walter Dorwin Teague), and engineering (Raymond Loewy); others were mavericks in architecture and engineering, like R. Buckminster Fuller.[24] Synthesizing these various fields, industrial design gave equal emphasis to product performance, dramatic presentation, and consumer appeal. Employed by both the fair's organizers and its corporate exhibitors, industrial designers created the startling syntheses of theatre, futuristic design, and technology that became the fair's most popular exhibits. The fair would demonstrate the beneficence of corporate control, from invention through mass production to merchandising, through aesthetically pleasing designs calculated to overwhelm the visitor.

Expositions, of course, are designed to do more than encourage competition and display goods and services. To succeed, an exposition also must meet the overarching goal of achieving consensus. Like political parades, exhibitions are collective social displays, defining the contours of the community. The fair not only had to attract a wide cross-section of the American public; it also had to make each visitor feel like part of the event. To this end, the techniques of promotion—well developed by 1939—were brought into play. States and groups were given honorary days. Organizations were invited to hold meetings on the grounds. Contests and lotteries sprinkled a few free tickets into almost every community in the country, and a well-oiled publicity machine flooded local papers with colorful stories. Railroad companies sold package tours, and the large exhibitors incorporated the fair into their advertising campaigns. All over the nation, virtually every citizen heard about the fair several times a week, saw souvenirs in local stores, and was invited to enter contests for free tickets.

Attendance, not profit, was the measure of success. Ethnic and racial groups were courted through a variety of publicity stunts. For example, in March of 1939 some 500 residents of Harlem were invited to come and dance at the groundbreaking ceremony for the fair's Savoy Theater. The *Washington Afro-American* reported that "with the thunder of flying feet set to the rhythms of swing . . . the army of cats from north of 125th Street went to work on the site with all the fury of a thousand steam shovels."[25] In all, more than 800,000 African-Americans were expected to attend. Even if African-Americans did not have a large role at the fair,[26] many wanted to see it. A three-day visit by 125 high school students from Washington, D.C., was arranged for late May.[27] In Chicago the Fortnightly Bridge Club, a middle-class African-American group, sold fair tickets.[28] Harlem's enthusiasm for "The World of Tomorrow" increased after special days at the fair honoring Crispus Attucks, "Colored Gold Star Mothers," John Brown, Frederick Douglass, and "the Negro press" were established.[29] A concert by Marian Anderson was announced,[30] and three jazz bands were employed on the exposition grounds. Later, *The Hot Mikado*, a WPA-sponsored swing version of Gilbert and Sullivan, played in a theater on the midway.[31] A Chicago newspaper sponsored a contest to choose five women who would receive an all-expenses-paid trip via "DeLuxe Pullman car."[32] Similar contests and promotions took place in ethnic communities, whose performers, bands, and special-interest groups were likewise invited.

Who came? Because amusement operators complained that the admission fee was too high, the fair's managers hired a survey team to find out.[33] Statistics gathered at the entrances, and more than 1000 interviews (each lasting from 10 to 20 minutes), revealed a solidly middle-class audience. A disproportionate number were professionals and white-collar employees. Attendees were evenly divided by sex. People between the ages of 20 and 44 were overrepresented. At a time when only 4 percent of the population had college degrees, 38 percent of the visitors did. This was a discerning audience, less interested in the amusement zone than in the foreign pavilions. Such individuals had the means to travel, and nearly half of the attendees came from more than 250 miles away. Only a fourth were from the New York area. More than 10 percent came from the upper Midwest, and almost as many from the Southeast. More than half reported having made between two and six visits, and many people reported having surveyed the entire fair. More than a quarter of the visitors were reported to have come from farms and towns of under 25,000. The most popular exhibits reflected the tastes and values of the educated Eastern middle class. Though one might expect such an audience to have been mobilized by public relations and advertising, in fact the surveyors found that word of mouth had been more influential than newspaper publicity.

Before arriving, visitors were already familiar with the Trylon and Perisphere from postage stamps, bumper stickers, buttons, and magazine covers.[34] People queued up to get into the central buildings on ramps suspended over an artificial lake, and from this elevated vantage point they got an awe-inspiring view of the entire fairgrounds.

To enter the Perisphere one took a 50-foot escalator up to one of two circular balconies which overlooked a miniature of "Democracity," a great metropolis of tomorrow. The official guidebook described this as "a perfectly integrated, futuristic metropolis pulsing with life and rhythm and music."[35] Robert Kohn describes the scene as follows: "In the near distance, yet separated by belts of green . . . satellite settlements devoted to manufacturing, mining, milling, shipping, and other industries, each of these with suitable housing. . . . In the further distance, almost at the horizon . . . areas of farm and cattle-raising country."[36] This was not the disorganized and sprawling city that Americans knew so well; it consisted instead of a geometrically neat commercial center for a quarter-million people, with a uniform architecture of streamlined contours and white surfaces, surrounded by five satellite towns with ample green spaces between them. Superhighways and transit lines integrated the parts. As the Perisphere's balconies began to rotate, the lights dimmed and stars shone in the dome overhead, completing the illusion that one was floating in space, looking down at the future. Once the visitors had contemplated this future world, they were presented with a powerful vision that one commentator compared to "a secular apocalypse."[37] Now the lights of the city dimmed. To create a devotional mood, a thousand-voice choir sang on a recording that André Kostelanetz had prepared for the display. Movies projected on the upper walls of the globe showed representatives of various professions working, marching, and singing together. The authoritative voice of the radio announcer H. V. Kaltenborn announced: "This march of men and women, singing their triumph, is the true symbol of the World of Tomorrow."[38]

Through lighting, music, film, and the model itself, the designer Henry Dreyfus had achieved the illusion of a future reality. His creation had precedents in the amusement zones of previous fairs, however. Both the Chicago Exposition of 1893 and the Pan-American Exposition of 1901 had featured scale models of the volcano Kilauea. The Creation and the Galveston flood had been on view at the St. Louis Fair of 1904.[39] Amusement parks had copied such exhibits and added many more, so some of the visitors to the World of Tomorrow had previously bought a trip to the moon or to Antarctica, witnessed the Chicago fire or the Klondike gold rush, or even circled the globe from the inside, looking out at the oceans and continents.

But whereas the earlier miniaturizations of floods and volcanic eruptions were aimed at evoking fear, at the New York fair the corporations wanted to suggest the banishment of disaster and the taming of powerful natural forces, announcing man's domination and control of nature. Eliminating Burkean terror, the corporate exhibitors expansively emphasized how their expertise would create a harmonious world. Their architecture of the geometrical sublime stressed infinity by placing the subject at the top of a tall building or by providing the illusion of flying high in the air. Not only was the sense of space enlarged, but time was accelerated—a day's journey through an immense landscape required only 15 minutes. These suspensions of ordinary spatial and temporal limits provided the sense of an olympian perspective.

Burke's sublime was tinged with terror, but this panoramic omniscience was sanitized and safe. Democracity contained no unsightly garbage dumps of the kind that the fair itself had been constructed on. It had no poor neighborhoods, no traffic jams, no polluted streams, no smog, no ruins, no large factories, no industrial blight, no signs of war, and no military installations. In contrast with the true situation at the fairgrounds, there were no racial tensions and no job discrimination. The technological sublime of the Perisphere nearly passed over into the realm of the beautiful, retaining from the sublime only a stunning shift in perspective that awed visitors as they entered.

At the end of the fair's first week, the *New York Times* reported that the Perisphere was among the most popular exhibits, together with Consolidated Edison's City of Light, General Electric's House of Magic, and General Motors' Futurama and City of Tomorrow.[40] Subsequent polls confirmed their popularity and added another favorite: the transportation exhibit in the Chrysler pavilion. What did these exhibits have in common? All were favored with central positions on the grounds, but this alone did not ensure their popularity. Nor did their success depend upon emphasizing how applied science could shape the future, since this was a theme throughout the fair. Rather, they succeeded because of the dramatic form of their presentations, which combined all conceivable kinds of visual and aural persuasion (as the multi-media show inside the Perisphere attested) to create a synthesis of sublime experiences: an olympian view of the future.

Consolidated Edison's City of Light, the world's largest diorama, reproduced the New York metropolitan area from Westchester to Coney Island inside a structure as long as a city block and as high as a three-story building. In this monumental miniaturization, designers Walter Dorwin Teague and Frank J. Roods combined the electrical and the geometrical sublime, using 130,000 miniature lights. They presented New York as a city whose form was superior to those of other major cities. The exhibitor's brochure proclaimed: "London, Berlin, Paris, Moscow and Rome seem like strange earthbound cities of an older age when compared to this. No other city has advanced nearly so far along the road that leads to the world of tomorrow." The diorama turned the city into an enormous aesthetic object, with some of the model skyscrapers about 18 feet high. The show, which began with nightfall, was a synchronized performance that simulated a 24-hour period. A "voice from the sky" painted a "dramatic word picture of the ever-changing city." This recorded text, and "the symphonic poem which serves as background music; the sound of thunder and all the city noises used in the diorama show," were "broadcast from loudspeakers in the sky above the diorama city."[41] The functioning of the city's lights and the operation of the miniature subways varied according to the time of day. During a picturesque cloudburst, all the lights came on. The City of Light was thus hardly a static model; it was an approximation of the functioning city. Its miniaturization showed New York as an already-realized utopia—a splendid human achievement rather than a city plagued by poverty, unemployment, racial tensions, crime, and class conflict.[42]

A glance at two less successful exhibits underscores the importance of combining miniaturization and utopia. The Hall of Pharmacy's feature attraction was a stage in the form of a huge medicine chest, with a mirror that could become transparent to reveal "products the size of human beings." Puppets acted out a short play in several acts, between which a "special method" was used "to make the mirror alternately reflecting and transparent." The Ford Motor Company made a similar mistake, building a huge revolving turntable that displayed 87 exhibits with puppets representing the "Cycle of Production" and "showing the progression of raw materials from earth to finished cars." Ford's show, which featured 100 puppets, was not particularly popular. It did not depict a recognizable landscape, as did the City of Light. It did not lend itself to visions of either the past or the future. Worst of all, the viewer could imagine no personal participation in the production process that was depicted. However, people flocked to Ford's other major exhibit, which had none of these defects. Here visitors could take a ride in a new car on a futuristic "Road of Tomorrow" which spiraled above and through the Ford building on a half-mile course. Here the visitors were active rather than passive, they inhabited the future rather than the present, and they could identify with consumption rather than attempt to understand production.[43]

Precisely these qualities characterized the most successful exhibits, which dramatized the future and presented it as a utopian landscape, usually in the form of a diorama or a miniaturized world. Such exhibits might be compared to the sacred places of tribal societies. Within them the individual temporarily annulled the anxieties of everyday life. Each inscribed cultural meanings in ritual. By being inducted into this ideal realm, the visitors left behind their anxieties about jobs, money, success, and the possible coming war in Europe. The successful exhibits did not address these concerns directly; rather, they immersed visitors in what Victor Turner calls a liminal state—a state characterized by "a relatively undifferentiated . . . community, or even communion of equal individuals who submit together to the general authority of the ritual elders."[44] And who but the corporations took the role of the ritual elders in making possible such a reassuring future, in exchange for submission? They had purchased the bonds, designed the fair, erected the pavilions, and employed scientists and designers to make their utopias possible.

General Electric's exhibit exemplified how corporations created forms of the sublime. Both Kant and Burke had identified thunderstorms as examples of the natural sublime, but here calm GE technicians produced thunder and lightning on demand. One visitor recalled: "The signal is a shout and then a tremendous bolt of violet flame shoots from the electrodes and there's a thunderous blast of sound. The flash is so quick . . . but the sight is beautiful, exciting, crowd-pleasing."[45] This highly popular exhibit emphasized not education but magical display, and most visitors left the building impressed but no wiser about electricity than when they had entered. A sizable portion of the building was devoted to a "House of Magic" in which scientific phenomena were packaged as tricks. Here corn was popped and corn fried with radio waves, and objects were suspended in mid-air by magnetism. This mingling of the marvelous and the real prepared visitors to see new consumer goods in terms of an inevitable march of progress toward a technological future. The apparent magic heightened the appeal of the technological sublime, investing the corporation with supernatural powers.[46] Roland Marchand and Michael Smith note: "Almost everybody was into the magic business. . . . AT&T and Westinghouse both involved their audiences in talking with robots and AT&T magically enabled 'auditioners' among their visitors to hear their own voices emerge from the mouths of dummies in a stage presentation."[47] Westinghouse featured a riderless bicycle and a 7-foot robot, Elektro, who was advertised as being able to see, talk, smell, sing, and count on his fingers. In 1940 Westinghouse went further, adding to its exhibit the Nimitron, a one-ton "electric brain" that answered questions and played simple games. The public also saw an electric light bulb that could be blown out and an "electric ear."[48] A few companies even hired magicians as presenters.

By simulating magic, these corporations pushed the audience over the line from ordinary reality toward a simulated dream world in which everything seemed possible. At the same time, the magical presentation of science made it seem to have a life of its own. The self-regulating machines that mimicked human abilities or seemed to defy natural laws made ordinary people appear helpless by comparison. The magicians who hosted these exhibits served a mediating function. As curators of the future, they demonstrated its potentialities and wonders, emphasizing the ease and abundance to come. They did not always explain the marvels presented, nor did the exhibit designers want them to. Though some corporate officers may have harbored a lingering desire to educate, the trend was clearly away from instruction and toward self-aggrandizing entertainment.

The focal exhibit of the Transportation Section, designed by Raymond Loewy and located in the Chrysler pavilion, was a historical compendium of the dynamic technological sublime. The *Times* described the exhibit as a dramatic narrative that "sought to show . . . man's progress from the footpath to the stratosphere. History is dramatized by a huge map with flashing lights which chart the changes in man's speed from the days of sandal-clad runners."[49] To tell this story, Loewy created an enormous semi-circular chamber. The visitor entered the room through doors beneath a map that covered the one flat wall. The floor space was mostly empty, to accommodate the largest possible crowd, but it did contain scale models of many forms of transportation along the curved outside walls. Once the audience was inside for a performance, the doors closed and the lights dimmed so that attention would automatically be focused on the brightly lighted map of the world and on an accompanying screen. A film was shown that recapitulated the transformative effect of the steamboat, the railway, and the automobile. The history of transportation was divided into three eras. The first part of the program carried the story from walking to the Viking ships; the second focused on clipper ships, American covered wagons, and the Pony Express. Then came the mechanical period of "the swift automobile, the streamlined train, the Zeppelin, and the modern plane that can encircle the globe— 25,000 miles—in less than one week."[50] Using speed as a shorthand for progress, the exhibit emphasized the most recent developments. It closed with a startling rocket launch from a large model just below the map. Amid the roar of motors, trains and taxis arrived at the "Rocketport," disgorging their passengers into elevators that carried them into the ship. The flight to London was to take an hour. "Warning sirens were heard, and a magnetic crane hoisted the Rocketship and lowered it into the breach of the Rocketgun. After a brief silence there was a flash, a muffled explosion, and the ship vanished into the night."[51] Skilled stagecraft made this climax seem almost real, and the increasing tempo of this multi-media show, from the slow plodding of a runner to the fiery rocket launch, suggested uninterrupted acceleration through the ages. It barely alluded to the military uses of transportation, and it did not intimate that rockets might carry bombs as easily as people. (All too soon, German rockets would begin to arrive in London.) Instead, the dramatic lighting in the darkened hall, the sound effects in the film, and the many realistic details of the launch urged the viewers to forget current events and to understand history in terms of science, speed, and the technological sublime.

An even more theatrical presentation of the same message underlay the most popular exhibit at the entire fair, General Motors' Futurama (also called "Highways and Horizons"). More than 5 million persons managed to see it. President Roosevelt was so impressed by it that he invited designer Norman Bel Geddes to the White House to discuss the feasibility of building it in reality.[52] *Life* ran a lavish photo essay on the exhibit, and in 1940 Bel Geddes published a book—*Magic Motorways*—illustrated with photographs of it. From the first week, Futurama was besieged by long lines— 27,000 people a day waited on the curved ramps that led into the building, which covered the equivalent of several city blocks. Instead of watching a parade go by, the visitor was transported through an all-encompassing display in one of 600 "comfortable moving sound-chairs." As publicity for the exhibit emphasized, they would "enjoy a thrilling scenic ride into the future, a tour through what seems to be many miles of landscape. Past them streams a realistic miniature countryside, with cities and towns, rivers and lakes, valleys and mountains, forests and fertile fields executed in perfect detail. Through this landscape run super-highways of the future, busy with moving traffic and complete with radio traffic control towers, safety intersections, and automatic lighting." Just as the City of Light exhibit was narrated by a "voice from the sky," this one was explained over tiny loudspeakers mounted inside the moving chairs. This was a minor technological marvel in itself: the sound system delivered the same talk at 150 slightly different times, so that the recorded narration was synchronized with each chair's movement through the exhibit. The narrative began by welcoming visitors to a "magic Aladdin-like flight through time and space" that was designed "to demonstrate in dramatic fashion that the world, far from being finished, is hardly yet begun."[53] This idea is, of course, an underlying presupposition of the technological sublime. As they heard these words, visitors began a 16-minute journey that simulated a cruise in a low-flying aircraft over the United States of 1960. Bel Geddes had built an ideal landscape that contained enough familiar elements to be easily comprehensible but was novel enough to excite admiration. By presenting it as seen from an airplane, he drew upon the tremendous prestige of air travel in the interwar years.

As visitors gazed down at this vast model, they began their "flight" up a beautiful valley, while far below factories sent up their smoke. The flight followed a fourteen-lane expressway through an entire day, starting in the morning. Spread below was a panorama covering 35,000 square feet that depicted the future even more fully than the Perisphere. The pastoral landscape, unmarred by suburban sprawl or industrial blight, contained pleasant recreation areas and model farms (purported to use scientific agriculture to produce more food on less acreage). A million miniature trees represented recovered forestland. Everywhere, teardrop-shaped motor vehicles—50,000 in all—rushed along expressways, their movements said to be regulated by operators in radio control towers. At "night" all the roads were illuminated, so they stood out as bands of light across the countryside. The landscape also contained half a million houses and buildings, including a working amusement park. In addition to standard elements of the natural sublime—mountains and a sunrise—it had elements of the technological sublime: a huge hydroelectric dam, bridges, and skyscrapers.

Jeffrey Meikle has noted that, except for the superhighways and the streamlined buildings, "the projected America of 1960 resembled that of 1939."[54] This is true of the world depicted, but not of the commentary. Like the Perisphere, Futurama deleted pollution, poverty, unemployment, the problems of minorities, and war. The narrator not only explained what lay visible below but also dilated on new scientific marvels. According to this script: "Behind this visible America of 1960, hidden in the laboratories, are the inventors and engineers. By the spring of 1939 they had cracked nearly every frontier of progress."[55] Visitors were urged to imagine that life expectancy had increased greatly, and that women "still had perfect skin" at age 75. Doctors had found cures for cancer and infantile paralysis. Automobiles cost only $200 (one-fourth the current price). Strong plastics having revolutionized architecture, houses were built of entirely new materials and could be "easily replaced." Radio beams transmitted power, eliminating costly and unsightly transmission lines. Much of the population lived in "one-factory farm villages producing one small industrial item and their own farm produce."

Nearing the end of the ride, visitors saw a virtual recapitulation of the development of the sublime. After crossing great snow-capped mountains at dawn, they immediately got a vision of technology integrated with nature: beyond the mountains and across a river lay a metropolis of streamlined skyscrapers with generous green spaces between them. Elevated skywalks permitted pedestrians to walk freely above traffic, which moved efficiently on one-way streets. The whole design was harmonious, clean, and uncluttered. It was a regulated world in which each car, "being out of its driver's control," was "safe against accident."[56] Residential and industrial areas were strictly zoned. Throughout, technology and nature merged under the control of experts.

The journey into the future did not end abruptly. Rather, the "plane" came in for a landing near the city. This portion of the program was described in the brochure "Highways and Horizons" as follows: "By a series of unique close-ups one intersection looms larger and larger until its four corner buildings rise several feet into the air. The scene is alive with people and motor cars, flashing signs and bright show windows. But before the spectator can take in all the details his chair turns and he is looking out on a full-sized street intersection, the same scene he has just viewed in miniature." The visitor emerged into this scene on a pedestrian skyway above moving traffic, looking at the futuristic landscape of the fairgrounds. Standing in what seemed the actual reality of 1960, the visitor received a button that said "I have seen the future." Aided by the narrator, the visitor had been "reading" a panorama of the future, but at the end of the trip he apparently landed in the world he had been seeing. Momentarily creating the illusion of time travel, the Futurama served as a historical conveyor belt, carrying the consumer effortlessly into a dreamscape of ease and abundance.

Such exhibits recontextualized new consumer goods. Unlike the isolated objects displayed at nineteenth-century fairs, these appliances were intentionally displayed as legible parts of a larger historical process. Their display came as the last act in a scientific drama. First, future marvels were suggested through demonstrations, models, or dioramas such as the Perisphere, the City of Light, the Chrysler Transportation Exhibit, and Futurama. Each of these used a smoothly controlled or automated presentation to induce passivity in the viewer, who was urged to accept an ideal landscape as the probable future. Once this acceptance was elicited, the fairgoer's social construction of reality had been transformed. The objects displayed in subsequent exhibits were no longer merely new consumer goods; they were talismanic parts of an ideal future. They proved that this future was already emerging.

The fair's life-size exhibits proved less popular than the miniaturized worlds of Futurama, the City of Light, or the Perisphere. One such exhibit was the Electrified Farm, which demonstrated how crops could be tended and harvested, with radio-controlled machines doing much of the work. Another was the Town of Tomorrow, which showed in detail how people presumably would live in the world depicted in the Perisphere.[57] In the sense that these full-scale displays were seen as little more than enlarged details of the vast miniaturized models of the future, they were subordinate to the great corporate exhibits.

Model landscapes such as Futurama effectively omitted human beings as visible parts of the future. Instead of people, the visitor saw new machines, futuristic cars, idealized landscapes, and streamlined buildings. The inhabitants appeared, if at all, as tiny stick figures, without faces, discernible clothing styles, or visible class status. Such a representational strategy permitted each visitor to imaginatively enter the scene without impediment, but it did so at the expense of any attempt to grasp future human relations. In this realm of pure property, the individual had no apparent obligations or social ties.

The omission of human beings had an international dimension as well. This constructed future took no account of other nations or of minority groups within the United States. Previous world's fairs had prominently featured exotic aspects of other countries[58]; however misguided or reprehensible these representations were, they at least recognized the existence of non-Western cultures. At the 1939 fair there were fewer signs of interest in other cultures. The British, French, and Dutch exhibits did represent those countries' colonies, but one had to go to the amusement park to see "Jungleland," the Inuit family at the soon-bankrupt Children's World, and a few Hollywood Indians. This virtual disappearance of the colonized "other" was not an accident; it was a direct consequence of the fair's emphasis on self-contained simulated worlds at the expense of comparisons of products and people from different nations. Yet exoticism had not been excised from the fair. Rather, it was redeployed, with the economically depressed 1930s as the underdeveloped "Third World" of the future. The America that lay outside the fairgrounds could now be understood as a colony to be developed in accord with the miniaturizations seen at the fairgrounds.

This relocation of the "other" redefined the American public, which the corporate exhibitors now addressed in more patronizing terms than in the past. Just as in the 1920s advertisers had conceived of the public as essentially feminine (i.e., supposedly irrational and impulsive), corporations now saw the public as a backward "other" that had to be cajoled into modernization. In this project, the technological sublime ceased to have any relation to the education of the senses. Unlike the natural sublime, which Kant defined as a source of value and as part of a process that led to a heightened awareness of transcendental reason, the corporate synthesis of various forms of the technological sublime sought not to enlighten but to impress and pacify. The impulse to wonder had been subverted in magic and spectacle. The spontaneous crowd, which had been one important element of a sublime event, had been turned into paying spectators, who were told in detail how to interpret the wonders presented to them

The most popular exhibits at the fair were built by rivals in a corporate potlatch. Collectively, they declared that the social problems of the 1930s could be solved by applied science, with little recourse to politics. Yet, while all the corporate exhibits had this intention, the most successful of them were those that took the form of dramas with covertly religious overtones. As a group they reversed the ritual of the eternal return—found in many traditional societies—in which a people ceremonially recreates a golden yesterday that it believes existed before a troubled and chaotic present. In contrast, the fair only cursorily looked back to the revolutionary past for renewal as it articulated a forward-looking technological republicanism. Corporations promised a cornucopia of goods and services that would eliminate social problems and usher in an age of universal ease and abundance.

All of the five most popular exhibits celebrated the future. Three of these (General Motors, the Perisphere, and Consolidated Edison) were synchronic representations of ideal landscapes during an exemplary day, not in the distant future, but at most a generation later. Visitors glimpsed a promised land that could be entered in their own lifetimes and bequeathed to their children.[59] The other two (General Electric and Chrysler) were diachronic and emphasized that man already controlled the technological means necessary to build the utopias represented everywhere at the fair.

In these five dramas, unseen narrators described the future in authoritative detail, and unnamed scientists were credited with making the future possible. In 1909, illumination engineers had created the special effects of the Hudson-Fulton Celebration; in 1939, industrial designers produced the spectacles of the world's fair. There were few visible actors. Instead, there were detailed models of the world that invited fairgoers in search of renewal to project themselves into symbolic landscapes. Such displays did more than unveil the artifacts of future life; they extended the historical continuum of the museum toward idealized spaces and times waiting to be born. The artifacts displayed at the fair seemed to have been retrieved from the future by scientists and engineers, and the designer-curators had found suitable dramatic forms to represent these future objects. The most popular dramatic enactments of the World of Tomorrow allowed visitors to enter a "future state" in a process that resembled religious ritual. Yet despite affinities with the ritual process that Turner describes, these exhibits lacked certain crucial features of the ceremonies he investigated. The visitors had none of what Turner called "the ritual powers of the weak" to humiliate and revile an ordinarily respected figure—here the scientist—before giving that figure a position of authority.[60] Nor did visitors undergo any ordeals and humiliations that destroyed their previous status and prepared them to take on another identity. They did not need to acquire a new social being before becoming temporary citizens of this frictionless future world, because the fair offered not a renewal of humanity but a transformation of property. Its exhibits promised not the recreation of social relations but economic restoration. They gave the visitor an olympian perspective that appeared to transcend politics, instead offering a detailed rendering of a new realm. Just as the fair had subdued and transformed the Great White Way into a static landscape of symbols, it promised to incorporate the dynamism of new technologies into a static, harmonious design.

The 1939 fair was a quasi-religious experience of escape into an ideal future equally accessible to all. It offered a new version of republicanism—one that retained only a few vestiges of the classical architecture that had once articulated that vision. No speeches given at the fair were as memorable as its technological exhibits, which were designed less to sell particular products than to synthesize the technological, electrical, and geometrical sublimes into one form that modeled the future. Within the fairgrounds, the worried Americans of the late 1930s could feel re-empowered as they gazed down on immense miniature utopias. The fair was a shrine of modernity, offering what seemed an achievable future. Fairgoers could step away from the contradictions of their own time into a monumental perfection that demanded less of them than earlier forms of the sublime had, since it provided an immediate explanation for each startling discontinuity as it was produced. One of the most revealing aspects of this transformation was that both the terror of the natural sublime and the sense of human weakness had been virtually eliminated. The synthesis reduced fear to a temporary disorientation that lasted only until the senses adjusted to the illusion of olympian vision. In the postwar years these techniques would reappear in a new form of permanent exposition: the theme park, where a synthesis of sublimes could be experienced in perpetuity.

Atomic Bomb and Apollo XI: New Forms of the Dynamic Sublime

The 1939 World's Fair, despite its emphasis on how technology could deliver Americans into a utopian future, failed to imagine how the rocket and the atomic bomb would reshape the real world of 1960. By the end of the war both were terrifying realities to civilians, bringing terror back to the technological object and erasing any illusions that science was intrinsically beneficent. Both of these new technologies, in quite different ways, manifested speed and power, and this made them potential forms of the dynamic sublime. Moreover, atomic weapons and rockets together were more frightening than either was by itself. The guided missile made the bomb an immediate threat, and each advance into space suggested more surveillance and more vulnerability.

In popular perception, the combination of the two technologies justified support for space research not on its own merits but as an element of national security. The U.S. government simultaneously developed atomic energy and rocketry for both war and peace. Convincing the public that atomic energy was friendly proved difficult, but the space program was popular.

Rockets had been developed and tested before World War II, but they became a terrifying reality when Germany fired them at London. The German V2 program suggested that flight outside the atmosphere was possible, and after the war ended this idea spread from a few scientists and readers of *Astounding Science Fiction* to the general public. Eight articles on space flight appeared in *Collier's* between 1952 and 1954, explaining rockets, space stations, weightlessness, and how it might be possible to make trips to the moon and to Mars. Wernher von Braun became a familiar figure through these articles and through his subsequent television appearances. The *Collier's* articles also provided background material for three

Walt Disney television programs: "Man in Space," "Man and the Moon," and "Mars and Beyond." The first two of these aired in 1955, the third in 1958.[1]

While *Collier's* and Disney presented space exploration as a peaceful, scientific activity, rockets were also widely understood as a means of delivering weapons. When the USSR launched Sputnik, in October 1957, Americans were not only startled to learn that a small orb was circling over the United States; they also perceived it as a military threat. Just a few weeks before orbiting Sputnik, the Soviets had claimed the first successful test of an intercontinental ballistic missile.[2] With a piece of Russian hardware circling overhead, people felt vulnerable. Suddenly, Americans found themselves in second place. *Life* put it this way: "Let us not pretend that Sputnik is anything but a defeat. . . ."[3] Rockets that could put satellites into orbit obviously could also carry atomic weapons to targets inside the United States. When the third Disney space program aired, in December 1958, the exploration of Mars it depicted appeared more probable than ever, but Americans feared that the Soviets might get there first.

Democrats in Congress seized on Sputnik as an issue, particularly after the Soviets orbited a second satellite a month later. Lyndon Johnson, chairman of the Preparedness Subcommittee of the Senate Armed Services Committee, held extensive hearings which gained him a good deal of publicity and which led to the creation of the National Aeronautics and Space Administration the following year.[4]

The fact that the American space program was born in reaction and fear added to its psychological importance, and the explosion of several early American rockets on the launch pad heightened the drama of competition. When the Russians sent the first man into space, in April 1961, Americans were certain that they had fallen desperately behind in what was dubbed "the space race," a contest that seemed clearly linked to the Soviet Union's rapid development of nuclear weapons and to the Cold War. At its inception, space exploration was less justified for itself alone than as a strategic response to a global challenge. Rockets were primarily military, with ICBMs a central part of the national "defense." Launching satellites involved somewhat different kinds of rockets, but the public perceived their success as a guarantee that ICBMs would work. Moreover, the satellites themselves had obvious military applications. All these factors gave psychological importance to the "space program" (as it came to be called during the 1960s) and made it a suitable repository for the sublime.

It is difficult today to recover the feelings that surrounded the first atomic test. We know so much about the dangers of atomic weapons, their fallout, and their potential for unprecedented destruction that characterizing the experiences of those who first witnessed the secret test at Alamogordo as sublime seems an abuse of the term. But consider the explosion in the sense in which the term has been developed here: it was a terrifying and irresistible force, like a hurricane or a volcano, which scientists believed they were observing in comparative safety.

Nothing human beings had made was more powerful than the atom bomb exploded on July 16, 1945. The men who had produced it were not certain that it would work at all, and they failed to predict how powerful it would be. They expected a massive steel tower half a mile from the explosion to be unharmed, but in fact it was twisted, flattened, and knocked to the ground. Most of the project scientists believed that a temperature inversion at 17,000 feet would stop the cloud from the blast, but in fact it "surged and billowed upward with tremendous power, reaching the substratosphere at an elevation of 41,000 feet."[5] Because the test's success had not been certain, and because the device's power exceeded all expectations, it made an extremely powerful impression on the men involved in the last-minute preparations, who were sheltered only 10,000 feet from the explosion. Some danced for joy; all were spellbound:

The effects could well be called unprecedented, magnificent, beautiful, stupendous and terrifying. No man-made phenomenon of such tremendous power had ever occurred before. The lighting effects beggared description. The whole country was lighted by a searing light with the intensity many times that of the midday sun. It was golden, purple, violet, gray and blue. It lighted every peak, crevasse, and ridge of the nearby mountain range with a clarity and beauty that cannot be described but must be seen to be imagined. It was that beauty the great poets dream about but describe most poorly and inadequately. Thirty seconds after the explosion came first, the air blast pressing hard against the people and things, to be followed almost immediately by the strong, sustained, awesome roar which warned of doomsday and made us feel that we puny things were blasphemous to dare tamper with the forces heretofore reserved to The Almighty. Words are inadequate. . . . It had to be witnessed to be realized.[6]

Isidor Rabi, who had received the Nobel Prize in physics in 1944, described the blast this way: "There was an enormous flash of light, the brightest light I have ever seen or that I think anyone has ever seen. It blasted; it pounced; it bored its way right through you. It was a vision which was seen with more than the eye."[7] After the enormous fireball came a column of smoke and dust that rose thousands of feet into the air and then mushroomed out into a thick, bright, roiling cloud.

The scientists had awed themselves with the forces they had unleashed. They felt triumphant at having made the bomb, but puny before its effects. They felt superior to poets, having created an awe and a beauty beyond imagination. Little wonder that other observers of this first blast reacted with "a hushed murmur bordering on reverence." Religious feelings and quotations welled up in most of the witnesses. One felt as if he had been "present at the moment of creation when God said, 'Let there be light.'" J. Robert Oppenheimer quoted Hindu scripture: "I am become Death, destroyer of worlds." After the initial emotions of sublime terror and celebration, however, came a somber feeling. Rabi thought "of my wooden house in Cambridge and my laboratory in New York, and of the millions of people living around there, and this power of nature which we had first understood it to be. . . ." George Kistiakowsky said, simply, "Now we are all sons of bitches."[8] Such a weapon undermined the self-justifications of the technological sublime, which had exalted the inventor and seen his material improvements as morally uplifting.

Burke had argued that "the sublime is an idea belonging to self-preservation."[9] But atomic weapons challenged the possibility of self-preservation and therefore transformed the conditions of consciousness. Edmund Husserl had defined the individual's sense of a "life-world" as "always already there, existing in advance for us, the ground of all praxis. . . . The world is pre-given to us. . . . Not occasionally but always and necessarily the universal field of all actual and possible praxis."[10] Instead of thinking that individual machines amplified an individual's life-world, this technological system had created the possibility of a "death-world." The atomic bomb undermined that sense of the world as "always already there." Nature and human existence ceased to be "pre-given" and became contingent. This shift in perspective had already begun with the death camps and the carpet bombing of civilians, but the invention of atomic weapons made it more nearly absolute. To describe this new, fragile sense of existence, Edith Wyschougrod has written of "the creation of death-worlds, a new and unique form of social existence in which vast populations are subjected to conditions of life simulating imagined conditions of death."[11] The phenomenology of the bomb undercut any sense of stability and continuity in the life-world for the ordinary citizen.[12]

Yet, terrible as the new weapons were, for the scientists and engineers who constructed them a successful test brought feelings of pride and relief and excellent prospects for promotion. General Leslie Groves, military head of the Manhattan Project, thought of "how Blondin had crossed Niagara Falls on a tightrope" and "thought to himself that his personal tightrope had been three years long."[13] As this statement reveals, one man's triumph over nature became a metaphor to represent another's success. Despite the radical difference in techniques, these were both acts of will, performed before an audience, in a situation where failure could prove catastrophic. Like Groves, the scientists and managers involved in building the first bombs knew that they had enhanced their personal reputations. The enormous Manhattan Project and the subsequent secret weapons laboratories, like all technological systems, developed with a logic of their own, which became expressed as a day-to-day routine in which the scientists could only intermittently recall the larger implications of what they were doing as they struggled to perfect one or another component of a new weapon.

The most common justifications were patriotism, the advancement of science, and the protection of the Free World. Others have seen less rational motives at work. Brian Easlea argues, in a book called *Fathering the Unthinkable*, that scientists repeatedly conceived of their work in terms of virility. A successful explosion was a "birth," and a powerful weapon was a "son" that they had procreated. In Easlea's words: "It seems entirely appropriate that the horrendously destructive end result of such masculine creativity—the ultimate outcome of masculine repression of and violence towards the feminine—should have been described . . . as the birth of a boy! Violence is male."[14] Though his tone is often this shrill, Easlea does point to a persistent pattern of metaphors and mental associations. Yet, if the bomb was a boy in the imagination of the scientists who dreamed it up, to the general public it was a faceless cloud of death.[15] Easlea is concerned with the psychology of invention, and he focuses attention on leading scientists, leaving the thousands of engineers and military personnel out of his story.

The explosion at Alamogordo was the culmination not only of abstract physics but also of applied science—the successful testing of a new product about to go into production. To make the atom bomb required $2.2 billion and the coordination of thousands of highly skilled workers and scientists working in Chicago, Tennessee, New Mexico, and Washington. As Thomas Hughes has emphasized, the Manhattan Project was not a unique but rather a characteristic combination of management, science, and engineering. The bomb was produced in a way that was analogous to the research and development process at large corporations such as General Electric, AT&T, and Du Pont.[16] As corporations had demonstrated at the 1939 World's Fair, they knew how to marshall inventors and engineers into teams that could produce the new. From the point of view of eventual production, Hughes notes, it seemed appropriate to General Groves to bring Du Pont into the project "to design, construct, and operate plutonium-production piles and separation plants."[17] Later General Electric and Westinghouse became involved in the building of U-235 plants at Oak Ridge, Tennessee.

Like the construction of the Eads Bridge, the Manhattan Project forced engineers to find solutions to problems that science had scarcely defined yet. However, in these two cases the relationship of the workforce to the project could scarcely have differed more. All of St. Louis watched and understood the construction of the Eads Bridge, discussing each new difficulty or delay. Many residents descended into the depths of the piers to watch the work proceed, and many mourned each death entailed by the construction. But the Manhattan Project was so carefully guarded that the public never heard of it, and most of the individuals working on it in various parts of the country did not know the ultimate goal. Though tight security was no doubt necessary because of the war, later such secrecy became typical of atomic research. As a result, the development work that led to the hydrogen bomb was not subjected to public debate. This secrecy, however necessary it may have appeared at the time, flouted most of the values of republicanism. The structure of atomic decision-making created a hierarchy of authority in which a small elite made policy in secret. After the war, on grounds of national security, ordinary citizens were denied the chance to discuss or to vote on the desirability of weapons of mass destruction. But republican government requires open public debate, as the American revolutionaries well understood from experience. Kant, like his American contemporaries, regarded full public disclosure of debates and decisions as essential to the functioning of a free society, going so far as to say that "all acts relating to the rights of other men, the maxims of which are incompatible with publicity, are unlawful."[18]

Lewis Mumford was one of the harshest early critics of both the invention of the atom bomb and the decision to use it. He wrote in the *Saturday Review of Literature*: "We are living among madmen. Madmen govern our affairs in the name of order and security. The chief madmen claim the titles of General, Admiral, Senator, scientist, administrator, Secretary of State, even President. And the fatal symptom of the madness is this: they have been carrying through a series of acts which will lead eventually to the destruction of mankind, under the solemn conviction that they are normal, responsible people, living sane lives, and working for reasonable ends."[19] This, one of the first clear formulations of opposition to atomic weapons, was articulated both as a matter of morality and as a matter of practical necessity. Mumford clearly sensed the emergence of the "death-world" of pure technique dedicated to destructive ends. As Everett Mendelsohn commented, "For Mumford there was no possibility of 'managing' nuclear weapons, and certainly no justification for integrating them into the military and strategic policies of the United States."[20]

The atomic bomb had come into being not as a result of open debate but as the result of a secret project that was never subjected to the normal controls of a democratic political process. The construction and dedication of the Eads Bridge and the Brooklyn Bridges were civic events. In contrast, the explosion of the first atom bomb was a spectacle only for the few who had created it. Because of wartime security, the members of Congress did not know about the Manhattan Project, and its funding was hidden in the military budget. The public heard about it as a *fait accompli*. When the bomb was first tested, in a remote area of New Mexico, the army created a disinformation campaign to explain the explosion away as a munitions-dump accident to civilians who happened to see it. In later years, many of the scientists who had created the bomb, including Oppenheimer, were eventually separated from it administratively, so that they could no longer influence its use or its further development. In short, neither the public nor their elected leaders nor the scientific community was in control of the weapon; control had largely passed into the hands of the military.

The first people to see the bomb were the citizens of Hiroshima, who had never heard of the weapon and did not apprehend the mortal danger posed by a single bomber flying over their city. When the blast came, they hardly could appreciate it with the detachment from immediate physical danger that is required for the experience of the sublime. For them, the bomb did not reveal a new landscape, nor was it the ultimate in spectacular lighting. Anyone within a mile of the detonation was killed instantly. In the museum dedicated to the memory of those who died, perhaps nothing is more eloquent than a child's wrist watch with the hands melted into the dial at 8:15—the time of the explosion. As Spencer Weart puts it, the bombings of Hiroshima and Nagasaki "seemed less like a military action than a rupture of the very order of nature."[21]

Only when an atomic explosion was viewed voluntarily from a distance, with some sense of safety (however illusory), could it conceivably be a sublime experience, as it had been for General Groves and the scientists who had created it. Between 1945 and 1962 more than 200 bombs were exploded above ground, in every case with witnesses present. Thousands of civilians lived downwind of the test sites, and for a decade they were subjected to repeated exposures. Continually reassured that there was no danger, many made a point of watching the blasts.[22] A quarter million servicemen were ordered to Nevada or to the Marshall Islands and subjected to atomic explosions. Most were less than 10 miles from "ground zero," and some were only 2 miles away, dressed in ordinary uniforms and crouching in dirt trenches. These men were not informed of the deadly radiation risk they were being exposed to. Like General Groves in 1945, many of them expressed awe and wonder at the power of the blast, the intensity of the light, the shock waves, the winds, and the unusual colors. Another common experience was seeing through one's bones. One marine recalled: ". . . with my eyes tightly closed, I could see the bones in my forearm as though I were examining a red x-ray. I learned many years later that I had been x-rayed by a force many times greater than a normal medical x-ray."[23] The experience was traumatic. The same man declared: "The event left an indelible impression on me, and to this day, I shudder and perspire when I recall my experience in that trench." For these men, almost as much as for the victims at Hiroshima and Nagasaki, the experience of the atom bomb went beyond sublimity to sheer terror, leaving trauma and a life of radiation poisoning in its wake.

Though it may seem implausible that nuclear tests could ever have been a tourist event, in fact there was public fascination with the new weapons. Some Americans made a point of seeing an atomic explosion. The Atomic Energy Commission once promoted the viewing of a test as an exciting holiday event. Fifty miles from the Nevada test site, local families and school groups often climbed Angel Peak to watch blasts, which at times were visible as far away as Los Angeles.[24] In Las Vegas, "a hotel filled its pool with two thousand mushrooms, . . . the Sands Hotel sponsored a Miss Atomic Blast beauty contest," and the piano player at the Desert Inn "featured a boogie-woogie number, 'Atomic Blast Bounce.'" A woman "timed her wedding ceremony to coincide with a nuclear test," and her wedding cake was shaped like "an atomic blast mushroom."[25] The gambling industry found the tests profitable. Postcards showing a mushroom cloud behind the famous casino strip were sold, and the Chamber of Commerce provided tourists with "shot calenders" and road maps to the best vantage points. Even the *New York Times'* travel section advised people on the fine points of "the honorable pastime of atom bomb watching."[26]

Open-air testing continued into the days of the Kennedy Administration. When a giant hydrogen bomb was detonated over the Pacific, tourists lined Hawaii's beaches to watch the blast, which

A postcard from Las Vegas. Courtesy of Special Collections, University of Nevada, Las Vegas.

like a fireworks display was thoughtfully staged after dark. *Life* reported the event to the millions so unfortunate as to miss the pyrotechnic display: "The blue-black tropical night suddenly turned into a hot lime green. It was brighter than noon. The green changed into a lemonade pink . . . and finally, terribly, blood red. It was as if someone had poured a bucket of blood on the sky."[27] In short, despite the government's attempts to make atomic energy seem acceptable, the bomb remained a source of sublime terror, as it had been from the first.[28]

But many scientists and a portion of the public greeted atomic energy as the harbinger of future prosperity. In doing so, they were only recalling what had been predicted for at least 20 years before World War II. For example, in 1934 Waldemar Kaempffert, science editor of the *New York Times,* declared that in the future "probably one building no larger than a small-town post office of our time will contain all the apparatus required to obtain enough atomic energy for the entire United States."[29] The Atomic Energy Commission, created immediately after the war, was dedicated to the proposition that atomic energy could be used for peaceful purposes. Proponents, echoing prewar predictions and the science fiction of the 1930s, proclaimed that energy would soon be too cheap to meter, and that Americans would use their unlimited supply of power to control the climate, increase productivity, travel cheaply, and create a social utopia. A science editor for the Scripps-Howard newspapers, David Dietz, wrote a long paean ("Atomic Energy in the Coming Era") in which he predicted that people would soon drive their cars for a year "on a pellet of atomic energy the size of a vitamin pill," that "summer resorts will be able to guarantee the weather," and that "artificial suns will make it as easy to grow corn and potatoes indoors as on the farm."[30] Such ideas were logical extensions of the technological sublime. As Michael Smith points out, by the end of the 1950s a government publicity campaign had converted the atom into "our friend" through a welter of books, comics, and films.[31] A 1965 film, *The Atom and Eve,* showed atomic power growing to maturity along with a girl, providing her with more and more power to run an ever-increasing array of appliances. This campaign domesticated atomic power, dissociating it from weapons and making it part of the more "familiar trappings of a consumer culture."[32]

The domestification of atomic energy underscores the shift away from terror and toward control that is a central characteristic of the technological sublime. Atomic power reaffirmed man's control over nature, both in the awesome explosions of the bomb and even more impressively (it was thought) in the use of this power for peaceful purposes. Yet it proved difficult to separate attitudes toward reactors from fear of the bomb. The relatively uncritical support the AEC enjoyed in the 1950s faded during the following decade, when controversies over the thermal pollution caused by nuclear plants turned many environmentalists against nuclear power. In the early 1960s the Sierra Club had favored atomic reactors as a "clean" form of energy, but such support soon evaporated.[33] In addition, the nuclear program had delivered less than it had seemed to promise at the end of World War II. Daniel Ford's assertion that in 1970 "nuclear plants provided the nation with less energy than firewood"[34] may overstate the case, but in fact only a few plants had actually gone into operation, supplying but 2 percent of the nation's electrical energy. However, a large number of plants had been approved after 1965, and many were nearing completion. In 1971, when Glenn Seaborg and William Corliss published *Man and Atom* (subtitled *Building a New World Through Nuclear Technology*), many new plants were about to go on line. Seaborg had been a presidential science advisor under Eisenhower and had headed the AEC from 1961 until 1971. His belief in the beneficence of atomic power remained unwavering, despite rising public fears. He declared that with nuclear power "not only will we be able to raise a greater part of the world's people to a decent standard of living, but we will be able to move all mankind ahead into an era of new human advancement—human advancement which takes place in harmony with the natural environment that must support it."[35] This new rhetoric of harmony, often contradicted by specific proposals in the same volume, sought to address the mounting critique by environmentalists. The imperial 'we' belied the claims of environmental awareness, however. Seaborg and Corliss accepted the logic of the technological sublime, which exalts not human reason in general but the expertise of an elite that proposes to uplift mankind.

As might be expected, Seaborg and Corliss described nuclear reactors as safe and nonpolluting. Indeed, their book reflects an assumption that by 1971 reactors had become institutionalized. They had new ideas on the drawing board, such as creating a million-degree "fusion torch" that would break down garbage—"container and all"—into a hot mixture of ions, which could then be decanted into "elements of high purity."[36] Policy makers also had been exploring the possibility of a nuclear-powered mechanical heart and the use of "cleaner" nuclear bombs to excavate harbors, to blast new canals in Suez and Panama, and to force a national waterway through the Rocky Mountains.[37] Such ideas were part of the Plowshare Project, which took its name from the well-known biblical passage. Implausible as such plans may appear today, the federal government spent more than $17 million on feasibility studies for a "Panatomic Canal" that would cross the Isthmus of Panama at sea level (thus requiring no locks). That this last project remained unrealized was due in part to its ecological implications: no one knew what the biological consequences would be if marine life from the Gulf of Mexico and the Pacific intermingled. A greater obstacle, however, was the 1963 agreement between the Soviet Union and the United States banning open-air tests.

Before the test ban, a number of "cratering experiments" were tried at the AEC's Nevada test site. On July 6, 1962, a 100-kiloton bomb was used to make a crater 1200 feet in diameter and 320 feet deep. Six years later, the "first nuclear row charge" dug a trench 860 feet long, 280 feet wide, and 68 feet deep. Seaborg confidently predicted that, instead of using conventional equipment to cut through the 1000-foot mountains along a proposed route in Panama, canal builders would use nuclear explosives to "make long terrestrial incisions in less than a minute, slicing down 200 feet below sea level."[38] After 1970, however, such "planetary engineering" became anathema to the public—not least because the government had systematically distorted and supressed evidence about the dangers of fallout. The "downwinders" of Arizona and Nevada lost thousands of farm animals to strange diseases, and a disproportionate number of them to cancer, while for two decades the government insisted that its tests were safe.[39]

Not only did above-ground nuclear explosions become unacceptable; after the accident at Three Mile Island in March 1979, so did nuclear reactors. The valve problems that caused the system failure at Three Mile Island were hardly unprecedented. The Nuclear Regulatory Commission's files document eleven incidents of valve malfunction at other sites between 1968 and 1976. Worse still, in some cases the components that failed were discovered to have had operating problems even before the reactor went on line. The "Our Friend the Atom" campaign had not prepared the public for a display of confusion and mismanagement by the NRC or for the intentional concealment of information by Metropolitan Edison (operator of Three Mile Island).[40] Later revelations of inadequate training of reactor service personnel frightened the public further, and in the 1980s it became virtually impossible to license a new nuclear plant.

Yet, in a curious irony, Three Mile Island became a popular tourist site during the 1980s. The cooling towers of the plant had become world famous, as identifiable as the Eiffel Tower or the Statue of Liberty. To accommodate the public and to put a positive gloss on the accident, Metropolitan Edison hired a large staff of full-time public relations people and opened a visitor center. It is surrounded by green grass, and there are picnic tables nearby. Visitors are told a reassuring story of a Three Mile Island where nuclear power is fully under control, where technological systems worked, and where the cleanup was flawless. The visit culminates in a trip around the plant in a minibus. Tourists might expect only to ride by the closed reactor—but, as Sharon O'Brien found, "the unexpected happens: the guide stops the bus by the disused cooling tower and tells us that you can enter. This is an extraordinary and seemingly spontaneous moment on the tour, the moment when the cooling tower—symbol of Three Mile Island, of nuclear power, and, after 1979, of the dangers of nuclear accidents—becomes invested with religious meaning as awed tourists enter. . . . "[41]

For Burke and Kant the sublime was a constant, but history has shown that it seeks new objects. Yesterday's technological wonder is today's banality, and, as at Three Mile Island, tourists are ever on the lookout for novelties. Today no one pays attention to a row of brilliant street lights, and even a city skyline may evoke only a polite murmur. So much of experience has become prepackaged and predictable that, as Daniel Boorstin once remarked, "we go not to test the image by the reality, but to test reality by the image."[42] How valuable, then, the few experiences that cannot be entirely prepackaged, that "must be seen to be believed"! One of these is the launch of a manned space vehicle, a public experience that Americans are willing to travel thousands of miles to see.

Since the 1960s, on the night before any countdown at Cape Canaveral, innumerable Americans in campers, cars, and four-wheel-drive vehicles occupy every spot with a clear view of the launch pad. The crowds grew larger throughout the early years. In 1962 about 50,000 came to see John Glenn become the first

American to orbit the earth. Twice as many turned out to greet his motorcade in Florida two days later, and 250,000 hailed his arrival in Washington. In New York, more than 4 million came to his tickertape parade.[43] In the following decade the astronauts remained popular heroes, though none of them, with the possible exceptions of Glenn and Armstrong, achieved Charles Lindbergh's level of recognition. As Robert Baehr has pointed out, Lindbergh was much better at communicating his experiences in nontechnical language than the astronauts, he represented no company or government organization, and he not only flew the *Spirit of St. Louis* but also had it built, tested it, and planned his own trip. In contrast: "The astronauts were not perceived to be acting independently like Lindbergh. They were seen to be protected by the limitless resources of budgetless financing, the entire American scientific community, and the whole of the federal bureaucracy. They were part of something immense, at the top of the pyramid to be sure, but not alone."[44] After the middle of the 1970s, most of the adulation for astronauts subsided. However, launches and landings increased in popularity. This is partly a case of the machine's displacing the hero; however, when viewed from the perspective of the technological sublime it is more.

Launches of spacecraft, like atomic explosions, represented a new stage in the historical shift from man to machine to process. In 1828 the artisans and craftsmen of Baltimore displayed themselves in a parade and celebrated the decision to build a railroad. The more than 5000 individuals marching in the procession were members of the same community as the spectators. By the 1850s artisans had been partially eliminated from some processions; by the time the Brooklyn Bridge was dedicated, in 1883, they had disappeared entirely. In theory, the tickertape parades for the first few American astronauts might have included scientists who had worked on the project, engineers who had created the hardware, technicians who had directed the flight, and representative workers from the many factories where the thousands of individual parts of the rockets had been made. Instead, only a few officials and the astronauts themselves appeared. While the complexity of technological projects and the number of spectators had increased, the performing cast had shrunk to a small number of celebrities.

These cultural choices had political ramifications. The decision to build the Baltimore and Ohio Railroad was arrived at with support from both the Maryland legislature and the elected officials of

Baltimore. It was rightly understood as a decision reached by the community as a whole. Consequently, the hero of July 4, 1828, was not an engineer but a representative of the political process: Charles Carroll, a signer of the Declaration of Independence and a founder of the railroad. In contrast, the space program was the product of America's competition with the Soviet Union. Its first heroes did not emerge from politics but from the space program itself. Presidents Kennedy, Johnson, and Nixon all astutely recognized the benefits of treating the astronauts as heroes whom they could use to add luster to their administrations.[45] In recent years, however, astronauts have been less valuable as political commodities, because the public has less interest in the individuals who go into space than in the spectacular technology that takes them there. Americans come by the millions to Cape Canaveral because each liftoff breaks through normal experience, creating the "disconcerting disproportion between inner and outer" that is the mark of the sublime moment.[46] Like the Eire Canal or the Union-Pacific Railroad, the technological sublime generates an enthusiasm which politicians seek to tap. The politically active citizen, so valued by republicanism, has been replaced by the mesmerized spectator.

At one launch of the shuttle *Columbia*, in April 1981, local police estimated the crowds at half a million, and there was bumper-to-bumper traffic for 15 miles in all directions. Most of those people were seeking an immediate, intense experience. After the launch, a 25-year-old engineer who had journeyed from New York declared to a *New York Times* reporter: "I didn't think my heart could take it, but it did. It was such an intense experience. I felt it in every bone in my body. It was an exalted feeling." A college student witnessing the same liftoff said: "I've been looking forward to this all my life, but I never thought it would be so powerful, so awesome." An 82-year-old retired insurance salesman from Michigan told the reporter: "Seeing that thing go up and hearing all the cheers was the most exciting thing I've ever seen."[47]

Launch spectators travel to the site voluntarily. Many wait patiently for days, having come early to secure a good observation point. A *Washington Post* reporter found the crowds along the coastline awaiting the 1981 launch mentioned above unusually happy: "There is a sense of enjoyment on this beach, of things being right, of the marvel of spectating that is not often seen. . . . It's the calm part that's different."[48] In fact, the event is less a matter of spectatorship than a pilgrimage to a shrine where a technological miracle is confidently expected. As Victor and Edith Turner have observed, "pilgrimage was the great liminal experience of religious life"—a form of "exteriorized mysticism" in which the ordinary person sought to "get out, go forth, to a far holy place approved by all."[49] In secular America, Cape Canaveral is such a place. And unlike a ritual of initiation, which is obligatory at a certain moment in life, the pilgrimage to Florida is voluntary. The pilgrim breaks out of his usual round of activities, leaving profane social structures behind.[50] Each manned space launch and each landing calls out a vast array of citizens from every walk of life. As they often say explicitly, they come to renew their belief in the powers of American technology and to reinforce their patriotism. One retired auto worker from Ohio characteristically effused, just after *Columbia* disappeared into the sky, "Doggone it! It's about time we showed somebody we could do something." A woman from Florida was even more explicit: "Great, great, absolutely great! There isn't another country in the world that's going to do this—you've got to say America's first."[51]

Some of the million people who went to see Apollo XI lift off, July 16, 1969. Courtesy of National Aeronautics and Space Administration.

In Kantian terms the sublime event is, of course, that which is absolutely great. But the witnesses quoted above differ from Kant in their interpretation of the sublime object. Whereas Kant postulated the realization of reason through the experience of the absolutely great in nature, the pilgrim to Cape Canaveral realizes patriotism through the experience of the absolutely great in technology. To reformulate Weiskel's analysis of the natural sublime: as people recover from the powerful shock of the launch, they interpret the indeterminacy of the event as symbolizing their relationship to the nation. Hence the conclusion that "you've got to say America's first." This statement carried with it, of course, an implicit comparison with the Soviet Union. The technological sublime is not absolute but comparative. Not only are its objects soon obsolescent; they are often consciously constructed and perceived as demonstrations of greater power and expertise than an adversary possesses. This competitive attitude was nowhere more evident than in the decision by the Kennedy administration to put a man on the moon before the Soviets. This political decision led to the paradigmatic liftoff experience: the Apollo XI launch of 1969.

The novelty and the daring of the Apollo XI mission, coupled with the political importance attached to it, literally called out the populace. Surveying the area, Norman Mailer discovered people waiting everywhere: "Through all that several hundred square miles of town and water and flat swampy waste of wilderness, through cultivated tropical gardens, and back roads lined with palms, through all the evening din of crickets, cicadas, beetles, bees, mosquitoes, grasshoppers and wasps some portion of a million people began to foregather on all the beaches and available islands and causeways and bridges and promontories which would give clear view of the flight from six miles and ten miles and fifteen miles away."[52] More than half of the members of Congress accepted invitations to be on the viewing stand, along with 69 ambassadors from foreign countries, various cabinet members, and other dignitaries. NASA chartered six airliners to bring them from Washington to Florida.[53] The *New York Times* reported that Apollo XI attracted "the largest crowd ever to witness a space launching," amounting to more than a million and including "former President Lyndon B. Johnson, members of the Poor People's Campaign, African and Asian diplomats, youths carrying Confederate flags, vacationing families, hippies, scientists and surfers, and students and salesmen."[54] The blastoff fascinated people from all walks of life, many of whom wore buttons that read "I Was At The Apollo XI." The sheer size of the crowd became part of the meaning of the event.

Yet this immense gathering did not represent a clear majority. In the year before the launch 49 percent of the American public did not support the "aim of landing a man on the moon."[55] Only 39 percent were in favor, with the rest undecided. The chief criticisms focused on the cost; a minority concurred with the proposition "God never intended us to go into space." The *Baltimore Afro-American* of July 19, 1969, put a photo of the three astronauts on the front page and called the flight "mankind's most thrilling adventure, a spectacular payoff to centuries-old dreams of reaching another planet," but this was even more a minority view in the black community than among whites. In stark contrast, the *Chicago Daily Defender* of July 12 ran an editorial-page cartoon that showed Uncle Sam standing on earth, observing a smoky stream of dollar bill symbols—$$$$$$$—that stretched from the earth to an orbit around the moon. A naked hand labeled "Humanity on Earth" reached out to him imploringly from one side. The caption read "What about the space between races of man?" A similar drawing appeared in the July 25 issue of the religious newspaper *Muhammad Speaks*. New York's *Amsterdam News* of July 12 had a more militant cartoon, depicting President Nixon sitting on a large sphere and smiling up at the moon. However, he sat not on the earth but on a bomb, with a lighted fuse, labeled "minority frustrations." The accompanying editorial attacked the "outlandish costs of the space race" and declared: "The Black man isn't too moved over this jaunt into space, this ambitious leap into the future, this fantastic achievement of interplanetary travel. . . . Man can conquer space, yes. But man has still to conquer his homeland. And that's where the real action is, brother." Interviews with Harlem residents, reported in the July 12 issue of the *Amsterdam News*, uncovered sentiments similar to those of whites. A saleslady said: "This is the age of miracles. I hope they are successful in landing on the moon and find it habitable. We are living in a miracle age and everything is possible." Others interpreted the event in religious terms: "If it's God's plan, man will get there; if not, he'll be wasting his time." The most typical response was that of a policeman, who agreed it was "a wonderful human achievement" but who nevertheless spoke of "a lot of money going down the drain." A lecturer at Columbia was quoted as saying: "It's folly at a time when the great needs of the country demand that we allocate resources to the poor." These responses were in marked contrast to the black community's enthusiasm for the training of young African-American aviators in 1939. Aviation, an earlier form of the technological sublime, had appealed to virtually all Americans.

As the moon shot approached, a Lou Harris survey found that the public favored the Apollo Program by a margin of 51 to 41 percent, although a majority still did not think the space program was "worth the $4 billion a year that has been spent on it."[56] On the night before the launch, some of the onlookers had come not to admire but to protest. Yet, as the launch drew near it had a magnetic effect on all who waited. Against the darkness, people could see the silvery Saturn rocket, tall as a skyscraper, standing on the launch pad in a blaze of lights. "In the distance," Norman Mailer wrote, "she glowed for all the world like some white stone Madonna in the mountains, welcoming footsore travelers at dusk. Perhaps it was an unforeseen game of the lighting, but America had not had its movie premieres for nothing."[57] Mailer might have traced that lighting back to the world's fairs, the vision of the white city, and the dramatic illumination of national monuments. He might have recalled that the electrical sublime had long been a central part of the American attempt to transcend the mere facts of this world.

Mailer noted that the vessel that was to carry men to the moon was surrounded with the aura of religion, and that it seemed a sacred object. Mailer knew his Henry Adams, and he realized that here was the ultimate machine of the twentieth century taking on the trappings of a shrine, with its pilgrims, its mysteries, and its terrible, controlled violence. The Saturn contained the equivalent of a million pounds of TNT, with the explosive force of a bombing raid by 1000 planes in World War II. These facts, in and of themselves, did not move Mailer, who was as recalcitrant a participant in the sublime experience as could have been found. Rather, he wondered to what extent the Nazi backgrounds of Wernher von Braun and other specialists signaled that some part of the ideology of the Third Reich lay buried in the interstices of the experience, and he feared that on the whole the space program might be something evil. Unlike the great mass of Americans who stood on the swampy roadsides, Mailer resisted the experience, tried to frame it with theory, and sought to tamp down any patriotism or excitement and to view the launch dispassionately. But he too was overwhelmed.

Later, Diane Ackerman described the launch of a space shuttle in a dizzying blend of metaphors, attempting to capture the synesthesia of the experience: "When floodlights die on the launch pad,

Apollo XII on the launch pad, October 29, 1969. Courtesy of National Aeronautics and Space Administration.

camera shutters and mental shutters all open in the same instant. The air feels loose and damp. A hundred thousand eyes rush to one spot, where a glint below the booster rocket flares into a pinwheel of fire, a sparkler held by hand on the Fourth of July. White clouds shoot out in all directions, in a dust storm of flame, a gritty, swirling Sahara, burning from gray-white to an incandescent platinum so raw it makes your eyes squint, to a radiant gold so narcotic you forget how to blink." This is more than a visual experience: "The air is full of bee stings, prickly and electric. Your pores start to itch. Hair stands up stiff on the back of your neck. It used to be that the launch pad would melt at lift-off, but now 300,000 gallons of water crash from aloft, burst from below. Steam clouds scent the air with a mineral ash. Crazed by reflection, the waterways turn the color of pounded brass. Thick cumulus clouds shimmy and build at ground level, where you don't expect to see thunderheads."[58] As the shuttle lunges into the air, trailing 700 feet of flame, many cry. Just as the visitors were struck dumb with amazement by the Pan-American Exposition's nightly illuminations, and just as crowds at the Hudson-Fulton Celebration gazed in silent admiration at the electrical fireworks, the throngs along the Florida beaches stare with aching eyes in open-mouthed delight.

How did Norman Mailer respond to the launch of Apollo XI? His account alternates between private thoughts and the broadcast commentary from the command center (a mix of technical jargon, reports from inside the module, and the countdown). Only a few moments before liftoff, Mailer "was still not properly ready for the spectacle." Yet in the final seconds he realized that his throat was dry. "A tiny part of him was like a penitent who had prayed in the wilderness for sixteen days, and was now expecting a sign." Yet, at the same time, "nothing in his mood was remotely ready for the experience he had been promised," which was that "the sight of a large rocket going up was unforgettable and the sound would be remarkable, the ground would shake." This prediction proved accurate. Mailer was mesmerized by the "two horns of orange fire" that burst from the Saturn rocket 9 seconds before liftoff. The rocket was in the air a full 6 seconds before the sound reached him. Because of the initial silence, "the lift-off itself seemed to partake more of a miracle than a mechanical phenomenon," not least because of the enormous flames. "And in the midst of it," Mailer continued, "white as a ghost, white as the white of Melville's Moby Dick, white as the shrine of Madonna in half the churches of the

world, this slim angelic mysterious ship of stages rose without sound
out of its incarnation of flame and began to ascend slowly into the
sky, slow as Melville's Leviathan might swim, slowly as we might
swim upward in a dream looking for air."[59] Here are the characteristics of the sublime: irresistible power, magnificence, complexity,
and a journey into the infinite reaches of space. And Mailer's experience was not yet complete, for after this silent vision came a
sound which he compared to "a million drops of oil crackling suddenly into combustion," to "the earsplitting bark of a thousand
machine guns," and to "the thunderous murmur of Niagaras of
flame." Burke had long before noted that "excessive loudness alone
is sufficient to overpower the soul, to suspend its action, and fill it
with terror,"[60] and the blastoff of Apollo XI was a case study. The
earth began to shake, and Mailer realized that he was repeating
over and over again, "Oh, my God! oh, my God! oh, my God! oh,
my God!"

After reading this epiphany, one can begin to understand why so
many want to see a launch. Those who come, unlike Mailer, usually
are not skeptical about technology, and they are patriotic in an
uncomplicated way that he could not be. Millions of Americans
have journeyed to Florida to see a rocket lunge into the heavens.
What they see cannot be transmitted by television. The sheer scale
of the event mocks the small frame of any camera; the blinding
brightness and subtlety of the colors cannot be broadcast any more
than one can transmit the violent roar of the engines, the smell of
the fuel mixed with that of the surrounding swampland, or the feel
of rocket's thrust shaking the earth. (The National Air and Space
Museum compensated for these deficiencies as much as possible,
using film made with frames ten times the usual 35 millimeters and
a special camera. Projected on a screen five stories high and accompanied by quadraphonic sound, the film still only began to suggest
the experience.[61]) It should not be surprising that the value of
launches as television events soon declined—the small screen simply could not convey the drama. Furthermore, a long countdown,
often interrupted by delays, had to be filled with commentary, and
it proved hard to hold an audience; people soon realized that the
launch took only a few moments and could be seen later on the
news.

After 1969 the space launch became a form of authentic experience, continuing to draw a large crowd well after it had ceased to
be attractive to television. At least 100,000 people are present at any

Apollo XIII lifts off, April 11, 1970. Courtesy of National Aeronautics and Space Administration.

launch (usually many more), and providing good seats has become a profitable business. In the 1980s the National Space Institute ran tours to all civilian launches, for $60 providing transportation to a vantage point "only four miles from the launch site." The advertising copy in the February 1985 issue of *Space World* said: "You've seen it on TV, but nothing can compare to being right there—inside the Kennedy Space Center gates—when six million pounds of thrust propel America's reusable spacecraft into orbit . . . and you feel the vibration . . . hear the roar . . . see the brilliance!"[62]

Some of these paying visitors are quite knowledgeable, having read a good deal about NASA's technology. Like Melville, who combined metaphysical musings on whales with detailed measurements and descriptions of their anatomy, these high-tech tourists often combine an engineer's knowledge with a sensuous appreciation of the sublime object. Mailer also sought to link his immediate reponse to a more scientific account of "that remarkable mechanical white whale which rose into the air at Cape Kennedy."[63] He did this not as a spectator mesmerized by the roar but as a technically knowledgeable insider, describing the hardware and its functioning during the takeoff and the procedures necessary to go to the moon. By providing this technical information, Mailer sought to lift the reader out of ignorance into a position simulating the understanding of someone who had worked on the project—someone who knew the complexity of the machine and still could see its poetry. For the insider, as Mailer early realized, language is merely functional; only the machine is sublime. His task was to find a language adequate to describe the technician's understanding.

After describing the launch as seen by an outsider, Mailer recapitulated it twice, once from the point of view of technicians on the ground and once from inside NASA. In each case he avoided resorting to the techno-language of the mission control center. To perform this feat, Mailer resorted to a duality. On the one side was the elegance of physics, which decrees the general principles of gravitation and the movements of bodies from which the trajectory of a flight must be constructed. On the other side were the thousand small compromises and approximations of engineering. Here Mailer achieved the same kind of double vision that characterized *The Naked and the Dead.* In that novel the action was seen both from the trenches and from the abstract viewpoint of the commander, who tracked the action through maps and aerial photographs. Like the Brooklyn Bridge, the moon shot had to be approached from multiple perspectives, and it eluded a single definition.

Of course, Apollo XI was not over after the successful takeoff. Half a billion people followed the astronauts' progress for days on television. To maximize the impact of this experience, some NASA officials had realized in the months before the event that they would need to provide the public with more information and more direct exposure to the astronauts than had been the practice with earlier flights, when there was "one stilted press conference with each network, for a limited time and in sparse surroundings." The old practice, they realized, "had presented the astronauts as stereotypes." NASA's public relations experts wanted to have each crew member spend a full day with the networks, perhaps even traveling back to his home town, in order to provide human-interest material that could be broadcast during the days and nights while the spacecraft was on its way to the moon. They believed that "the public would be better able to share in the ventures of these men on the moon if it knew who they were, why they were there, and what they were doing, a knowledge that could be achieved only through more time with the men and better training documentation, films, and taped reports of the progress of the launch."[64] But such ideas did not square with the rigorous training schedule and with the need to quarantine the three men; instead, they were interviewed by the press, eleven days before liftoff, through a protective plastic shield.

Apollo XI was a highly mediated event. Its planners spent considerable time worrying that if the first step onto the moon happened to be into the shadows it would photograph badly. Likewise, public relations people inside NASA complained about the decision of scientists to use black-and-white rather than color film for photographing the lunar surface. Committees discussed what objects should be left behind on the moon, and they canvassed the Smithsonian, the Library of Congress, and many other organizations for suggestions. The suggestions included carrying miniature flags of all the United Nations states, but in the end the American flag alone was unfurled. This decision was appropriate, in view of the political impetus for space flight and the intensity of the public's desire to see America win the "space race."

To get the most political "mileage" out the Apollo XI mission, Nixon spoke with the astronauts from the White House by telephone shortly after the landing. And when they splashed down in the Pacific, Nixon was there to greet them on the aircraft carrier

Hornet. Looking through the window of their decontamination trailer, he declared: "This is the greatest week in the history of the world since Creation."[65]

Apollo XI was the high point of the public's interest in the space program. Yet even before this mission many had begun to question the usefulness of a program that would leave more than a million dollars' worth of high-tech litter on the moon at a time when Nixon was cutting back on social-welfare programs. The decision to land a man on the moon had been made during the economic upsurge of the early 1960s, but NASA would find it increasingly difficult to obtain funding in subsequent years. The Apollo Program alone cost $24 billion, and in 1968 President Lyndon Johnson had cut its yearly budget by $370 million. On the day of the Apollo XI launch, Mike Mansfield, the leader of the Democrats in the Senate, rejected calls for a program to land a man on Mars, instead emphasizing the needs of people on earth. While public support briefly surged in the heady aftermath of the successful landing, in August 47 percent of Americans still believed that the space program was "not worth

Tickertape parade for Apollo XIII astronauts, Chicago, May 1, 1970. Courtesy of National Aeronautics and Space Administration.

it."[66] Twenty years later, however, three out of four Americans felt it had been worthwhile to send men to the moon.[67]

Large crowds continued to attend launches long after 1969, testifying to continuing excitement about them. Yet a launch was no longer automatically front-page news. By the early 1980s general interest had begun to wane, though interest in "being there" remained high. One journalist wrote in 1981: "The space shuttle does not generate the same enthusiasm as the old Mercury and Apollo shots, it is true, but there is still something fascinating about seeing a ball of fire take human pilots into space and a shiny metal sculpture bring them back safely."[68] The landing of the shuttle *Columbia* drew no less than 750,000 people to the White Sands Missile Range in southern New Mexico. This enormous crowd appeared despite the fact that the landing had been shifted to White Sands from a rained-out Edwards Air Force Base (in California) on only four days' notice. Although television coverage lessened and the general response to space flight weakened, seeing the space shuttle "in person" remained popular. The landings drew huge crowds, who waited patiently in unpleasant conditions. A journalist tried to describe the experience they were waiting for: "High above the desert one hundred miles north of Los Angeles, the space shuttle *Columbia* heralds its return from earth orbit by cracking the stratosphere with resounding sonic booms. The cannon-like blasts send waves of excitement through the crowds . . . an excitement that swells to cheers when the shuttle lands there just minutes later."[69] The crowds that turn out for such landings often number 400,000, despite the fact that they must view the event at a distance of several miles. Watchers need binoculars and telephoto lenses. They wait for hours in dry heat, 35 miles from the nearest services, and they must bring plenty of food and water for the vigil.

The *Challenger* disaster enhanced the excitement, however morbid, of witnessing a launch. The crew of the *Challenger* had been given the media exposure that had been recommended for the astronauts of Apollo XI. This had had the intended effect of whipping up public interest in the flight, particularly because a young, attractive schoolteacher had been chosen to take part. Because of her participation, millions of the nation's schoolchildren watched the launch and saw the rocket blow to pieces shortly afterward. Those who did not witness it had ample opportunity to see the replays that came with every news broadcast for the next 24 hours. The *Challenger* disaster dramatized the danger of the launches and

rescued them from the banality of uninterrupted success. It ful-
filled the dark promise of unimaginable violence that had always
been an unconscious part of the experience of witnessing a launch.
As one NASA official had admitted in 1981, "in many minds, this is
like a race track to which spectators come to watch crashes."[70]
Paradoxically, the news media swarmed to an event whose majesty
and sublime power could not be represented on the slight chance
that it would fail. A television technician remarked: "God forbid
that anything should go wrong, but that's why we are here."[71]

In the late 1980s, while NASA was in disarray, military expenditures
on space technology increased. The full budget is difficult to deter-
mine, since many projects remained secret, but the Reagan administration pushed defense appropriations to the highest level ever
attained during peacetime. Not only did the military continue to
launch satellites and use the space shuttle; it also began to develop
a new superweapon. The new project, known officially as the
Strategic Defense Initiative, was nicknamed "Star Wars" by oppo-
nents and came to be popularly known by that name. As others had
often done in the past, President Ronald Reagan offered Americans
a "technical fix" for the problem of military preparedness. Since
the seventeenth century many Americans had been prone to
believe that they lived on the brink of apocalypse. This attitude,
clearly present in many Puritan sermons and in the American literature's first best-seller, *The Day of Doom*, resurfaced in the periodic
religious revivals of the eighteenth and early nineteenth centuries.
During the later nineteenth century a whole genre of books
described the future destruction of the country by better-armed
foreign nations, and the theme likewise infuses much of the twentieth century's popular literature. H. Bruce Franklin has analyzed
the American imagination of superweapons and has shown that
Americans often have been fascinated with weapons of mass
destruction, confidant that each new weapon would prove so devastating that war would become unthinkable. However terrifying each
new weapon, Franklin notes, it could be explained as a moral
good.[72] In 1798 Robert Fulton argued in support of his submarine
that "[when] warships [are] destroyed by a means so new, so secret,
and so incalculable, the confidence of the sailors is destroyed, and
the fleet rendered worthless." Offensive war, he argued, would then
become impossible. Similar arguments were later used to justify
machine guns, strategic bombers, germ warfare, and nuclear arms

as defensive. The Strategic Defense Initiative, too, was described as defensive.[73]

The SDI program was thought to be necessary because the atomic bomb had clearly not made war unthinkable. Yet SDI was justified in the same way as the bomb and other superweapons always had been. It too would be the ultimate defensive weapon, permanently frustrating attack and ushering in universal peace. As Franklin notes: "Like the submarine, the torpedo, the steam warship, the strategic bomber, and thermonuclear deterrence, [the laser weapons of SDI] would be purely defensive and intended only to guarantee peace."[74] Yet even in its first use the atomic bomb was not a defensive weapon. It had been developed because of Germany's research along the same lines. The leaders of the Manhattan Project had never considered Japan to be in a position to make such a weapon.[75] Nevertheless, the scientists at Alamogordo worked feverishly to complete the bomb after Germany's surrender. As with earlier weapons, the justifications for making the atom bomb were forgotten once it could be used.

The public has made few attempts to celebrate atomic weapons in the way the Brooklyn Bridge or the Baltimore & Ohio Railroad was celebrated. An orator could confidently say that no one can look upon the Brooklyn Bridge and not feel proud, but who can make this universal claim for the hydrogen bomb or the nuclear reactor? The classic form of the technological sublime has broken down not because the objects of our contemplation have ceased to be fearful but because terror has become their principal characteristic, and we have no sense that we can observe them in safety. The natural sublime impressed the observer with man's insignificance and weakness. Even the older technological sublime had always proved ephemeral—the railroad, bridge and the light bulb, which once seemed sublime, soon became commonplace. But the Bomb could not become commonplace despite the government's best public relations efforts to domesticate it, nor could it be effectively surpassed by greater weapons that might diminish it through force of contrast. To anyone who contemplates them, nuclear weapons can only be a permanent, invisible terror that offers no moral enlightenment. The 1939 New York World's Fair had promised regular one-hour flights to London in the 1960s, but instead the next generation lived in fear of nuclear-armed missiles. In the nineteenth century the technological sublime had exalted human reason and made heroic the inventor, the builder, the entrepreneur, and the solo pilot. But by the end of the twentieth century technologies had become so complex and inhuman that they could make a mockery of the individual. Whereas a space launch awakened the will to believe, the bomb evoked uncertainty and dread. A technology so terrifying ceased to seem sublime. It could no longer claim to be an engine of moral enlightenment, to be contemplated in the same spirit in which one would approach natural scenery.

Rob Wilson has called the shift from the affirmative nineteenth-century experience of the sublime to the overwhelming terror of some forms of contemporary technology "the postmodern sublime." He coined this term to refer to the "worrying, mind-quelling force released at Los Alamos [*sic*] and Hiroshima, the Apollo 11 moon landing, nuclear winter, or superpower explorations of black holes and the ozone layer on Mars."[76] Wilson's list conflates two forms of technology, just as policy makers did for a generation. The rocket launch and the exploration of distant planets retain their fascination, because they represent a fundamentally different kind of enterprise from the building of nuclear weapons. The public quite rightly avoids atomic explosions, which have ceased to be tourist sights, but the triumphant blastoff into the heavens retains its attraction long after it has ceased to serve the ends of policy makers. Though the space race had no clear goal after Americans reached the moon and the USSR dropped out of the competition, there remained the existential pleasures of observing a rocket launch—perhaps the final avatar of the dynamic, technological sublime after the steamship, the railroad, and the airplane. Here remain possibilities for surpassing the achievements of the present hour. Here the individual can be lifted out of quotidian experience in a fundamental rupture of the usual sensory impressions, to be overwhelmed by technological spectacle. Here millions crowd together seeking an experience that can powerfully represent national greatness.

And here the tradition of republicanism confronts the paradox of a machine that does not serve as an engine of moral enlightenment but rather reduces citizens to awed spectators before enormous systems or to terrified dreamers of a nuclear apocalypse. John Adams once warned Thomas Jefferson that "every one of the fine arts from the earliest times [had] been enlisted in the service of Superstition and Despotism."[77] As John Kasson has shown at length, such fears were characteristic of the early days of the United States, when many hoped that the nation would concentrate on the useful arts on the ground that they were more suitable to a democratic nation than the fine arts. The steam engine and the spinning jenny, this line of thinking held, were inherently moral objects both because they were based on natural laws and because they improved the condition of mankind. Throughout the early development of the railroad, the contemplation of complex machinery had seemed to be an act of sublime meditation that ennobled the observer.

By the end of the twentieth century, however, the inherent virtue of concentrating on "the useful arts" was no longer apparent. The civilian population in the atomic age sensed the death-world just beneath the surface of consciousness, and a few prescient observers, such as Mumford, became obsessed with the fundamental shift in human affairs. The habitual perception of the sustaining life-world that is "always already there" retains its strength, but intimations of the "death-world" emerge in moments of international tension, such as the Cuban Missile Crisis or the 1991 Gulf War. There are periodic awakenings to the life-world's insubstantial underpinning—for example, the upsurge of student awareness during the early 1980s that invigorated the American "nuclear freeze" movement. There is an uneasy oscillation between the habitual admiration for large technological projects and the fear of nuclear holocaust.

At the deepest level, the existence of atomic weapons has undermined the possibility of the sublime relationship to both natural and technological objects. The experience of the natural sublime rests both on the sense of human weakness and limitation and on the power of human reason to comprehend the infinitely large and powerful. But when human beings themselves create something infinitely powerful that can annihilate nature, the exaltation of the classic sublime seems impossible. The Kantian relationship to the object required a sense of personal security. One was exposed to the power of the hurricane, but nevertheless one saw it in relative safety. This necessary precondition evaporates in the superheated wind of an atomic blast. The technological sublime, in which the observer identifies with the power of a man-made object, becomes absurd. Who identifies with the bomb? The collective sense of achievement, another hallmark of the technological sublime, is radically undercut and destroyed. Just as important, contemplation of the bomb transforms admiration for inventors, engineers, and scientists into fear and mistrust. Viewed in this perspective, the early linkage of the space program with national defense served to obscure the radically different implications of the arms race and manned space flight. The sublimity of a manned launch could easily overwhelm reflection on the military uses of rockets.

Public enthusiasm for the space program represents a nostalgic return to the technological sublime, a turning away from the abyss of the nuclear holocaust seen all to clearly after 1945. In the 1960s a launch allowed many to recuperate the sense of technological sublimity. The affirmative sense of achievement that followed the moon landing made government high-tech programs attractive at the very moment when the nuclear stockpile was large enough to destroy every living thing on earth. In an atomic age, the pilgrimage to the Kennedy Space Center promised a sublime experience that renewed faith in America and in the ultimate beneficence of advanced industrialization. This final avatar of the technological sublime is a literal escape from the threatened life-world.

Rededicating the Statue of Liberty

It was during the 1909 Hudson-Fulton Celebration that the first air-plane flew over New York—piloted by Wilbur Wright, and with a canoe tied underneath just in case a water landing might be neces-sary. Wright flew down the harbor and circled the Statue of Liberty, observed by approximately a million people. A photograph juxta-posing the plane and the statue appeared on the cover of *Harper's Weekly*. The following year, Glenn Curtiss made the first flight from Albany to New York, retracing the route of Robert Fulton's *Clermont*. Curtiss too circled the Statue of Liberty. Since that time there have been many photographs of airplanes next to the statue.[1] In the early images the aircraft of Wright and Curtiss are lilliputian, almost like small birds in comparison to a human body. In later photos the planes are larger and they fly higher, no longer circling Liberty's head but now looking down on her. One might read this shift in scale and altitude as a metaphor for the increasing domi-nance of technology over political liberty, but in fact the relation-ship is considerably more complex. The Statue of Liberty and the airplane are, each in its own way, sublime. Their juxtaposition con-trasts the dynamic and the mathematical sublimes. As a succession of airplanes soar toward infinite space, the statue provides an olympian perspective on New York City.

Burke argued that certain abstract words, notably 'liberty', were capable of arousing human emotions, even if they called up no immediate association with a tangible object.[2] In fact, this lack of clear referentiality makes it possible for a culture to attach the word 'liberty' and the emotion it evokes to a statue that monumentalizes the idea into a powerful symbol. The statue functions at once as a visualization of an abstraction and as an object of grandeur, impres-sive because of its sheer scale. Its creator, Frederic-Auguste

Bartholdi, understood this relationship well, writing of his statue that "the immensity of form should be filled with the immensity of thought, and the spectator, at the sight of the great proportions of the work, should be impressed before all things else, with the greatness of the idea of which these ample forms are the envelope."[3]

The Statue of Liberty is one of the great works of the geometrical sublime, equally impressive when seen from a distance and when used as an observation point. Furthermore, since it has been illuminated more and more elaborately over the years, the statue has been appropriated to the electrical sublime as well. Not only does it combine the idea of liberty, monumentality, spectacular illumination, and the sublime, but it has also been used as the focal point of a great national ceremony (on the occasion of its rededication by Presidents Reagan and Mitterand in 1986). It is thus a fitting subject with which to conclude this survey of the American appropriation and redefinition of the sublime.

Some critics have mistakenly focused on how the statue appears when one is standing near it, on Bedloe's Island; they have attacked the form of the drapery, the sharpness and rigidity of the crown, the almost harsh expression of the face, and so forth. Marina Warner, for example, complains that the statue is "remarkably hideous for a public sculpture of its date." She continues: "The twist on the figure as she steps forward forcefully to brandish her torch provides the only movement in the statue, but it is visible, given Liberty's size, from the right side only, in profile, and there it looks frozen, exaggerated and slightly vulgar, a soubette's over-emphatic gesture."[4] Such comments assume that one is meant to stand beside the statue when viewing it, as though it were a close-up in a film, when in practice it is almost invariably seen from a great distance, like a gigantic actor on the stage of New York Harbor. From a distance of half a mile or more, the drapery seems to flow more and is less obviously metallic, and because the features are sharp they can still be distinguished. Details that seem overemphatic at close range are necessary to give the statue character when viewed from a passing ship or from the shore. Marvin Trachtenberg perceptively points out that, although Liberty faces more south than east, her position gives the spectator on land the impression that "she seems to face outward towards the Atlantic and across to the Old World— the world she enlightens." At the same time, he notes, "to the shipboard observer Liberty presents an illusion of forward movement— of striding into the path to New York (and America)."[5]

Whatever faults the statue may have cannot be seen from the base, because of the massive pedestal. But those who come to the island, like visitors to a skyscraper, do so less to inspect the outside than to go inside. They usually climb up at least to the top of the pedestal, and preferably up the winding stairs and into the head. The statue affords a fine view of the skyscrapers of Manhattan, one form of the geometrical sublime saluting another.

Warner has pointed out that the association between women and the abstract names of the virtues permeates European languages. In Latin and Greek the words for virtue, justice, faith, hope, charity, and the seven liberal arts are feminine. Warner argues that such associations may have prompted a tendency to identify virtues with the female form. Certainly in classical sculpture women were often made to stand for virtuous abstractions, while men were not usually treated in this way. This pattern of association was passed on to the Renaissance and, principally by way of France, to the United States—not least by Bartholdi, whose work was steeped in the classical tradition.[6] As Pierre Provoyeur has noted: "Liberty has been a woman since the third century B.C. In Rome, in the temple dedicated to her on the Aventine, she is robed, wears a Phrygian cap of a freed slave, and holds a scepter denoting her independence."[7] Both Provoyeur and Trachtenberg trace the lineage of imagery and associations from that time until the 1870s, when Bartholdi assimilated them in the pose, dress, and ancillary symbolism of his statue.[8]

Bartholdi had once planned a colossal statue to be erected in Egypt at the mouth of the Suez Canal, but that project fell through owing to a lack of funding. Later, when the Statue of Liberty had come to New York, he sought to disassociate the two projects, telling a reporter: "The Egyptian affair would have been a purely financial transaction," whereas the New York statue was "a pure work of love."[9] Surviving drawings show certain similarities between the two colossi, however. Each was conceived as bestriding the entrance to an important harbor, bearing a flame. The Egyptian statue was specifically planned as a lighthouse, not only sublime in scale but also exploiting dramatic lighting to increase its impressiveness. In this sense, it anticipated spectacular electrical illumination. The double function seems to have been carried over, as the Statue of Liberty at first was conceived as both a symbol and a lighthouse. The text accompanying Bartholdi's 1875 model says that "by night a halo of light will shine from her forehead far out across the boundless sea."[10] Yet as Provoyeur notes, Bartholdi abandoned this

The arm and the torch of the Statue of Liberty at the Philadelphia Exposition of 1876. Courtesy of Collections of Library of Congress.

scheme—and for a revealing reason: "Familiar as he was with Universal Expositions, the sculptor knew the public delight in climbing to a great height; the most spectacular ascents were those made in a balloon. . . . Bartholdi twice offered people a good climb: once into the torch at Philadelphia in 1876 [at the Centennial Exposition], and then into the head, at the Universal Exposition in Paris, in 1878. . . . The tourists won; the light was relegated to the torch."[11]

Bartholdi had to choose between the certainty of the geometrical sublime, which was bound to be impressive, and the less certain possibility of creating a powerful light. In 1875 the arc light was still in the experimental stage. Only a few lighthouses had special forms of electric light, and gas light was too faint to be impressive at any distance. Bartholdi undoubtedly knew that the theaters and opera houses of Paris sometimes used elaborate lighting techniques, such as limelight, the *laterna magica*, and arc lights, but these were "impractical" techniques that lasted only a few minutes.[12] While considering the uncertainties of finding an impressive and reliable light source, Bartholdi could observe the pleasure of spectators who climbed up inside the already-built arm, and he could imagine how spectacular the view would be from atop the entire statue when it was mounted on a high pedestal in New York Harbor.

Trachtenberg notes that, although the statue appears to be solid, its "skin" is not self-supporting but consists of many overlapping pieces of hammered copper, only 2.5 centimeters thick, each fastened to an iron skeleton. The structure combines the external appearance of traditional sculpture with Gustave Eiffel's iron trusswork frame, which drew on his ingenious bridges and looked forward to the Eiffel Tower (which he began almost as soon as the statue was completed). To build Liberty, Bartholdi had to work on an unprecedented scale, direct a large work force, and confront new problems of perspective; Eiffel had to find a skeletal form strong enough to bear the weight of the copper cladding and resilient enough to withstand wind shear and temperature changes. Trachtenberg notes that "Liberty is an archetypal illustration of the aesthetic tension of its time—when technology had already attained great advances and power and a hold over the mind, but when the conscious eye was still dominated by traditional imagery."[13] Yet the contrast between the neoclassical exterior and the technological complexity of the interior was not regarded as a contradiction at the time. Rather, it was understood as an ingenious use of materials to achieve something classical and familiar on a colossal scale.

The Statue of Liberty eventually became an all-encompassing national symbol, but it did not always have that status. After receiving the gift from France, Americans were slow to appreciate it, and did not immediately raise sufficient funds to erect the pedestal. The New York committee charged with this task did virtually nothing from 1877 until 1882. When the campaign finally began in earnest, wealthy New Yorkers at first disdained to contribute generously, and efforts elsewhere proved disappointing because the rest of the country did not identify with the project. New York State refused to contribute, and a committee of notables struggled to raise $200,000 over a four-year period. Only a determined campaign by Joseph Pulitzer's *New York World* spurred the general public to subscribe a desperately needed $100,000, permitting the pedestal to be completed and the statue erected.[14] By the fall of 1886 the statue was in place, a little more than ten years after its upraised torch had appeared at the Philadelphia Centennial.

On the statue's dedication day, October 28, 1886, the celebrations included all the traditional elements of democratic ritual: a parade, speeches, and fireworks. Despite rain and mist, a million people turned out for the ceremonies, crowding Fifth Avenue to watch a procession that included more than 100 bands. Civil War veterans (some of them African-American) and troops from France, Italy, Belgium, and Switzerland were among the 20,000 marchers.[15] The dedication ceremony itself was audible only to the comparative few who stood on Bedloe's Island, while the rest of the crowd waited in boats and along the shore for the unveiling. The speakers of the day made no memorable remarks, and, as John Higham notes, their rhetoric "concentrated almost exclusively on two subjects: the beneficient effect on other countries of American ideas, and the desirability of international friendship and peace."[16] (Almost entirely unspoken was an important French motive for this colossal gift: the need for American approbation of the French plan to cut a canal through Panama. That project, if completed, would decisively reassert France's presence in the Americas and give France control over a key military site.) During the speeches, Bartholdi, who stood inside the head, inadvertently unmasked Liberty's face in the middle of a senator's address. The spectators, who could not hear the speaker, immediately began to shout, and cannons roared from the flotilla. The pandemonium lasted 15 minutes. For most of the citi-

The inauguration of the Statue of Liberty, October 28, 1886. Photograph by H. O'Neil; courtesy of Collections of Library of Congress.

zenry the unveiling derived its meaning not from the words spoken but from the visual elements of the spectacle. The effacement of the orator had begun, anticipating the Hudson-Fulton Celebration's wordless text and the emergence of the Great White Way. But the new electric lamp of Liberty was feeble on the evening of the inauguration, all but invisible in the rain and fog. The *World* reported that it looked like a glowworm.[17]

As has already been mentioned, the Statue of Liberty did not instantly become a symbol of the United States. It was a gift from France, after all, and it was meant to embody an abstract idea. Although a steady stream of tourists visited it, for decades the statue was neglected by the federal government, which skimped on maintenance and which had to be prodded constantly by private associations to protect the monument and develop the site. For decades Bedloe's Island was ill-kept. There were many ramshackle warehouses, and the dock where tourists landed was an eyesore. Despite such negligence, however, the statue grew in symbolic significance, helped by successively brighter electrical illuminations that made it especially noticeable at night. It was second only to the Brooklyn Bridge as a prominent sight along New York's waterfront.

Chief among the early associations that clustered around the statue was the New World's promise for immigrants. But this was not part of Bartholdi's intended meaning, which was to commemorate the friendship of France and the United States and their common political commitment to democracy. However, the proximity to Ellis Island's immigration center made it easy to associate the statue with immigrants. As Werner Sollors emphasizes, some writers made only negative associations. Nativists such Thomas Bailey Aldrich and Herbert Croly imagined Liberty as a "white goddess" who "should guard freedom against the menace of the rather beastly invaders."[18] Emma Lazarus's poem, written before the statue was erected, was neither read nor mentioned at the dedication ceremony in 1886. A proposal that a new receiving station for immigrants be located on Bedloe's Island provoked many complaints—Bartholdi himself called it a "monstrous plan."[19] Yet by 1903 the link between the Statue of Liberty and immigrants was officially recognized. The words of Emma Lazarus were inscribed on a plaque, bidding the world to send America its huddled masses, tired, poor, and yearning to breathe free. The importance then attached to this plaque should not be overemphasized, however. The plaque was installed not by the front door, where it is today, but on the second floor, inside, where it attracted little attention. There was no official installation ceremony, and until the 1930s the poem received little comment. In 1933, when the National Park Service took over the statue, it did not emphasize immigration as part of the statue's meaning. Rather, as John Higham notes, it "clung to the traditional motifs—Franco-American friendship and liberty as an abstract idea." And "on the Statue's fiftieth anniversary, in 1936, patriotic organizations and public schools promoted a nationwide celebration which kept the usual themes steadily at the fore."[20]

The statue remained just short of being a national symbol for at least the first quarter-century of its existence. H. G. Wells found it unimpressive and not worth a visit when he came to New York in 1906. To Wells, the statue was "meant to dominate and fails absolutely to dominate the scene." He continued:

It gets to three hundred feet about, by standing on a pedestal of a hundred and fifty; and the uplifted torch, seen against the sky, suggests an

An immigrant family looking at the Statue of Liberty. Courtesy of Collections of Library of Congress.

arm straining upward, straining in hopeless competition with the fierce commercial altitudes ahead. Poor liberating Lady of the American ideal! One passes her and forgets.[21]

Although most other travelers and most immigrants were impressed, the Statue of Liberty lacked international stature as a symbol. Yet it was coming into its own as a symbol within American culture, and by World War I it had begun to displace Columbia as the female personification of the nation. The decision to call war bonds "Liberty Bonds" unleashed a tide of posters bearing the statue's image, and the press began to link the statue to the struggle for victory over Germany and to Woodrow Wilson's declared intention to make the world safe for democracy. One characteristic poster, clearly directed at people with immigrant backgrounds, depicted the statue seen from a ship and carried this text: "Remember Your First Thrill of American Liberty: Your Duty—Buy United States Government Bonds." By the 1920s the statue had become such a part of the American imagination that it was routinely associated with any heroic circumstance, such as Lindbergh's flight.[22]

As is true of any important symbol, the Statue of Liberty's meanings are not unitary, nor are they what was originally intended by the artist and his contemporaries. Rather, successive generations of interpreters have ascribed meanings to it, and a cluster of associations hover around it.

The continual reinterpretation of the Statue of Liberty's meaning is particularly evident in the celebration of the centennial of its dedication.[23] The preparations for this celebration began with a meticulous examination of the statue's physical structure. After a year's diagnostic work, the statue was surrounded by free-standing scaffolding that was never allowed to rest upon the skin. Hundreds of workers restored the copper sheathing and reinforced the iron frame, which had been weakened by metal fatigue. The restorers found that the upraised arm and the head had been installed incorrectly and were as much as 2 feet out of alignment, causing unintended strains on the skeleton. Both were repositioned, and the torch was replaced.[24] Many of the restorations are visible only to specialists, but overall there was one unforgettable improvement. During the previous century much of the interior had been subdivided by concrete floor slabs and other additions. Now, however, the visitor "discovers the beauty and mystery of the statue's interior:

a magnificent volume—a 100-foot-high 'room' of copper and metal, of light and shadows."²⁵ Dramatic lighting and viewing points along the stairways were added to heighten awareness of the unusual space. These efforts made the statue more striking both from a distance and from inside. The new gilded torch was better lighted than before, and it gleamed more noticeably at night. The walkway leading to the entrance was improved, and massive new doors and better stairways were installed. The restoration strengthened the statue's appeal as a form of the mathematical and geometrical sublimes.

What kind of event was constructed to celebrate the centennial? Significantly, the rededication was not held on October 28, the date of the first unveiling. Rather, like the dedications of the Erie Canal, the Baltimore and Ohio Railroad, the Eads Bridge, and many other projects, it was held on the Fourth of July. On Friday, July 3, the statue was relighted, and the following day it was reopened to the public with a massive fireworks display. Many of the weekend's elements were uncannily similar to the Hudson-Fulton Celebration, held precisely 77 years earlier. As in 1909, there were spectacular lighting displays in the harbor and feats of airmanship. The U.S. Navy's Blue Angels flew over the harbor, and eight jets of the French air force put on an "aerial ballet."²⁶ There was a naval procession, as in 1909; but, whereas the Hudson-Fulton event had celebrated the then-new technology of the ironclad battleship, by 1986 such behemoths had become familiar and an interesting reversal had taken place. Now the public was charmed by tall sailing ships, and such ships had been invited from all over the world to take part in the celebration. Nostalgia for the age of sail and aesthetic appreciation of the graceful lines of wooden ships had become stronger than the desire to see new technologies.

And in 1986 there were no parades, mass spectatorship having become largely the province of television. Rather than a line of floats passing down Fifth Avenue, or even a line of electric advertisements on Broadway, most Americans now saw a TV spectacle punctuated with patriotic commercials, including 23 made as a special series for Prudential Life Insurance. (The organizers of 1909 had explicitly eschewed commercialism and advertising.) The ABC network broadcast the evening news from Governors Island, using shots of the statue as a lead-in to a story on affirmative action and reminding viewers that full coverage of the unveiling ceremony was coming up later that evening.²⁷

Heading the team that organized the Liberty weekend was David
L. Wolper, a Los Angeles television producer whose understanding
of mass taste had been evidenced by his success with the miniseries
Roots, *North and South*, and *The Thorn Birds*. Wolper also had produced live public spectacles, notably the opening and closing ceremonies of the 1984 Summer Olympic Games. Wolper and his staff
approached the Liberty event as a semi-commercial venture—a
four-day miniseries, to be paid for by the sale of rights. Tickets for
choice viewing positions on Governors Island went for $10,000 a
pair, and raised more than $15 million, but did not ensure comfort.
The *Daily News* reported that "guests' feet sank into soggy, hastily
laid sod and crossed wooden walks still wet with paint," that "the
sound system was too loud," and that "a giant television screen was
out of focus."

In line with the commodification of the event, the last official
ceremony was held a good distance away from the statue in Giants
Stadium, where 50,000 people paid admission to see big-name performers such as the Temptations.[28] Wolper maintained that the
unveiling was essentially entertainment, and that this justified selling ABC the exclusive rights to broadcast the proceedings for $10
million. The other networks protested vigorously, pointing out that
the appearance of the president at the event and the national significance of the statue made the proceedings news rather than
entertainment, and that it was their duty to cover the news, which
was not a commodity. Though a compromise was worked out permitting all the networks to cover the opening ceremony,[29] ABC's
more complete coverage helped that network make a profit of $16
million from its programming, which topped the weekend ratings,
drawing 60 million viewers.

Newspapers around the country decried the commercialism that
surrounded the event. The *New York Daily News* complained that
"Wolper of Hollywood, whom we get rid of in a few days, came to
town and turned Lady Liberty into a Vegas chorus girl." The *Dayton
Daily News* editorialized: "The restoration of the statue turned into
one big commercial hype by sponsors who helped pay for repairs. . . .
The hype should be no surprise in a culture that has managed to
transform holy days into holidays and holidays into sales." The
Pittsburgh Press worried "whether American culture does not have its
dark side if the only way we can celebrate meaningful moments in
the nation's life is through show-biz overkill." The *Lexington Herald-
Leader* praised two judges who had refused to swear in new citizens

in conjunction with the ceremony because they knew "the difference between a patriotic event and crass commercialism." A U.S. District Judge in Washington canceled a swearing-in ceremony that was to be held at the Jefferson Memorial after he learned that "the proceedings would be followed immediately by commercials."[30]

Despite Wolper's efforts to wring every possible dollar out of the venture, the statue's harbor location permitted many to view the event free. Any boat owner could sail into the harbor and see the show for nothing—as did the wealthy magazine publisher Malcolm Forbes, who invited assorted Rockefellers, Henry Kissinger, and Frank Sinatra to his yacht. Lee Iacocca borrowed a $10 million, 143-foot yacht.[31] There were an estimated 40,000 boats of all sizes in the harbor, and it was almost impossible to sail through them. One company offered a half-day cruise for $125 per person, and the Circle Line (which ordinarily runs sightseeing boats around Manhattan Island) offered rides to an anchorage near the statue for the relighting ceremony for $100 a head. To see the fireworks from the same location the following night cost $195, including a box supper. For really big spenders, penthouses with a view were available for $15,000. The *Amsterdam News* headlined "Liberty Weekend, grand spectacle for super rich." Yet one of the best observation points was the Staten Island Ferry, which sailed right by the statue and charged only the usual fare, and it cost nothing to stake out a good point on the George Washington Bridge or at Liberty State Park on the New Jersey shore.

In 1909, for the Hudson-Fulton Celebration, all the hotels in New York were overbooked, and scalpers sold tickets to viewing stands along the parade route for outrageous prices. However, in 1986 many hotels found themselves with empty rooms even a few days before the event, and a company that had rented Windows on the World—a restaurant with a spectacular view of the harbor—still had 300 unreserved places on July 2.[32] In short, while an estimated 15 million New Yorkers and their out-of-town guests thronged the shoreline, most of America stayed home and watched the events on television as a complement to local celebrations.

St. Louis had a 41-foot replica of the statue near the Gateway Arch, and the Elks Ladies' Auxiliary Club of Longmont, Colorado, baked a 3000-pound Miss Liberty cake.[33] Vendors sold millions of tiny statues as souvenirs. A replica in chocolate stood in a hotel lobby. The Carnegie Delicatessen made a statue out of chopped liver and presented it to an official who had deplored the overcommercialization of the event. A Hollywood agency specializing in celebrity lookalikes searched for the woman who most resembled the statue.[34] There was considerable good-natured irreverence. John Gross noted in the *New York Times* that "in the land of liberty you are allowed to take liberties with liberty." He continued: "In the course of two minutes' window shopping this morning, I noted down a Cabbage Patch Miss Liberty, a bespectacled Miss Liberty, a make your own model of Miss Liberty (all cubes and cylinders, like a figure in a painting by Chiroco), a Mae West Miss Liberty . . . a Miss Liberty carved out of soap. . . ."[35] Walden Books sold a videotape of the statue's history for $24.95, apparently the same program that Liberty Mutual Insurance Company had funded for the Public Broadcasting Network station WNET.

Yet, while many viewed the rededication of the statue as an opportunity to cash in, New York City officials took a much different view, seeing the event as a weekend of free activities available to everyone. While promoters put together expensive hotel packages and rented the roofs of warehouses near the harbor for high prices, the city made available thousands of inexpensive camping places.

The destitute were not entirely forgotten. A National Coalition for the Homeless set up several shanties in Battery Park to call attention to the contradiction between the Statue of Liberty's promise and the lives of thousands of New Yorkers. Jesse Jackson appeared in their support, declaring "This should not be a party just for the affluent."[36] The San Francisco *Sun Reporter* complained on July 9 that, despite the public celebration, "there are millions of people living in the United States who are not free, particularly black Americans, Hispanic Americans, and, to some degree, Asian Americans." The *Baltimore Afro-American* took a similar stance: ". . . the promise offered by the statue was not offered to the millions of black people whose parents came to these shores in the hole of a ship, chained to each other. . . . America's immigrant policy is racist and has been for many years. People of color still have lower quotas than people from North and West Europe and this tradition of racist immigration policies coupled with shabby treatment that Haitian immigrants recently received dampens the enthusiasm that many black people feel for the Statue of Liberty celebrations." The *Chicago Defender* ignored Liberty weekend altogether, merely listing the ABC special in the TV section. But at the top of the front page of the July 12 issue was a story about a black South African teenager who had sought asylum in the United States "after three months at sea as a stowaway on two merchant ships" but had been put in an Oakland prison on July 4.[37]

New York's *Amsterdam News* of July 5 editorialized on its front page: "Black Americans gaze helplessly at this spectacle wondering whether to damn or cheer this Cecil D. DeMille production out of Hollywood, underwritten by the POWER ELITE of American industry, starring a president who can willingly give 100 million dollars to the Contras attempting to overthrow the legitimate government of Nicaragua while at the same time supporting the racist apartheid regime of South Africa. What kind of celebration of LIBERTY is that?" Beside these words was a large cartoon showing Ronald Reagan and New York Mayor Edward Koch lounging on a yacht with a smaller craft chained to it. The smaller vessel, crowded with ordinary Americans, carried a replica of the Statue of Liberty; its passengers shouted "You ain't goin' nowhere without us." Another article in the same paper ("Is Lady Liberty really symbol for Blacks too?") argued that the statue had been inspired by Bartholdi's visit to Egypt and noted that one of Bartholdi's acquaintances, the French historian Edourd de Laboulaye, had been chairman of the French Anti-Slavery Society. In this version of Liberty's origin, Laboylaye had "conceived of a monument to send as a gift to abolitionists in recognition of the end of slavery."[38] This interpretation could have been grounded directly in American history as well: John Greenleaf Whittier, in a poem read at the statue's inauguration, had barely mentioned the immigrants but had celebrated the abolition of slavery.

Having determined to repossess the statue as a symbol for African-Americans, the *Amsterdam News* adopted a fervent tone in an editorial-page piece titled "Turn Around, Ms. Liberty":

We truly love what you stand for and adhere to what you represent, it's just that no one has given them to us. When you arrived 100 years ago, Miss Liberty, we were already here for over 300 years. . . . Please if only for a second, turn around, you'll see 60 percent Black youth unemployed, high rents in gutted buildings, poor education services, drugs at an all time high. . . . TURN AROUND MISS LIBERTY! You'll see the President, who will light you up, is packing the Supreme Court with reaction. . . . Oh, we believe in you Miss Liberty but you can't see us, that's why they pointed your face to the sea. The immigrants you welcomed and set free thrive on a land we developed, yet we are locked out of real participation.

The festivities began on Thursday, July 3, when President Reagan hosted the unveiling of the refurbished monument. While Reagan bathed in the reflected glow of the event, his administration had, in fact, refused to provide federal funding for the restoration, which had cost just under $70 million. Instead, Reagan had named a committee to raise money privately. Headed by Chrysler's chief executive officer, Lee Iacocca, the committee had done so well that, as the event neared, Republicans had begun to fear that the popular Iacocca might run for president as a Democrat in 1988. He seemed to embody the American dream that the statue had come to stand for. Son of Italian immigrants, educated at Lehigh and Princeton, he was now the head of one of the nation's largest corporations. His recent autobiography had sold 5 million copies. In the run-up to the ceremony, the Reagan administration had done what it could to diminish Iacocca's prestige. Secretary of the Interior Donald Hodel had dismissed him from the commission charged with refurbishing the statue and Ellis Island and had feuded with him in the press, even attacking him for raising more money than was necessary for the restoration and for suggesting that the $12 million surplus be used to defray any losses sustained in mounting the celebration.[39]

But it was Reagan, not Iacocca, who held the limelight on July 3, acting as master of ceremonies in an evening of festivities that included music, dance, speeches, and the dramatic swearing-in as American citizens of 276 immigrants from 100 different countries by Chief Justice William Rehnquist.[40] As many commentators have noted, President Reagan habitually not only drew upon acting skills but also reused lines from his films. The Wolper extravaganza was a perfect setting for him.[41] Hollywood invaded the content, the style, and the technology. There were few speeches; instead, the proceedings were dominated by aging stars, including Frank Sinatra, Elizabeth Taylor, and Shirley MacLaine. The structure of the event insisted that citizenship required little more than the consumption of images and entertainment. Reagan's rhetoric emphasized the metaphor of the torch: "We are the keepers of the flame of liberty; we hold it high tonight for all the world to see, a beacon of hope, a light unto the nations."[42] The few rhetorical flourishes were designed as sound bites for radio and television; there was no attempt to make a historic speech.

The main event began with the Navy Band and a Marine drill squad. Then came the singing of "The Star-Spangled Banner" by Kenneth Mack, a 13-year-old black boy from Louisiana, followed by a brief film on the statue's history, introduced by Gregory Peck. The film was clearly more effective for the millions of television viewers than for those at the site, and the use of this device suggests the complexity of the staging: A former-movie-star president stood in front of the actual Statue of Liberty, acting as the host of a made-for-television film that represented the Statue of Liberty as an image. Of course, the American public had no difficulty assimilating this mix of presiding, acting, image, reality, and simulation, which marked the event as a typical postmodern extravaganza.

Then came the highlight of the festivities: the unveiling of the statue after three years of restoration work. Standing on Governors Island with the French president, François Mitterand, at his side, Reagan prepared to press the button that would light up the dark silhouette that loomed in the harbor. Just as Theodore Roosevelt had illuminated the St. Louis World's Fair, just as Woodrow Wilson had pressed a button to illuminate the Woolworth Building, just as Calvin Coolidge had lighted Chicago's White Way, and just as Herbert Hoover had pushed a button to light the Empire State Building, it now seemed quite natural that Reagan should push a button in a ceremony where light would be used for symbolic purposes.

There had been technological improvements in lighting technology, of course. The electrical sublime no longer consisted of powerful searchlights working in the medium of steam, as it had at the Hudson-Fulton Celebration, nor would strings of lights in the rigging of ships or along the superstructure of bridges have had the impact in 1986 they had had in 1909. Over the years, the illumination of the Statue of Liberty had been increased several times, often at public instigation. For example, in 1915 the *New York World* raised $30,000 to improve the lighting. The intensity of the illumination was heightened again in 1931, when 96 1000-watt lamps were installed.[43] But these intensifications had not been coherent. For the restoration, a fully coordinated lighting design was created that increased the intensity of light as the eye traveled from the pedestal up to the torch. Yet even this new system was not deemed adequate for the rededication ceremony—that was felt to require special effects. When Reagan pushed the button, he activated a series of powerful colored laser beams that bathed the statue with pinpoint accuracy, lighting it in distinct stages. According to the *Times*, "beams of bright red caught the base of the lady first, then changed to red amber as they caught the pedestal." After a few seconds, "a steel-blue wash revealed the full, 151 foot monument." After another brief pause, powerful "floodlights splashed the full statue in an intense white light."[44] The climax came when Reagan pushed a button that relighted the torch and set off 5 minutes of fireworks. "As the Statue of Liberty was bathed in new red, white, and blue lights last night, lower Manhattan was awash in the spirit of spontaneous friendship," the next day's *Daily News* reported. "Strangers embraced one another. Some were so charged up that they jumped on top of benches in the Battery Park City complex and broke into 'God Bless America' and then 'The Star Spangled Banner.'" The technological sublime "charged up" normally disparate individuals, creating a communal bond.

July 4 dawned cool and clear. Millions had come to see ceremonies which they might have seen on television. They came both for patriotic reasons and because they knew that, as one put it, "You can watch it on TV all you want, but you won't get the thrill of really being here."[45] The collective technological sublime had become naturalized and was now "real" in comparison with the televised version. Whereas Kant had implicitly assumed that the sublime required a solitary communion with nature, here were several million people crowding and jostling together to contemplate an orchestrated public event. Part of the "thrill" was to be part of an enormous crowd that stretched as far as the eye could see, and to enter into collective expectation and enjoyment of a national festival. However, the authenticity of being there in person was highly mediated. Lawrence H. Estrin, Liberty Weekend's sound and communications director, was responsible for wiring not only Governor's Island but also parts of lower Manhattan and Giants Stadium, both of which were used in the celebrations. The system of microphones and loudspeakers was coordinated by technicians in constant contact through walkie-talkies. (Estrin, the *Times* reported, "usually wears two walkie-talkies, a beeper, and a cellular phone strapped to his waist.") Huge speakers all around the harbor carried music that was coordinated with the fireworks. For example, as the "Blue Danube Waltz" was played, blue Roman candles kept the beat while blue and gold shells streaked across the sky. Not surprisingly, the man chosen to coordinate these effects had previously designed sound systems for the Grammy and Emmy ceremonies and the Tournament of Roses parade.[46]

In keeping with the commercialization of the event, the man in overall charge of the fireworks show (which lasted 28 minutes and involved more than 40,000 projectiles and 100,000 bursts) was

Tommy Walker of the Disney organization. The Grucci family firm, winner of the fireworks contract, had erected thirty launching sites around the harbor, all linked to a central command computer on Governor's Island. The next day, reporters strained to find metaphors strong enough to suggest how stunning the display had been. Dirk Johnson wrote in the *Times*: "Against a slate black sky, Fourth of July fireballs soared higher than they have soared before [to 1500 feet, three times the normal height], then exploded into a million tentacles that reached toward the ground before becoming wisps of smoke. As millions gazed skyward—some flinching with each apocalyptic boom—fireworks were set ablaze in every direction, so that lower Manhattan seemed under a roof of color and glitter. The fireworks were stars, then glittery trees, then rubies exploding into diamonds."

Such an elaborate performance could never have been staged by manually lighting individual rockets and keeping time by approximation. For this event Felix Grucci had prepared new computer software, adapted from dBase III and the Lotus 1-2-3 spreadsheet program. Grucci's software took account of the weight and the propellant thrust of each rocket. It automatically withdrew each device from the inventory as it was added to the script. It warned the designer if he was developing an imbalance between different kinds of projectiles.[47]

While no one denied the impressiveness of the pyrotechnics, some criticized the event severely. A *Baltimore Sun* editorial declared that the centennial "captures the distinctive mood of the Reagan era: unabashed patriotism, unflinching optimism, unapologetic indulgence in the good life" and that "glitz and glamor are on display in New York harbor and imprinted on videotapes for the edification of the ages." While most other newspapers wrote as if the past had been innocent of commercialism, the *Sun* reminded its readers that the Statue of Liberty "is not just a beacon of hope to huddled masses waiting to breathe free. She is the product of an earlier gilded age that gloried in grandiosity and in flinging up technological marvels and in charitable enterprises financed by robber barons." (This was only half right, of course—the robber barons had conspicuously refused to contribute to the pedestal, and a populist newspaper campaign had rallied the citizenry to pay for it, just as many ordinary Frenchmen had given small sums to pay for the statue.) The *Sun* further recalled that this gift of a statue to represent liberty had come from France during the "height of its colonial expansion," when the "voices of Manifest Destiny were nudging the United States toward the Caribbean and the far Pacific." In quite a different spirit, President Reagan made a similar connection in a radio broadcast on July 5, as he sought to use the Statue of Liberty as a prop for his administration's Central America policy. He compared France's assistance to the United States during its revolution to latter-day American aid to "the pre-democratic freedom fighters in Nicaragua."[48]

The centennial of the Statue of Liberty demonstrates how, by the late twentieth century, the sublime had become an achievable part of a public spectacle. What had occurred in this movement from awed contemplation of natural scenery to the rapt enthusiasm for technological display? The emotion once experienced in the presence of Niagara Falls or the Grand Canyon had been transformed. In the new sublime, the subject seemed to glimpse not divinity but national greatness, and the meaning of the nation was expressed less in words than in powerful images. The orator's role, once central to the Fourth of July, had gradually been upstaged and superseded by technological display. The process had already begun in 1886 at the original dedication of the Statue of Liberty. Yet, if the meaning of the event was not articulated, it was inscribed within it nonetheless. This was done not through great parades, once so central to manifesting the structure of the community, but through a barrage of television coverage, accompanied by parades of products that replaced the processions of producers once common to every festive day in America.

And yet television was widely understood to be a poor medium for viewing the rededication. The relighting of the statue and the fireworks were too large and complex to be captured on a small screen, as was equally true of the space launches. Participation in an immense crowd apparently had become a precondition for the American technological sublime. The very size of the audience had become an important part of the meaning of the event. Just as important, this crowd was not violent, agitated, or demanding. It was not the dangerous anonymous crowd that modern people have learned to fear and shun. Rather, at these public moments, another kind of public community briefly becomes visible. As at the rocket launches in Florida, the crowds at the rededication of the Statue of Liberty were much more peaceful than had been anticipated. The New York police had 15,000 men on duty, and many apartment buildings had hired extra security guards, but there was little for them to do.[49] Expectations of violence or a terrorist attack proved unfounded. The *Daily News* headlined "Nightmare Just Never Happened" and concluded that "the start of Liberty Weekend was a pure delight."[50] The *Times* reported that the crowds were "notably serene, almost reverent, as if paying homage to something right-eous and inviolable," that "people moved about more lazily," and that "faces were less taut and many wore radiant smiles."[51] One member of the crowd told the latter reporter "There's a brother-hood and everybody's friendly."[52]

This freely circulating public is not to be confused with the audi-ence for the finale, which consisted of 50,000 paying customers at Giants Stadium in New Jersey. That crowd had filed into tiers of seats, to sit passively and be entertained by Elizabeth Taylor, Willie Nelson, the Four Tops, and several thousand other entertainers. However fine the show, it had no more to do with the sublime than the phenomenon of the rock concert in general, which fore-grounds the performer and is heavily textualized.

In contrast, the history of the technological sublime is that of the movement from word to spectacle, from individual to crowd, from nature to the machine, from substance to electric image. Its history records a shift in emphasis from natural to artificial landscapes, a shift that simultaneously transformed the position of the subject in relation to the sublime object. Whereas Americans had once made pilgrimages to natural wonders in order to sense their place in the natural order, in 1986 they treated technological achievements as signs of political stability and dominance over nature. That attitude had already begun to emerge in the public response to the railroad and the telegraph, but by the late twentieth century the omnipo-tence of engineering had been internalized. It was no longer neces-sary to declare that machines were sublime. Indeed, it was hardly necessary to say anything at all, as President Reagan's short speech suggests. He could simply turn on the lasers and watch the fire-works.

It was not only, as so many commentators pointed out, that show business and commercialization had intruded into the political arena. American politics had so merged with the technological sub-lime that its power now had much to do with special effects. Mark Twain had glimpsed this possibility in the 1880s in *A Connecticut Yankee in King Arthur's Court*, when he had described "acres of human beings" awed and overwhelmed by the technological artistry of Hank Morgan. Morgan's control over technology was translated directly into control over the English nation, not because of physical might but because of the awe induced by his displays of fireworks. Unlike Morgan, Reagan did not need to pretend to magical powers. His public had long since come to depend upon technology for periodic demonstrations of America's greatness and invincibility.

The Americans of 1986 wanted to be awed by means that were not the mysterious proof of wizardry but rather triumphs of the human will. The atomic bomb had failed as a source of sublime because it had short-circuited the sense of transcendence, threatening the ego with annihilation. In contrast, the experience offered in New York Harbor on July 4 was not a terrifying extension of the dynamic technological sublime to its limits; it was a revamped version of the geometrical sublime. It combined the monumentality of the Statue of Liberty with the spectacular effects of a sound and light show produced by Hollywood. The result was an experience of that peculiar hyperspace of the electrical sublime in which objects are highlighted, blacked out, flattened, dematerialized, recolored, and presented as outlines, creating a new vision of the world seen by daylight. When the scene that is transformed contains both natural and man-made landscapes, the two are conflated into an impossible middle landscape. The geometrical sublime turned objects into maps or panoramas of themselves. When electrified, these panoramas lost their three-dimensionality and became a glittering sea of lights no longer amenable to the laws of perspective, requiring that the spectator mentally compare the recollection of the world by daylight with the electrified landscape. When this ceremonial transformation is invested with patriotic music, leaders, celebrities, and vast multitudes, and consummated with fireworks, it becomes an overdetermined experience in which the individual embraces the nation.

The Statue of Liberty ceremony was most affecting for those who were immediately present. Much of its force was lost once it was channeled through television, which left viewers with a miniaturized and tamed version of the spectacle. As a result, the mediation of the experience gave considerable scope for the egotistical sublime, because the subject's expectations could easily outrun the experience of watching TV. Those who observed the technological sublime on television could be unimpressed by a space launch or the rededication of the Statue of Liberty; those who attended in person, like Norman Mailer at the blastoff of Apollo XI, could be impressed and perhaps overwhelmed even if they intended to remain aloof.

The Americans in New York Harbor on July 4, 1986, appreciated the rededication of the Statue of Liberty in much the same spirit in which they had enjoyed the space program. At such events they had become spectators rather than participants in the creation of political culture, and they focused less on the explicit ideology articulated by presidents than on physical sensations induced by an unveiling or a launching. The great crowds lining New York Harbor, like those who journeyed to the Kennedy Space Center, were seeking renewal. Secular pilgrims, they sought a rupture with ordinary experience which they could interpret as a fundamental break with daily life, objectified in the moment of liftoff or the relighting of Liberty's torch. In the politics of perception, such events are far more important than the shopworn phrases of speeches or the limited square of the televised image. A powerful technological synthesis creates a temporary community, investing the spectators with a sense of personal and national transcendence. These new formations of the sublime articulated a distinct political and social relationship to technology. Each event was organized according to a timetable that built toward a planned emotional release. Each induced the sense of unity and the sense of future possibility that are essential ingredients in the achievement of political hegemony. Each encouraged human beings to temporarily disregard divisions in the community. Each merged the individual into the crowd waiting expectantly for a technological transformation. Each appeared to renew the "life-world."

What was missing from this scene? Gone was the power of the word that Jefferson had known to rouse the national spirit. Gone was the citizen who, in Baltimore or New York in the 1820s, actively helped to form the parade or watched friends and neighbors display the cohesion of the community. Gone were the visible links between work and product, between commerce and politics, between technology and human agency. To the public, the technologies that Ronald Reagan put into play by pushing a button at the 1986 event were anonymously spectacular. The rededication of the Statue of Liberty, like the launch of Apollo XI, was nearly empty of the contents of political life required by republicanism.[53] Neither made any reference to a virtuous citizenry, and neither made the once-common claim that a new technology was a moral machine that would elevate the people. Instead, each was a massive display of organizational and technical power. Each encouraged the belief that democracy and state control of advanced technology were compatible. Each event presented a technical achievement as a sign of national greatness, encouraging the citizen to introject this vision of power and make it a fundamental part of personal identity. In this way, each enhanced the technological sublime as a category of American political experience.

The Consumer's Sublime

In 1845 an American visitor to Niagara Falls stood watch on the piazza of his hotel during a thunderstorm. He later wrote: "The finest thing we have seen yet—and one of the grandest I ever saw— was a thunder storm among the waters . . . the other night, which lighted up the two cascades, as seen from our piazzas, with most magnificent effect. They had a spectral look, as they came out of the darkness and were again swallowed up in it, that defies all description and all imagination."[1] This spectacle combined the mathematical sublime of the vast waterfall with the dynamic sublime of thunder and lightning, creating a still more powerful effect.

In a few years, the gorge at Niagara was spanned by Roebling's railway bridge, which many thought as wonderful as the falls. Two generations later, the falls were illuminated by searchlights. These lights—major attractions of the electrical sublime—received their power from Niagara's new hydroelectric power stations, which themselves became tourist attractions and which provided the theme for the 1901 Pan-American Exposition in Buffalo. The industrial sublime also emerged here, for the factories attracted to the site by cheap electricity drew crowds to witness the wonder of mass production. Thus, a succession of natural and technological sublimes appeared at this one site.

The sublime has persisted as a preferred American trope through two centuries. Aside from the specific characteristics of each form, Americans seem to have a particular affinity for sharp discontinuities in sensory experience, for sudden shifts of perspective, for a broken figure of thought in which the quotidian is ruptured. The psychology of the sublime has long been a recurrent figure of thought as Americans have established their relationship to new landscapes and new technologies. This underlying pattern persists in the descriptions of objects as disparate as Niagara Falls, the

Grand Canyon, the Natural Bridge of Virginia, the first railroads, suspension bridges, skyscrapers, city skylines, spectacular lighting, electric advertising, world's fairs, atomic explosions, and the rockets of the space program.

Europeans neither invented nor embraced the vertical city of the skyscraper. Europeans banned or restricted electric signs, and rightly saw the landscape of Times Square as peculiarly American. Europeans did not see atomic explosions as tourist sites. Europeans seldom journeyed to see rockets go into space, but Americans went by the millions.[2] There is a persistent American attraction to the technological sublime.

Not only is the sublime a recurrent figure in American thought; potential sublimity has also justified the creation of new technologies in the first place. This was not merely a matter of the rationalization of new projects, or a simple form of class domination in which an overawed populace acceded to new displays of technological power. There is an American penchant for thinking of the subject as a consciousness that can stand apart from the world and project its will upon it. In this mode of thought, the subject elides Kantian transcendental reason with technological reason and sees new structures and inventions as continuations of nature. Those operating within this logic embrace the reconstruction of the lifeworld by machinery, experience the dislocations and perceptual disorientations caused by this reconstruction in terms of awe and wonder, and, in their excitement, feel insulated from immediate danger. New technologies become self-justifying parts of a national destiny, just as the natural sublime once undergirded the rhetoric of manifest destiny. Fundamental changes in the landscape paradoxically seem part of an inevitable process in harmony with nature.

After centuries of neglect, the sublime—first described in classical antiquity—reemerged in the eighteenth century in tandem with the apotheosis of reason and the advent of industrialization. This broken figure of thought, which permitted both the imagination of an ineffable surplus of emotion and its recontainment, was not based on a perceived opposition between nature and culture in the ancient world or in Enlightenment England. Nor was there an absolute opposition between the natural and the technological sublime in Jacksonian America. Although there is an undoubted tension between what Leo Marx terms "the machine" and "the garden," Americans looked for sublimity in both realms.[3] Each provided a disruption of ordinary sense perception, and each was interpreted as a sign of national greatness.

At first the sublime was understood to be an emotion almost exclusive to the male sex, women being relegated to the realm of the beautiful, as in Kant's early writings. The mature formulation of Kant's aesthetic held, however, that the sublime was a universal emotion, accessible to all. Yet, in practice, the scenes that elicited the sublime were usually more accessible, and accessible at an earlier time, to men than to women. The explorers who first saw Yosemite and the Grand Canyon were men. The engineers who built the railways, the architects that designed the skyscrapers, and the technicians who created the electrical displays were men. Men created virtually all the central objects of the technological sublime that have been the focus of this study. It was the male gaze of domination that looked out from the railway engine, the skyscraper, and the factory manager's office, surveying an orderly domain. Though women were not absolutely excluded, they were marginal.

Yet women played a vital part in the incorporation of the technological object into ordinary life. Once the initial shock of the sublime object had passed, it was domesticated and made familiar through a process of feminization. The railway engine soon became a "she," and train crews spoke of getting "her" to the station on time. The steel mill became an industrial mother, giving birth to molten ingots. Electricity was usually represented as a woman in statuary and paintings.[4] In the 1950s there was a "Miss Atom Bomb" beauty queen. In 1986 the sophisticated engineering of Eiffel and the copper cladding of Bartholdi merged with laser beams and computer-designed fireworks in the rededication of the Statue of Liberty. Feminization transformed the alien into the familiar and implied the emergence of a new synthetic realm in which the lines between nature, technology, and culture were blurred if not erased.

Feminization tamed technology, preparing the subject to experience new forms of the sublime. Despite its power, the technological sublime always implies its own rapid obsolescence, making room for the wonders of the next generation. The railway of 1835 hardly amazed in 1870, and most Americans eventually lost interest in trains (though that particular "romance" lasted longer than most). During each generation the radically new disappeared into ordinary experience. The shape of the life-world, the envelope of techniques and customs that appeared natural to a child, kept changing.[5] Travelers on the first trains, in the 1830s, felt that their disruptive experience was sublime, and remained in a state of heightened awareness, but the sense of an extraordinary break in experience soon disappeared. Similarly, the wonder of 1855 was to see a freight train cross the suspension bridge over Niagara Gorge. In the 1880s came the Brooklyn Bridge; in the 1930s came the Golden Gate Bridge. Each of these structures was longer than its predecessors, sustaining more weight over a greater span.

The logic of the technological sublime demanded that each object exceed its precursors. Similarly, each world's fair erected more elaborate lighting displays than its predecessors, creating more powerful illusions to satisfy the viewers' increasingly voracious appetite. The first skyscrapers, which had fewer than 15 stories, soon were overshadowed by far larger structures. The once dizzying and disorienting view from atop the Flatiron Building became familiar, and higher towers were needed to upset the sense of "normal" spatial relations. Likewise, the first airplanes to appear over Chicago or New York drew millions out into the streets; yet within 20 years daredevil stunts were necessary to hold the crowd's interest, and by 1960 aviation had passed into the ordinary. Each form of the technological sublime became a "natural" part of the world and ceased to amaze, though the capacity and the desire for amazement persisted.

Although the wonder of one generation scarcely drew a glance in the next, each loss of sublimity prepared the subject for more radical shifts in perspective yet to come. In the early 1950s Las Vegas newspapers complained that a recent round of open-air atomic explosions had been too small: "'Bigger bombs, that's what we're waiting for,' said one Las Vegas nightclub owner. 'Americans have to have their kicks.'"[6]

Sublime experience is not merely a matter of vision; all the senses are engaged. Burke noted that, although the eye was often dominant, movement, noise, smell, and touch were also important. A city sounds much different at the top of a skyscraper than on the street below. The wind makes one feel more vulnerable out on the open span of a long bridge. The steam locomotive shook the ground and filled the air with an alien smell of steam, smoke, and sparks; the Saturn rocket did much the same thing on a larger scale. The strong contrast between the silence of a rocket's liftoff and the sudden roar that follows a few seconds later is also a vital element in making that spectacle sublime. The sheer size of the crowd attracted to a technological display further arouses the emotions. In each event, the human subject feels that the familiar envelope of sensory experience has been rent asunder.

Most Americans have not interpreted these powerful experiences according to Kant. In the case of the natural sublime, while some well-read individuals deduced from the magnificence of Niagara Falls and the Natural Bridge the existence of a universal moral sense, many early visitors to these sites felt that they proved the existence of God. Technological marvels were even less likely to be "read" in Kantian terms. Most Americans thought that a man-made rupture in ordinary experience proved the potential omnipotence of humanity. A railroad, a skyscraper, or a hydroelectric dam proclaimed the ever-increasing power of technicians, demonstrating their ability to disrupt what had become normal perception and creating the discontinuity that Americans seemed to crave. This radical break in experience became a necessary epiphany; it reinforced the sense of progress. At the same time, the "Americanness" of this epiphany invigorated nationalistic sentiment.

There were occasional dissenters. Since the 1820s some have greeted new technologies with skepticism. Thoreau suggested that the railroad, for all its admirable celerity, nevertheless exemplified a more general development, in which (in Emerson's words) "things are in the saddle and ride mankind."[7] Later, others felt the skyscraper destroyed the scale of the city and reduced the citizen to a stick figure, a featureless pedestrian seen from a great distance. This opposition suggests a contradiction at the heart of the technological sublime that invites the observer to interpret a sudden expansion of perceptual experience as the corollary to an expansion of human power and yet simultaneously evokes the sense of individual insignificance and powerlessness. One is both the all-seeing observer in a high tower and the ant-like pedestrian inching along the pavement below. One can either be outside, terrified by the speed and noise of the railway, or riding triumphantly over the landscape. One may (like the young Lewis Mumford) experience the electrified skyline as an epiphany, or one may (like Claude McKay) feel overwhelmed by the swirling lights of Times Square. Technologies can be perceived as an extension and affirmation of reason or as the expression of a crushing, omnipotent force outside the self. In this bifurcation, those who have the greatest political, economic, and social power are more likely to find themselves inside the panopticon, surveying the vast surround.

Such a division was less evident during the Jacksonian period, when new technologies were embraced by a cross-section of the body politic. In the 1828 parade that celebrated the construction of the Baltimore and Ohio Railroad, representatives of virtually all trades and every social class took part. (Such events contrasted with unruly, spontaneous parades of the unskilled and the working poor.) During the remainder of the nineteenth century, formal parades organized by local elites gradually deleted the skilled worker and the ordinary citizen, focusing instead on engineers, entrepreneurs, and a few elected officials. Even more important, attention shifted away from human beings to technological objects. The huge Corliss Engine became the icon of the 1876 Centennial Exposition. A decade later, few of the millions who saw the inauguration of Brooklyn Bridge or that of the Statue of Liberty heard the speeches. The word was increasingly replaced by massive representations, a process advanced by the electrical sublime at world's fairs and the Hudson-Fulton Celebration of 1909. Gradually the word ceased to be the carrier of sublimity. Reenactments and displays became important before film and television, which proved ill suited to convey the full force of the technological sublime. It had to be seen, heard, and felt in person. The citizen had become less an active participant than a tourist and a consumer—a fact well understood by corporate exhibitors at the New York World's Fair of 1939 and self-evident by the rededication of the Statue of Liberty in 1986.

The classic location of such experiences was the urban metropolis. It possessed the magnificent railway stations, the artificial cliffs and ravines of the new skyline, the skyscraper panoramas that encouraged fantasies of domain, the de-natured industrial districts, and the phantasmagoria of the Great White Way. Each of these offered the visitor a powerful demonstration of what human beings had accomplished. In the view from a skyscraper, in the artificial totality of the factory zone, or in the special effects of the Great White Way, nature had disappeared. The visitor was by turns terrified, exalted, and reduced to insignificance. The new skyscrapers reduced the individual to an insect or a shadow. The new factories contained vast interiors filled with workers whose sheer numbers demonstrated the insignificance and interchangeability of individuals. Electric advertising did not emphasize producers, engineers, architects, or inventors; rather, it paraded the brand names of mass-produced products. Whereas Independence Day had been a once-a-year commemoration, the factory tour and the skyscraper observation platform were open each day and the Great White Way shone every night.

The paradox of the technological sublime is that it pretends to present a legible materialization of the unrepresentable—as though a physical construction could be infinite, as though the boundless could be bounded, as though the shapeless could be shaped. Slavoj Zizek has called the sublime "a unique point in Kant's system, a point at which the fissures, the gap between phenomenon and Thing-in-itself, is abolished in a negative way, because in it the phenomenon's very inability to represent the Thing adequately is inscribed in the phenomenon itself."[8] For Kant the hurricane was not a concretization of reason, but Americans have long thought that new machines are nothing less than that. For Kant the natural sublime object evokes the feeling of enthusiasm, or a pleasure of a purely negative kind, for it finally concerns a failure of representation. In contrast, the American technological sublime is built on a pleasure of a positive kind, for it concerns an apparently successful representation of man's ability to construct an infinite and perfect world.

In the search for this positive pleasure, a "consumer's sublime" has emerged as Americans shop for new sensations of empowerment. Icons of the natural sublime, such as the Grand Canyon, are subjected to this process, and indeed the national parks have far more visitors today than in the early twentieth century. But increasingly they are appreciated not as signs of nature's immeasurable power and sublimity but as contrasts to a civilization that threatens to overwhelm them. The national park thus becomes a species of theme park, making available a relatively unspoiled nature which society has spared from development. In this reconfiguration, part of the Grand Canyon's attraction and interest is that it is threatened by an array of technologies and therefore should be seen immediately before it is further defiled by the smog blowing in from the west, by acid rain, and by the "light pollution" that makes it hard to see the stars at night. The tourist is told to feel pleased if the weather is clear. A fortunate few fill the quotas for river raft trips, although the rapids are more predictable now than in the nineteenth century because the waters of the Colorado are dammed to the immediate north and south of the park. Glen Canyon Dam impounds the silt and rock that once would have been carried through the gorge and replenished its sand banks. It releases water at an even rate, thus slowing the erosion that carved the canyon in the first place. The moderate flow permits the growth of dense foliage where once the

An NBC camera crew at the Grand Canyon, 1955. Courtesy of National Park Service.

plants would have been scoured off the banks by flash floods. The water also is much colder now, since it comes from the deep lake behind the dam, and several species of fish unique to the area, such as the humpback chub, have been driven back to the warmer tributary streams. At the other end of the canyon, a 1984 government report found, unsupervised "use of the remote portions of the Colorado River from Diamond Creek to Lake Mead is increasing dramatically. Boating upriver from Lake Mead is increasingly popular and difficult to regulate."[9] (Environmental protests protected the canyon from another dam within its boundaries.) Noise pollution is also a problem. Overhead, planes and helicopters carrying tourists from Las Vegas and Phoenix can be heard during half of the daylight hours.[10] Five million people came to the park in 1993, and the visitation rate has doubled every decade since 1919. This massive influx greatly exceeds the park's hotel capacity of 3628 hotel rooms and 329 campsites. One feels fortunate to get a room inside the park, fortunate to have good visibility, fortunate if there are not too many planes overhead. In short, the Grand Canyon, if one is lucky, may appear undisturbed and "natural."

Yet the environmental threats to the canyon and the increased visitation are only one side of the problem. In the past the Grand Canyon invited reflection on human insignificance, but today much of the public sees it through a cultural lens shaped by advanced technology. The characteristic questions about the canyon reported by Park Service employees assume that humans dug the canyon or that they could improve it so that it might be viewed quickly and easily. Rangers report repeated queries for directions to the road, the elevator, the train, the bus, or the trolley to the bottom. Other visitors request that the canyon be lighted at night. Many assume that the canyon was produced either by one of the New Deal dam-building programs or by the Indians—"What tools did they use?" is a common question.[11] These remarks reveal a common belief that the canyon is a colossal human achievement that ought to be improved. Visitors enjoy it in its present form, but some see room for improvements: light shows, elevators, roads, tramways, river-boats, luxury hotels at the bottom. The assumption of human omnipotence has become so common that the natural world seems an extension of ourselves, rather than vice versa. These are by no means the opinions of all visitors, of course. Many want no "improvements" in the Grand Canyon, and some hope to see 90 percent of it set aside as a designated wilderness. Yet even the Park Service, which does not want further development, has asked the managers of Disney World for advice on how to deal with the enormous influx of visitors.

Paradigmatic sites of the natural sublime have long been made more accessible by improved transportation, beginning with the Erie Canal. But increasingly advanced technologies not only get people to the site; they also provide alternatives to seeing it. A maximum of 92 people per night are allowed to stay at the Phantom Ranch, at the bottom of the Grand Canyon; but outside the park 525 people per hour can sit in an IMAX theater and watch *The Grand Canyon—The Hidden Secrets*, a 34-minute film shown on a 70-foot screen with six-track Dolby sound. As an additional attraction, the theater advertises "Native Americans in traditional dress on the staff." During the performance guests are encouraged to "enjoy our fast food, popcorn, ice cream" and other snacks. Why bother to hike into the canyon when all the highlights have been prepack-aged? Similar theaters now serve Niagara Falls and Yosemite. Their gift shops sell videocassettes, postcards, photographs, slides, maps, guides, and illustrated books. These representations cannot extirpate the sublimity of these sites, but they can make it difficult for tourists, who usually remain for only one day, to rid themselves of their pre-visualizations. Many people experience a natural wonder as incomplete, disappointing, or overrated, and even when the Grand Canyon satisfies it often does so because it lives up to its image.

This egotistical sublime is hardly new. In the nineteenth century it was familiar to Margaret Fuller, to Nathaniel Hawthorne, and to the geologist Clarence Dutton. But this response has become pervasive, signaling the triumph of the consumer's vision, which asks that each object be clearly labeled, neatly packaged, and easily appropriated. Each year several people fall into the Grand Canyon because they venture too close to the edge, and their relatives then sue the Park Service because there were no fences or guard rails—as if a terrifying immensity that people journey great distances to see ought to have hand rails and warning signs; as if the "thing in itself" ought to be equipped with a snack bar; as if the unrepresentable could be projected on a 70-foot screen. The notion of a consumer's sublime is ultimately a contradiction in terms. Not only does it deny the existence of transcendental values that might be deduced from the natural object (that denial was always latent within the technological sublime): it also denies the exaltation, the danger, the difficulty, the immensity, and the otherness of the wild and the nonhuman.

The American fascination with the trope of the technological sublime implies repeated efforts at more and more powerful physical manifestations of human reason. For more than a century it was possible to incorporate each new technology into the life-world. The succession of sublime forms—dynamic technological, geometrical, industrial, electrical—fostered a sense of human control and domination that was radically at odds with the natural sublime. From c. 1820 to c. 1945 this incompatibility, rooted in the difference between Kantian reason and technological reason, was muted. However, the atomic bomb manifested a purely material negativity; it was the ultimate dead end of any attempted representation of the technological "thing in itself." The bomb materialized the end of human technique even as it dematerialized its target. Here the shaped and the bounded works of man merged with the shapeless and the boundless, and Americans first glimpsed the death-world that the technological sublime might portend.

Yet history is not a philosopher's argument. It records not logical developments but a mixture of well-reasoned acts, unintended consequences, accidents, shifting enthusiasms, and delusions. Though by the logic of argument the technological sublime "ought to" have gone into terminal decline after Hiroshima, something more complex occurred instead. While most people dreaded atomic weapons, many embraced the space program and most enjoyed spectacles such as the rededication of the Statue of Liberty. At the same time, Disney transformed the exposition, merging the amusement park with magical corporate displays. These changes marked a nostalgic return to the nineteenth-century technological object and to a sanitized recollection of the city before the diaspora into suburbia. Like the Jacksonians who mingled their awe for natural and man-made wonders, late-twentieth-century Americans seem oblivious to the logical impasses posed by the technological sublime, as their leisure travel increasingly demonstrates. They flock to the Kennedy Space Center, the Empire State Building and other skyscrapers, the Gateway Arch, the Smithsonian Air and Space Museum, Disneyland, and, most recently, Las Vegas.

Unlike the Ford assembly line or Hoover Dam, Disneyland and Las Vegas have no use value. Their representations of sublimity and special effects are created solely for entertainment. Their epiphanies have no referents; they reveal not the existence of God, not the power of nature, not the majesty of human reason, but the titillation of representation itself. The genuine ceases to have any special status; the faked, the artificial, and the copy are the stock-in-trade.

Las Vegas, with more annual visitors than Yellowstone, the Grand Canyon, Niagara Falls, the Empire State Building, and Cape Kennedy combined, exemplifies the full development of the consumer's sublime, which first emerged prominently at the 1939 New York World's Fair. The traditional underpinnings of a city do not exist in Las Vegas. Not the forces of trade nor those of industry nor those of religion nor those of government explain its emergence as a metropolitan area with a population of more than 800,000. One-third of all its jobs are in the hotel and gaming industries, and a good many of the inhabitants who do not work in those industries benefit from tourism indirectly.

Las Vegas's popularity is recent. Begun as a watering hole for westward-bound migrants traveling by horse and wagon, it emerged as a raw railroad town in the early twentieth century. Its spectacular growth began after 1931, when the State of Nevada legalized gambling (which was already prevalent).[12] The construction of Hoover

Dam and the establishment of military bases brought in new customers for the casinos. After World War II, the mushrooming population of California and the rest of the Southwest enlarged the
clientele further. As casinos grew more spectacular,[13] the city invented its "strip," a long row of lavish hotels decked with the most elaborate lighting west of Times Square and distinguished by an
approach to architecture later celebrated by Robert Venturi.[14] In
the 1960s air travel made the city accessible to the rest of the
nation, and it became a convention and entertainment center as
well as a gambling mecca. More than 2100 conventions were held
there in 1992, when Las Vegas had more hotel rooms and a higher
occupancy rate than London. The more than 21 million visitors
stayed an average of four days, gambled more than $430 per per-
son, and produced more than $14.6 billion in revenues.[15] Measured
in terms of time or money, Las Vegas has become one of America's
most important tourist sites. Built not on production but consumption, not on industry but play, not on the sacred but the profane,
not on law but chance, Las Vegas is that rupture in economic and
social life where fantasy and play reign supreme, the anti-structure
that reveals the structure it opposes.

Neither the slow economy of the 1970s nor the stagnation of the
early 1990s affected Las Vegas's surging growth. Long one of the
most spectacular sites of the electrical sublime, the city has undergone a multi-billion-dollar development program that includes several skyscrapers and most of the elements once associated with
world's fairs and amusement parks. The new casinos combine
hotels, gaming rooms, and "Disney-quality" theme parks. The 3000-
room Mirage Hotel (1990) cost $630 million. Despite dire predic-
tions that it would never be profitable, it has proved a "cash
machine." The Mirage's $30 million, five-story volcano, which
erupts several times each hour in front of the hotel, was such a success that in 1993 the owner opened, right next to it, another casino
hotel: Treasure Island, where battles between full-size replicas of a
pirate ship and a British frigate are staged. Nearby the 4000-room
Excalibur (1990), an ersatz castle with hotel towers attached, stress-
es a pseudo-Arthurian motif, with jousting matches every evening.
The Luxor (1993) is a 2500-room pyramid-shaped hotel with a
replica of the Sphinx, King Tutankhamen's tomb ("Tut's Hut"),
talking robot camels, and a "Nile River" boat ride under the casino,
"past murals illustrating the rich history of the Egyptian empire."
Circus Circus, which looks like an enormous tent, stages acrobatic

The entrance to the MGM Grand Hotel in Las Vegas.

acts every hour through the night and contains a $75 million climate-controlled amusement park called Grand Slam Canyon. The film *The Wizard of Oz* provides the motif for the billion-dollar MGM Grand Hotel (1993), the largest in the world. The visitor passes between the golden paws of a 109-foot lion and along a yellow brick road to a 60-foot-high Emerald City, beyond which lie a 33-acre theme park, a 15,200-seat sports area, and eight restaurants.[16] Theme casinos and their amusement parks attract whole families rather than only individuals and couples.[17] They combine fantasies and legends with simulated sublime experiences, including volcanoes, waterfalls, Grand Canyon raft trips, and skyscrapers.[18]

The transformation of Las Vegas into a family theme park in the desert makes it the premiere postmodern landscape—a fantasy world for the middle class. It develops the disembodied illusions of the Great White Way into a multidimensional hyperreality, juxtaposing scenes from fairy tales, history, advertising, novels, and movies to create a dreamscape of disconnected signifiers. At the upscale Forum Shops a cornucopia of designer goods are for sale on a simulated Roman street where the vaulted sky passes from morning to evening to night and "Roman gods come to life in the form of 'animatronic' robots."[19] For those with less cash there are ordinary malls near other casinos and factory outlet stores at the end of the strip. Likewise, Las Vegas stages entertainment for every taste, including Broadway musicals, magic shows, dancing girls, tennis tournaments, world championship boxing matches, demonstrations of virtual reality, and rock concerts. It even recycles elements of the zoo, displaying caged white tigers and tanks full of sharks. This is the landscape of capitalist surrealism, where a man-made order seems to replace the natural order entirely. Visitors experience a potpourri of the technological sublime in a synesthesia of lights, heights, illusions, and fantastic representations. All of these enhance the visceral excitement of gambling, which contains the terror of financial catastrophe and the allure of a huge jackpot.

Whereas the older forms of the technological sublime embodied the values of production, and literally embodied the gaze of the businessman as he surveyed a city from the top of a skyscraper or appreciated steel mills from the window of a passing train, Las Vegas validates the gaze of the consumer, who wants not the rational order of work but the irrational disorder of play. Las Vegas manifests the dream world of consumption. The steamboat, the train, or the Corliss Engine has no role here. The city presents itself as a world entirely without infrastructure and beyond the limitations of nature.[20] The dislocation and the "incoherence" of this landscape are its strongest attractions, suggesting that technology faces no natural or economic constraints. Las Vegas offers visitors an intensification of experience, speeding up time and extending itself round the clock. Its lavishly mirrored interiors, like its landscape of electric signs, destroy normal spatial relations. The aesthetic of Times Square merges here with the automotive strip as the consumer moves along the conveyor belt of images from one fantasy to another. Las Vegas fuses the theme park, the shopping mall, and the casino, presenting itself as the central attraction and other nearby attractions—the Grand Canyon, Hoover Dam, the atomic testing site—as secondary.

The consumer's sublime on offer in Las Vegas anthologizes sublime effects. These effects emphasize not solitary communion with nature but experiences organized for the crowd—not the mountaintop but man-made towers; not three-dimensional order but spatial and temporal disorientation. Kant had reasoned that the awe inspired by sublime objects would make men aware of their moral worth despite their frailties. The nineteenth-century technological sublime had encouraged men to believe in their power to manipulate and control the world. Those enthralled by the dynamic technological, geometrical, electrical, or industrial sublime felt omnipotence and exaltation, counterpointed by fears of individual powerlessness and insignificance. Railways, skyscrapers, bridges, lighting systems, factories, expositions, rockets, and bombs were all extensions of the world of production. Through them, Americans inscribed technological systems on consciousness so that the unrepresentable seemed manifest in human constructions.

But in the consumer's sublime of Las Vegas or Disneyland, technology is put to the service of enacting fantasy. The technological sublime had exhorted the observer to dominate and control nature. It celebrated rationality, substituting technique for transcendental reason, celebrating work and achievement. But the consumer's sublime erases production and excites fantasy. It privileges irrationality, chance, and discontinuity. There had always been an element of escapism in the tours to sublime objects, but there was an uplifting lesson to be drawn from Niagara Falls (God's handiwork) or from a railway, a skyscraper, a factory, or a moon rocket (man's accomplishments). The electrical sublime elided the line between nature and culture and, in the process, dissolved the scale and the three-dimensionality of the landscape; yet it could still serve a vision focused on progress. In contrast, the billions of dollars spent in Las Vegas represent a financial and psychic investment in play for its own sake. The epiphany has been reduced to a rush of simulations, in an escape from the very work, rationality, and domination that once were embodied in the American technological sublime.

Notes

Introduction

1. Technical details from E. Cromwell Mensch, *The Golden Gate Bridge: A Technical Description in Ordinary Language* (Government Printing Office, 1935).

2. The first walk over the bridge took place on May 27, 1937. See Allen Brown, *Golden Gate: Biography of a Bridge* (Doubleday, 1965), p. 121.

3. *San Francisco Examiner*, special souvenir edition, May 25, 1987.

4. Émile Durkheim, "The Elementary Forms of the Religious Life," in *Durkheim on Religion*, ed. W. S. G. Pickering (Routledge & Kegan Paul, 1975), p. 153. Durkheim was generally correct in this conclusion, though one certainly must grant that his separation of magic and religion was overdrawn and that he over-stated the power of religion *vis-à-vis* the state. See Robert A. Nisbet, *Émile Durkheim* (Greenwood, 1965), pp. 73–89, 137–141.

5. David I. Kertzer, *Ritual, Politics, and Power* (Yale University Press, 1988), p. 67.

6. The centrality of the technological sublime appears to me to be a distinctive feature of American culture. There obviously remains a great deal of work to be done on how racial and ethnic minorities view the objects that have been central to this tradition in the United States. Nevertheless, I attempt in this book to reawaken discussion of the idea of national character, and I use the term 'American' fully aware of the recent emphasis on multiculturalism in the American academy. A decade of living in Europe, with frequent visits to the United States, has convinced me that a core American culture exists, even if it is not fashionable to admit that this is so. The commonalities of this national culture are rooted in material life, notably in large systems such as the form of cities, the dependence on the automobile, the massive distribution of hand-guns, and the centrality of television. The denial that a core American culture exists prevents effective analysis of racial injustice, economic inequality, and other social problems.

7. On the power of the crowd, see also Émile Durkheim, *The Rules of Sociological Method* (Free Press, 1938), pp. 4–5.

8. These earlier uses of 'technological sublime' occur in the following: Perry Miller, *The Life of the Mind in America from the Revolution to the Civil War* (Harcourt, Brace, and World, 1965), pp. 295–306; Leo Marx, *The Machine in the Garden: Technology and the Pastoral Ideal in America* (Oxford University Press,

1965), pp. 195–207, 230–231; John Kasson, *Civilizing the Machine: Technology and Republican Values in America, 1776-1900* (Penguin, 1977), pp. 162–172; Barbara Novak, *Nature and Culture* (Oxford University Press, 1980), pp. 166–167; Roland Marchand, *Advertising the American Dream* (University of California Press, 1985), p. 280; John F. Sears, *Sacred Places American Tourist Attractions in the Nineteenth Century* (Oxford University Press, 1989), pp. 182, 192, 201.

9. My efforts to expand the investigation began in my previous book, *Electrifying America: Social Meanings of a New Technology* (MIT Press, 1990). Four other recent studies have also been useful. Two deal with art and landscape: Raymond O'Brien, *American Sublime: Landscape and Scenery of the Lower Hudson Valley* (Columbia University Press, 1981); Elizabeth McKinsey, *Niagara Falls: Icon of the American Sublime* (Cambridge University Press, 1985). Two concern poetry: Mary Arensberg, ed., *The American Sublime* (State University of New York Press, 1986); Rob Wilson, *American Sublime: Genealogy of a Poetic Genre* (University of Wisconsin Press, 1991).

10. The sights and sounds when thousands of men do battle rank among the most awesome and terrifying, and they deserve separate investigation. Warfare on the whole would appear to belong within the dynamic sublime, although it does not permit the safety that Kant assumes in his analysis. The experience of particular battles is not repeatable, as is a visit to the Grand Canyon or Niagara Falls, but it might be compared to a powerful storm with thunder and lightning. Separate analysis would have to be given to bombing runs, to fire fights at night, to war at sea, to the experience of seeing a great number of soldiers, horses, cannon, etc., and to invasions (particularly D-Day on the beaches at Normandy).

11. Edmund Burke, *A Philosophical Enquiry into the Origin of Our Ideas of the Sublime and Beautiful* (Oxford University Press, 1990), p. 160. (Henceforth this work will be abbreviated as *Enquiry*.)

12. Durkheim, "Elementary Forms of the Religious Life," p. 154.

13. O'Brien, *American Sublime*, pp. 279–280.

14. Zoja Pavlovskis, *Man in an Artificial Landscape: The Marvels of Civilization in Imperial Roman Literature* (Leiden: Mnemosyne, Biblioteca Classica Batava, 1973), p. 1.

15. Ibid., pp. 33, 20 and passim.

16. See p. 436 of Nicholas Taylor, "The Awful Sublimity of the Victorian City: Its Aesthetic and Architectural Origins," in *The Victorian City, Images and Realities*, vol. 2, ed. H. J. Dyos and M. Wolff (Routledge & Kegan Paul, 1973).

17. "The Addition to the Capitol, July 4, 1851," in *The Papers of Daniel Webster, Speeches and Formal Writings*, vol 2, ed. C. M. Wiltse (University Press of New England, 1989).

18. Walt Whitman, "Song of the Exposition," in *Complete Poetry and Selected Prose by Walt Whitman*, ed. J. E. Miller, Jr. (Houghton Mifflin, 1959).

19. In adjoining passages Lyotard speaks not of terror induced by a natural object but of the terror produced by a police state, which demands a certain

form of realism. See Jean-François Lyotard, *The Postmodern Condition* (University of Minnesota Press, 1984), pp. 75–78.

20. His 'sublime' merits another name. Herbert Grabes has put forward an explanation that clarifies the distinction between Kant's aesthetic and Lyotard's. Grabes finds the term 'sublime' inappropriate to name the viewer's experience of contemporary art, and suggests that Lyotard is dealing with an aesthetic problem that he ingeniously names the "aesthetic of the strange." The effort to "make strange" has been a preoccupation of the avant-garde for most of the twentieth century and is a central characteristic of postmodern art. Grabes suggests "an aesthetics of the strange" as the appropriate name for Lyotard's 'sublime'. Grabes points out that in selecting this name he is "following Kant, who names each aesthetic according to what induces the particular aesthetic experience through sense perception whilst at the same time being a subjective phenomenon. Like the beautiful and the sublime, the strange is perceived 'out there' by the senses, and at the same time is clearly the result of an 'inward' subjective act of cognition. This is quite important, for it is the subjective element which also in this case may allow a 'transcendental deduction', that is, justify a judgement of taste with a claim to general validity." (Herbert Grabes, "Pleasure and the Praxis of Processes: A Transcendentalist Aesthetic of the Strange," paper delivered at Würzburg University, 1991)

21. In his later writings Lyotard takes more interest in the role of technology in "making strange" the world, and in this sense his concerns and those of American Studies have converged. Nevertheless, postmodernist writing on technology focuses on recent times and takes a simplified and schematic view of the past that has little analytic force or precision. One can defend Lyotard's work on other grounds, but it has little usefulness in analyzing the interplay between the imagination and material culture, particularly before 1920.

22. Editorial, *San Francisco Examiner*, May 23, 1987.

Chapter 1

1. For a useful short introduction to the sublime see Rasario Assunto, "The Concept of the Sublime," in *Encyclopedia of World Art*, vol. 14 (McGraw-Hill, 1967). One might sample the literary debate about the sublime by reading the following: Steven Knapp, *Personification and the Sublime: Milton to Coleridge* (Harvard University Press, 1985); Frances Ferguson, "The Sublime of Edmund Burke, or the Bathos of Experience," *Glyph* 8: 62–78; Hugh J. Silverman and Gary E. Aylesworth, eds., *The Textual Sublime. Deconstruction and Its Differences* (State University of New York Press, 1990).

2. David B. Morris, *The Religious Sublime: Christian Poetry and Critical Tradition* (University of Kentucky Press, 1972), p. 1.

3. Marjorie Hope Nicolson, *Mountain Gloom and Mountain Glory: The Development of the Aesthetics of the Infinite* (Cornell University Press, 1959), pp. 321–323.

4. Ibid., p. 32.

5. Howard Mumford Jones, *O Strange New World* (Viking, 1967), p. 70.

6. Ibid., p. 251.

7. Raymond O'Brien, *American Sublime: Landscape and Scenery of the Lower Hudson Valley* (Columbia University Press, 1981), p. 106.

8. S. Dorsh, ed., *Aristotle, Horace, Longinus: Classical Literary Criticism* (Penguin, 1965), pp. 107–108. I will refer to the author as Longinus in order to avoid clumsiness.

9. This conception of the universality of the sublime was later fundamental to Kant's definition, as he argued that everyone was capable of recognizing the sublimity of certain natural scenes. As applied to the arts, however, the sublime combines a democratic idea of its universality with an elitist view of its origin. Many can discern and appreciate a sublime discourse, but only a few can create one.

10. See Charles L. Sanford, *The Quest for Paradise: Europe and the American Moral Imagination* (University of Illinois Press, 1961), p. 137.

11. Alchemy was not, of course, the mere search for a method to produce gold from base metals, but a mystical activity. Its practitioners were bent on penetrating the secrets of matter in a project of self-perfection. For discussion see Mircea Eliade, *The Forge and the Crucible* (Harper, 1962). The sublime might also be understood as a secular version of sacred space, as discussed in Eliade's books *The Sacred and the Profane* (Harcourt, Brace, 1959) and *The Myth of the Eternal Return* (Princeton University Press, 1954). Eliade elaborated the ideas expressed in these two basic works in the following two decades.

12. Mulford Q. Sibley, *Political Ideas and Ideologies: A History of Political Thought* (Harper, 1970), p. 347.

13. For example, Burke (*A Philosophical Enquiry into the Origin of our Ideas of the Sublime and Beautiful* [Oxford University Press, 1990], p. 136) writes: "When we have before us such objects as excite love and complacency, the body is affected, so far as I could observe, much in the following manner. The head reclines something on one side; the eyelids are more closed than usual, and the eyes roll gently. . . . All this is accompanied with an inward sense of melting and languor. These appearances are always proportioned to the degree of beauty in the object." Burke goes on to cite Newton's work on optics. For other examples see ibid., pp. 137–138 and passim.

14. Nicolson, *Mountain Gloom and Mountain Glory*, p. 62.

15. Ibid., pp. 97–99. The new, sublime view of nature was not static but potentially millenial. If the earth was not a fallen world but a developing and changing landscape that reflected the creator, then the same complex of ideas that led to the revaluation of mountains could imply human progress and even the perfectibility of man. Calvin hardly drew this conclusion, but it emerged later as part of the shift in sensibility during the Enlightenment.

16. Cited in Samuel Holt Monk, *The Sublime: A Study of Critical Theories in Eighteenth Century England* (University of Michigan Press, 1960), p. 87.

17. See John Watson, *The Philosophy of Kant Explained* (James Maclehose and Sons, 1908; Garland, 1976), pp. 490–491. The translation quoted is from book II of the *Critique of Aesthetic Judgement.*

18. John T. Goldthwait, translator's introduction to Immanuuel Kant, *Observations on the Feeling of the Beautiful and Sublime* (University of California

Press, 1960), p. 37. Except where explicitly noted, all references to Kant's philosophy of the sublime refer to his *Critique of [Aesthetic] Judgement* (see n. 21 below) rather than to this early work.

19. Watson (n. 17 above) summarized Kant's view this way: "The sublime is not to be found in nature, but only in our ideas. This may be expressed by saying that the sublime is that in comparison with which all else is small. It is plain that, since nature is simply the sum of sensible phenomena, nothing in it can be judged to be infinitely great or infinitely small."

20. Kant, in Watson, pp. 498–499.

21. Immanuel Kant, *Critique of Judgement*, [1790] (Oxford University Press, 1952), p. 504.

22. Elizabeth McKinsey, *Niagara Falls: Icon of the American Sublime* (Cambridge University Press, 1985), p. 251.

23. Burke, *Enquiry*, p. 53. He continues: "Astonishment, as I have said, is the effect of the sublime in its highest degree; the inferior effects are admiration, reverence and respect." Cody's views, undersigned by thirteen members of his party, were printed in Captain John Hance's *Personal Impressions of the Grand Canyon of the Colorado River* (Whitaker and Ray, 1899), p. 63. The word 'sublime' is used throughout this volume of reactions, and in many cases the travelers make a point to say that they have been all over Europe and seen nothing remotely comparable. For example, one Mrs. G. Peters wrote: "I have visited the whole world. I travel nine months in the year. I have never seen anything so grand as a sunset view of the Grand Canyon. . . ." (p. 52) Nationalism is a recurrent theme. Another writer declared: "Doubtless, God might have made something more wonderful or more magnificent, but doubtless he never did. America for Americans." (p. 90)

24. Joseph Wood Krutch, *The Grand Canyon* (William Sloane Associates, 1958), pp. 4–5.

25. The same reaction is common in descriptions of Niagara Falls. For example, Charles Bigot, on p. 155 of *De Paris au Niagara: Journal de voyage d'une delegation* (Paris, 1887), says: "La Niagara n'est pas seulement grand, imposant, magnifique, il est terrible, il est formidable, il est effroyable. . . . C'est une puissance de la nature déchâinée, aupres de laquelle l'homme n'est rien."

26. Burke, *Enquiry*, pp. 58–59.

27. Ibid., p. 72.

28. Colin Fletcher, *The Man Who Walked Through Time* (Knopf, 1968), p. 5.

29. Frank Waters, *The Colorado* (Rinehart, 1946), pp. 373–374.

30. Nevertheless, not everyone acts according to Kant's theory. At the rim of the Grand Canyon some visitors declare it must be man-made. A few connect it to the dam building or the WPA projects of the Roosevelt Administration. The assumption seems to be that if human begins can dig the Panama Canal or build the pyramids they can do anything. In fact, the great pyramids would disappear entirely if placed inside the Grand Canyon, which represents a greater excavation than all the world's canals combined.

31. Clarence E. Dutton, *Tertiary History of the Grand Cañon District* (Government Printing Office, 1882; Peregrine Smith, 1977), p. 140.

32. "Ash Cloud," *Yakima Herald-Republic,* May 21, 1980.

33. "Escape: Photographer Flees Ash Cloud," *Seattle Times,* May 23, 1980.

34. Editorial, *Oregonian,* May 25, 1980; "Survivors Recall Scenes of Horror," *Columbian,* May 19, 1990.

35. "Pilot Tells of 'Balls of Fire,'" *Yakima Herald-Republic,* May 19, 1980.

36. Editorial, *Rocky Mountain News* (Denver), May 21, 1980.

37. Photographs reproduced in *Yakima Herald-Republic,* May 22, 1980; reminiscence in same paper, May 24, 1980. For other first-person accounts see "Explosive Plume Spreads Terror" and "Survivors Recall Scenes of Horror," *Columbian,* May 19, 1990.

38. Thomas Weiskel, *The Romantic Sublime: Studies in the Structure and Psychology of Transcendence* (Johns Hopkins University Press, 1976), pp. 24, 49.

39. *Yakima Herald-Republic,* May 19, 1980.

40. Charles Mason Dow, *Anthology and Bibliography of Niagara Falls,* vol. 1 (State of New York, 1921), p. 152.

41. Sarah Margaret Fuller Ossoli, *Summer on the Lakes, in 1843* (Little, Brown, 1844), pp. 1–13.

42. McKinsey, p. 152.

43. Dutton, p. 142. From my own observation as an American living abroad, some cultures seem to value the sublime more than others. To make a bald generalization that would require a chapter to develop and refine, I suggest that Danish society values the beautiful, the small, the cozy, and the comfortable. It may well be that I became aware of the American sublime as a topic for investigation because of the absence of something equivalent in Denmark.

Chapter 2

1. For documentation of this point see Lewis O. Saum, *The Popular Mood of Pre-Civil War America* (Greenwood, 1980), pp. 193–194.

2. Saul K. Padover, ed., *The Complete Jefferson* (Books for Libraries Press, 1969), p. 579. For discussion see Gary Wills, *Inventing America* (Doubleday, 1978). Cuvier, born in 1769, was only 16 when Jefferson's book was published in Paris. Jefferson later knew of Cuvier's scientific work but did not adopt his nomenclature or his system of classification in natural history, preferring that of Linnæus.

3. Padover, pp. 581–582.

4. Sarah Margaret Fuller Ossoli, *Summer on the Lakes, in 1843* (Little, Brown, 1844), p. 3.

5. Reprinted in Charles Mason Dow, *Anthology and Bibliography of Niagara Falls,* vol. 1 (State of New York, 1921), pp. 233–234.

6. Ibid., p. 161. Original source: James Flint, *Letters from America* (W. and C. Tait, 1822).

7. *Table Rock Album and Sketches of the Falls and Scenery Adjacent* (E. R. Jewett, 1859), pp. 12, 14, 21, 39.

8. Ibid., pp. 44, 27.

9. See Ernest Lee Tuveson, *Redeemer Nation* (University of Chicago Press, 1968), pp. 149–171.

10. One tourist said it cost him $8.00 in fees to see the falls during the course of two days. See Dow, pp. 1132–1133.

11. The area remains something of a "Coney Island" to this day. See Rob Shields, *Places on the Margin* (Routledge, 1991), pp. 132–134.

12. Speech reproduced in Dow, *Anthology and Bibliography of Niagara Falls*, vol. 1, p. 1120.

13. Cited by Saum, *Popular Mood of Pre-Civil War America*, p. 177.

14. Ibid., p. 197.

15. "Miscellanies," *New England Magazine*, April 1832, pp. 357–358.

16. Cited in Louis Legrand Noble, *The Course of Empire and Other Pictures of Thomas Cole, N. A.: Life and Works* (New York, 1853), p. 94.

17. Charles L. Sanford, *The Quest for Paradise: Europe and the American Moral Imagination* (University of Illinois Press, 1961), pp. 138, 145.

18. John F. Sears, *Sacred Places: American Tourist Attractions in the Nineteenth Century* (Oxford University Press, 1989), p. 4.

19. Wilbur Zelinsky, *Nation Into State: The Shifting Symbolic Foundations of American Nationalism* (University of North Carolina Press, 1988), p. 186.

20. Earl Pomery, *In Search of the Golden West* (Knopf, 1957), pp. 51–60; Alfred Runte, *National Parks: The American Experience* (University of Nebraska, 1990), pp. 11–12; Victor and Edith Turner, *Image and Pilgrimage in Christian Culture* (Columbia University Press, 1978), p. 241. In 1992 more than 250 million people visited the national parks.

21. From Samuel Kercheval, *The History of the Valley of Virginia* (first published in 1833, second edition 1850; reprinted in Edmund Pendleton Tompkins and J. Lee Davis, *The Natural Bridge and Its Historical Surroundings* (Natural Bridge of Virginia, Inc., 1939).

22. Clifton Johnson, *Highways and Byways of the South* (Macmillan, 1904), p. 225.

23. Ibid., pp. 226–227.

24. In 1987 3. 5 million people visited the Grand Canyon—about 10,000 per day (*Statistical Abstract of the United States, 1989*). When I visited the park in 1993 the projected attendance for the year was 5 million.

25. Cited in Tompkins and Davis, *The Natural Bridge and Its Historical Surroundings*, pp. 25–26.

26. Cited in ibid., p. 139. For more examples see pp. 9–10, 140–146.

27. Perry Miller, *The Life of the Mind in America from the Revolution to the Civil War* (Harcourt, Brace and World, 1965), p. 65. Miller died after completing little more than a quarter of the projected work, which was to begin with a preface entitled "The Sublime in America."

28. Cited in Elizabeth McKinsey, *Niagara Falls: Icon of the American Sublime* (Cambridge University Press, 1985), p. 170.

29. Annette Kolodny, *The Land Before Her: Fantasy and Experience of the American Frontiers, 1630–1860* (University of North Carolina Press, 1984).

30. Immanuel Kant, *Observations on the Feeling of the Beautiful and Sublime* (University of California Press, 1960), pp. 78–79.

31. Cited in Michael L. Smith, *Pacific Visions: California Scientists and the Environment, 1850–1915* (Yale University Press, 1987), p. 77.

32. Ibid., p. 84, 87.

33. Leo Marx, *The Pilot and the Passenger: Essays on Literature, Technology, and Culture in the United States* (Oxford University Press, 1988), pp. 30–31, n. 17.

34. "Celebration of the Completion of the Erie Canal," *Albany Daily Advertiser,* October 11, 1825.

35. "Celebration" (October 15, 1825) and "The Grand Canal Celebration" (November 8, 1825), *Utica Sentinel.*

36. Ronald E. Shaw, *Erie Water West: A History of the Erie Canal, 1792–1854* (University of Kentucky Press, 1966), p. 186.

37. Cadwallader David Colden, *Memoir, prepared at the request of the City of New York at the Celebration of the Completion of the New York Canals* (Corporation of New York, 1825), p. 87.

38. Shaw, *Erie Water West,* pp. 189–190.

39. This American form of republicanism was not a blind devotion to country: "The questioning of what America stands for is not un-American; it is, paradoxically, part of the very core of what it means to be a patriotic American. To a degree rarely seen in history we are asked to love our country while at the same time purifying or ratifying our ardor by cultivating an awareness that our country may not be the best, certainly not the best conceivable, political order." (Thomas Pangle, *The Spirit of Modern Republicanism: The Moral Vision of the American Founders and the Philosophy of Locke* [University of Chicago Press, 1988]) See also Sean Wilentz, *Chants Democratic: New York City and the Rise of the American Working Class, 1788–1850* (Oxford University Press, 1984), p. 92.

40. "The Grand Canal Celebration," *Utica Sentinel,* November 8, 1825.

41. See J. B. Jackson, *Landscapes* (University of Massachusetts Press, 1970), p. 2.

42. Thomas Jefferson, *Notes on the State of Virginia,* ed. W. Peden (University of North Carolina Press, 1955), p. 175. In his view, nature was essentially good, and the farmer who owned his land was comparatively immune to corruption—in contrast to city dwellers.

43. Bryan Wolf, "When Is a Painting Most Like a Whale? Ishmael, Moby Dick, and the Sublime," in *New Essays on Moby Dick,* ed. R. H. Brodhead (Cambridge University Press, 1986), p. 155.

44. Leo Marx, *The Machine in the Garden* (Oxford University Press, 1965), passim.

45. Saum, *Popular Mood of Pre-Civil War America,* p. 181.

46. Jackson's second annual message, cited by Roderick Nash (*Wilderness and the American Mind,* third edition [Yale University Press, 1982], p. 41). Jackson was reformulating the common view of settlers in previous generations. See

Raymond O'Brien, *American Sublime: Landscape and Scenery of the Lower Hudson Valley* (Columbia University Press, 1981), p. 80. One of Jackson's contemporaries, Emily Lewis, wrote that Americans could "behold this thickly settled country, towns, villages, and even cities, where but a few years ago, were spread forests, and where the Aborigines wandered unmolested with here and there a wigwam to tell that human beings inhabited the wilds of America" (Saum, p. 187).

47. Marx, *Machine in the Garden*, p. 220.

48. "American Steam Navigation," *Hunt's Merchants' Magazine*, February, 1841, p. 14.

49. Donald Pease, "Sublime Politics," in *The American Sublime*, ed. M. Arensberg (State University of New York Press, 1986), p. 46.

50. Albert Boime, *The Magisterial Gaze: Manifest Destiny and American Landscape Painting, c. 1830–1865* (Smithsonian Institution, 1991), p. 75.

51. Cited in Smith, *Pacific Visions*, p. 84.

52. Jefferson has often been considered anti-industrial, but on closer inspection he fits into the general pattern. Like many of his contemporaries, he feared that the factory system, which he had seen firsthand in England, would have a pernicious effect on democracy. Yet his penchant for machinery and gadgets is evident in his home, Monticello, where he ran, in addition to a plantation, a small textile mill and a nail factory that produced 10,000 nails a day. Like many of his contemporaries, Jefferson hoped that American industrialization would lead not to increased urbanization but to decentralization.

53. John Kasson, *Civilizing the Machine: Technology and Republican Values in America, 1776–1900* (Penguin, 1977), pp. 24, 50.

54. George S. White, *Memoir of Samuel Slater, Philadelphia, 1836*, reprinted in *The New England Mill Village, 1790–1860*, ed. G. Kulik et al. (MIT Press, 1982), p. 355.

55. G. A. Pocock, "Civic Humanism and its Role in Anglo-American Thought," in Pocock, *Politics, Language, and Time* (Atheneum, 1973). Pocock's assumption that one political paradigm dominated the eighteenth century has in turn come under attack—it was an age of political debate, after all—but his work has undermined the simple whig model, in which Lockean theory leads to political revolution and economic progress. Instead, republicanism forged an ideology of social stability, which proposed a model of the good citizen grounded not in economic self-interest but in public virtue. In the United States, Lockean individualism was by no means as absent as Pocock and his school would have it. As Issac Kramnick (*Republicanism and Bourgeois Radicalism: Political Ideology in Late Eighteenth-Century England and America* [Cornell University Press, 1990], p. 168) has put it, "The revisionists have informed us of the continuity and hold of older political and cultural ideals, competing with a Lockean emphasis on natural rights and individualism, on the eighteenth century mind. But in its efforts to free the entire eighteenth century of Locke. . . . This broom has also swept away much that is truth." Middle-class reformers in England, like Locke, "divided society into an industrious, enterprising middle beset by two idle extremes," and they found legitimate differences in wealth that originated in "a difference in industry" (ibid., pp. 192–194).

56. Joyce Appleby, "Introduction: Republicanism and Ideology," *American Quarterly* 37 (1985), no. 4, p. 467.

57. Shaw, *Erie Water West*, p. 187.

58. In Boston, for example, the anniversary of the Boston Massacre, March 5, was the chief holiday commemorating the revolution until 1783, when the city decided to shift the emphasis to July 4. The choice of that date was debatable, but it was eventually accepted as "the" day, even by Adams, Jefferson, and other participants in the event (Daniel Boorstin, *The Americans: The National Experience* [Random House, 1965], pp. 377–380). On July 2, 1776, the Continental Congress had decided by voice vote that the country ought to be independent. Two days later they approved the text of the Declaration, but it was not signed by the members until August 2.

59. Cited by Boorstin, *The Americans*, p. 390.

60. Cited by Harry Alpert,"Durkheim's Functional Theory of Ritual," in *Émile Durkheim*, ed. R. A. Nisbet (Greenwood, 1965), p. 138.

61. Adams, cited by David I. Kertzer, *Ritual, Politics, and Power* (Yale University Press, 1988), p. 163. See Susan G. Davis, *Parades and Power: Street Theatre in Nineteenth-Century Philadelphia* (University of California Press, 1988).

62. "The Addition to the Capitol, July 4, 1851," in *The Papers of Daniel Webster, Speeches and Formal Writings*, vol, 2, ed. C. M. Wiltse (University Press of New England, 1989). Webster began his career as an opponent of protective tariffs, but by the 1820s, as New England industrialized, he had begun to reverse his position. See Clarence Mondale,"Daniel Webster and Technology," in *The American Culture*, ed. H. Cohen (Houghton Mifflin, 1968).

63. Cited by George William Douglas, *The American Book of Days* (H. W. Wilson, 1948), p. 374.

Chapter 3

1. Jacob Bigelow, *Elements of Technology* (Hilliard, Gray, Little and Wilkins, 1828). Quotations from p. 4 of second edition (1831).

2. On the engineer as hero at the Erie Canal celebration, see the notes to chapter 2 above. Also see Raymond H. Merritt, *Engineering in American Society, 1850–1875* (University of Kentucky Press, 1969). For the late nineteenth century, see Elizabeth Ammons, "The Engineer as Cultural Hero and Willa Cather's First Novel, *Alexander's Bridge*," *American Quarterly* 38 (winter 1986), p. 5. For the twentieth century, see Cecelia Tichi, *Shifting Gears: Technology, Literature, Culture in Modernist America* (University of North Carolina Press, 1987).

3. On the ideology of progress, see "A Sketch of the Rise, Progress, and Influence of the Useful Arts," *American Review* 5 (January 1847), pp. 87–95, and Leo Marx, *The Machine in the Garden* (Oxford University Press, 1965), p. 150. The argument of this chapter is obviously related to Marx's well-known discussion of the development of pastoralism in American literature. His definition of the ideal of the middle landscape between nature and civilization is a fertile and important idea, but it is not my point here.

4. Timothy Sullivan, *North American Review* 33 (January 1831), pp. 122–136.

5. *The Railroad Jubilee, An Account of the Celebration Commemorative of the Opening of Railroad Communication between Boston and Canada* (J. H. Eastburn, 1852), pp. 171–172.

6. Michel Chevalier, *Society, Manners, and Politics in the United States*, ed. J. W. Ward (Doubleday, 1961), pp. 71, 73–74.

7. Stewart H. Holbrook, *The Story of American Railroads* (Crown, 1947), p. 43.

8. Frederick C. Gamst, "America's First Railroad, on Boston's Beacon Hill," *Technology and Culture* 33 (1992), no. 1, p. 79.

9. Gamst argues that steam railroads ought to be seen as part of a continuous stream of innovations that began with medieval mining and continue in our own time. In this view, the motive force employed is only one aspect of the system. While this is an entirely defensible view when one is considering railroads as a technological system, the American public's perception of railroads is quite another matter. In 1828 there were fewer than 25 miles of non-steam railway (usually horse-drawn or gravity-powered) in the entire United States, so the American experience of railroads was extremely limited before the Charleston & Hamburg operated the first steam engine that drew cars over a track in 1830. Other lines under construction at the same time as the Baltimore & Ohio were the Delaware & Hudson, the Mohawk & Hudson, the Boston & Lowell, the Chesterfield, and the New York & Harlem. See Holbrook, *Story of American Railroads*, p. 23.

10. U.S. Bureau of the Census, *Sixteenth Census, 1940* (Government Printing Office, 1942), pp. 32–33.

11. See John Kasson, *Civilizing the Machine* (Penguin, 1977), pp. 50–51.

12. Like Durkheim, Burke (*Enquiry*, pp. 75–76) noted the powerful collective effect of a crowd, which others have since observed during strikes, athletic events, and political rallies: "The shouting of multitudes . . . by the sole strength of the sound, so amazes and confounds the imagination, that in this staggering, and hurry of the mind, the best established tempers can scarcely forbear being bourn down, and joining in the common cry, and common resolution of the crowd."

13. Col. J. Thomas Scharf, *The Chronicles of Baltimore; Being a Complete History of Baltimore Town and Baltimore City from the Earliest Period to the Present Time* (Turnbull Brothers, 1874), p. 426.

14. For an account of the first non-steam railroad in the United States, built in 1805, see Frederick C. Gamst, "America's First Railroad, on Boston's Beacon Hill," *Technology and Culture.* 33 (1992), no. 1.

15. *Niles' Weekly Register,* July 5, 1828, p. 297.

16. For the full text see Edward Hungerford, *Story of the Baltimore and Ohio Railroad, 1827–1927* (Putnam, 1928), p. 46. For a more recent history of the line, see John F. Sower, *History of the Baltimore and Ohio Railroad* (Purdue University Press, 1987).

17. James A. Ward, *Railroads and the Character of America, 1820–1887* (University of Tennessee Press, 1986), pp. 14–15.

18. Ibid., p. 17.

19. Ibid.

20. Scharf, *Chronicles of Baltimore*, p. 426.

21. *Niles' Weekly Register*, July 12, 1828, p. 324. The initial dialogue may also be found on p. 42 of Hungerford's *Story of the Baltimore and Ohio Railroad*.

22. That the B&O could heal sectionalism was a central theme in the speeches of the great railway celebration of 1857, held to mark the connection of Baltimore with New Orleans and the Mississippi River.

23. Scharf, *Chronicles of Baltimore*, p. 427.

24. *Niles' Weekly Register*, May 15 and 22 and June 11, 1830.

25. B&O passenger statistics from John F. Stover, *American Railroads* (University of Chicago, 1961), p. 32.

26. *Letters of a Man of the Times to the Citizens of Baltimore* (Sands and Nielson, 1836), p. 4.

27. The citizens of Pittsburgh tried to have the B&O's charter extended, as no other eastern railroad had reached them; however, Philadelphia investors won a complicated battle, the outcome of which was that the B&O's charter would be renewed only if the Pennsylvania Railroad failed to build 30 miles of track toward Pittsburgh by July 30, 1847. See Philip S. Klein and Ari Hoogenboom, *A History of Pennsylvania* (McGraw-Hill, 1973), pp. 151–152.

28. There was also an outdoor line operating from a coal mine to the Delaware & Hudson Canal.

29. See John F. Sears, *Sacred Places: American Tourist Attractions in the Nineteenth Century* (Oxford University Press, 1989), p. 194.

30. Even though a tunnel completed in 1870 eliminated the need for the gravity line to haul coal, the switchback railroad continued to carry tourists until 1933.

31. S. Henry, *Trains* (Bobbs-Merrill, 1947), p. 9.

32. *Niles' Weekly Register*, April 23, 1831, p. 136.

33. Ibid.

34. Ralph Waldo Emerson, "Nature," in *Complete Works*, vol. 1 (Houghton Mifflin, 1903), p. 51.

35. Cited in Oliver Jensen, *Railroads in America* (American Heritage, 1975), p. 22. Many of the first passengers on other lines burned their clothing as well. Lewis Saum (*The Popular Mood of Pre-Civil War America* [Greenwood, 1980], p. 185) notes the enthusiasm of a traveler from Maine to Cleveland who wrote home that "this flying along at the rate of 30 to 50 miles an hour is a lively business to what a wagon with a lazy horse is."

36. Chevalier, *Society, Manners, and Politics in the United States*, p. 135.

37. Jensen, *Railroads in America*, p. 25.

38. John F. Stover, *The Life and Decline of the American* Railroad (Oxford University Press, 1970), p. 16.

39. A. Howland, *Steamboat Disasters and Rail-Road Accidents in the United States* (Worcester, 1840).

40. Cited in Francis Klingender, *Art and the Industrial Revolution*, ed. A. Elton (Augustus M. Kelley, 1968), p. 147.

41. On the American "moral view of railroads" see Ward, *Railroads and the Character of America*, pp. 60–63.

42. Charles Dickens, *American Notes* (1842; Fromm, 1985), pp. 63–64.

43. Both examples are cited on p. 172 of Kasson's *Civilizing the Machine*. Chevalier (*Society, Manners, and Politics in the United States*, p. 136) used the dragon metaphor in describing a train seen after sundown: "The chimney yawned out at the top like the gaping mouth of some great beast; it threw out thousands of sparks. Although still at a distance, the noise of the quick breathing of the cylinders could be heard distinctly. In the darkness, in so wild a place, in the bosom of a vast wilderness and the midst of a profound silence, it was necessary either to be acquainted with mechanics, or to . . . believe this flying, panting, flaming machine was a winged dragon vomiting forth fire and flame."

44. Kasson, *Civilizing the Machine*, p. 256.

45. Such laughter is an important cultural marking device, and there were jokes of varying sophistication. Although virtually anyone could understand what a railroad or a steamboat was in a general sense, only the emerging class of engineers understood the technical fine points of its operation. Engineers had private jokes and puns using the jargon of their trade that were unintelligible to most people. In the nineteenth century engineers used such rhetorical devices to underline their separation from the public (see Carolyn Marvin, *When Old Technologies Were New* [Oxford University Press, 1988], pp. 18–31).

46. Perry Miller, *The Life of the Mind in America from the Revolution to the Civil War* (Harcourt, Brace and World, 1965, p. 305.

47. Russel B. Nye, *Society and Culture in America, 1830–1860* (Harper & Row, 1974), p. 260.

48. Walt Whitman, "To A Locomotive in Winter," in *Complete Poetry and Selected Prose by Walt Whitman*, ed. J. E. Miller, Jr. (Houghton Mifflin, 1959).

49. See *The Philosophy of Kant Explained* (Garland, 1976), pp. 498–499.

50. Robert Fulton was the fifth man to construct a steamboat in the United States. In 1790 John Fitch ran a steam ferry on the Delaware River. David Wilkinson briefly operated a steamboat on the Providence River, Oliver Evans launched a paddlewheel scow on the Schuylkill, and John Stevens of Hoboken ran a small steamboat, all before Fulton. See David Wilkinson,"Reminiscences," reprinted in *The New England Mill Village*, ed. G. Kulik et al. (MIT Press, 1982). On Evans see *Niles' Weekly Register*, April 23, 1831, p. 132. There are also solid European claims, but these need not concern us here.

51. Cited in Miller, *Life of the Mind in America*, p. 295.

52. Ibid., pp. 295–296.

53. Ibid., p. 294.

54. Cited in Holbrook, *Story of American Railroads*, p. 45.

55. Cited in Stover, *Life and Decline of the American Railroad*, p. 16. See also Holbrook, p. 25.

56. Nathaniel Hawthorne, "The Celestial Railroad," in *Selected Tales and Sketches* (Rinehart, 1960).

57. Kasson, *Civilizing the Machine*, p. 172.

58. Cited in Marx, *Machine in the Garden*, p. 195.

59. Ibid., p. 197.

60. C. Aiken, "Moral View of Railroads," *Hunt's Merchant Magazine* 26 (1852), no. 5, p. 579. Aiken had also preached in New York State on the occasion of the opening of the Erie Canal.

61. Ralph Waldo Emerson, "The Young American," in *Complete Works*, vol. 1 (Houghton Mifflin, 1903), p. 364.

62. Horace Greeley et al., *The Great Industries of the United States* (Garland, 1974), p. 1032. After the Civil War, the emerging profession of engineering tended to see nature as "a neutral, inert reality, without promise of either benevolence or felicity." Thus, a "creative struggle between knowledge and environment . . . replaced the transcendental identification with nature." Engineers were often "non-romantic manipulators of nature" who believed that a dam against spring floods, a bridge across the Mississippi, or a drained swamp turned into farmland was an improvement on nature and a benefit to mankind. The engineer saw mechanical achievements in a more complex way than the general public. The crowd may have been left speechless by a new machine, but the engineer could translate the machine back into charts, diagrams, and technical descriptions. (Raymond H. Merritt, *Engineering in American Society, 1850–1875* [University of Kentucky Press, 1969], pp. 133, 135)

63. Barbara Novak, *Nature and Culture* (Oxford University Press, 1980), pp. 167–179. See, e.g., Novak's discussions of T. P. Rossiter's "Opening of the Wilderness" (1846–1850) and Robert Havell's "Two Artists in a Landscape" (c. 1850). In the latter, the train is far away in the valley below and merges into the landscape.

64. On railway scenery see *Niles' Weekly Register*, April 23, 1831, p. 134.

65. See Novak, *Nature and Culture*, pp. 157–202. In addition to the changing perceptions of Niagara Falls, Elizabeth McKinsey (*Niagara Falls: Icon of the American Sublime* [Cambridge University Press, 1985]) discusses their technological conquest.

66. Klein and Hoogenboom, *History of Pennsylvania*, p. 184. See also Thomas C. Cochran, *Pennsylvania: A Bicentennial History* (Norton, 1978), pp. 90–91.

67. "Railroads of the United States," *Hunt's Merchants Magazine*, October 1840, p. 273.

68. "Effects of Machinery," *North American Review* 34, January 1832, pp. 245–246.

69. See David E. Nye, *The Invented Self* (Odense University Press, 1983), pp. 104–106.

70. Thomas Weiskel, *The Romantic Sublime: Studies in the Structure and Psychology of Transcendence* (Johns Hopkins University Press, 1976), p. 44.

71. Cited in Marx, *Machine in the Garden*, p. 231.

72. Ralph Waldo Emerson, final paragraph of "Art," in *Essays: First Series*. in *Selected Writings of Emerson*, ed. B. Atkinson (Modern Library, 1950), p. 315

73. Emerson, "The Poet," in *Essays: Second Series* (see preceding note), p. 328.

74. Miller, *Life of the Mind in America*, p. 306.

75. Cited in Jensen, *Railroads in America*, p. 25.

76. See Daniel Czitrom, *Media and the American Mind* (University of North Carolina Press, 1982), pp. 4–8.

77. *The Railroad Jubilee* (note 5 above), p. 162.

78. William Wordsworth, "On the Projected Kendall and Windermere Railway," in *The Poetical Works of William Wordsworth*. vol. 3 (Oxford University Press, 1946).

79. Everett, in *The Railroad Jubilee*, p. 174.

80. Ibid., p. 189.

81. Ibid., p. 175.

82. Chevalier, *Society, Manners, and Politics*, p. 136.

83. Susan G. Davis, *Parades and Power: Street Theatre in Nineteenth-Century Philadelphia* (University of California Press, 1988), p. 5.

84. Cited in Bernard Beckerman, *Theatrical Presentation: Performer, Audience and Act* (Routledge, 1990), pp. x–xi.

85. Alan Trachtenberg, *Brooklyn Bridge: Fact and Symbol* (University of Chicago Press, 1979), p. 116.

86. These details are drawn from the July 5 and 12, 1828, issues of *Niles' Weekly Register*, from the *Baltimore American and Commercial Advertiser* of July 3, 1828, and from Hungerford, *Story of the Baltimore and Ohio*.

87. Keith Melder, "Mighty Masses of Freemen: The Young Whigs of Baltimore, 1844," presented at American Studies National Convention, Baltimore, 1991.

88. William M. Ferraro, "The Baltimore Railway Celebration of 1857: Transportation, Economics, and the Question of Union," presented at American Studies National Convention, Baltimore, 1991.

89. Trachtenberg, *Brooklyn Bridge*, p. 116.

90. Davis, *Parades and Power*, p. 43.

91. *Railroad Jubilee*, pp. 10–11.

92. Ibid., pp. 63–64.

93. Ibid., p. 74.

94. Ibid., pp. 118–121.

95. Ibid., p. 119.

96. Ibid., pp. 140–147.

97. Ralph Waldo Emerson, "Progress of Culture," in *Complete Works*, vol. 8 (Houghton Mifflin, 1903), p. 219.

98. Alfred D. Chandler, *The Visible Hand* (Harvard University Press, 1977), passim.

99. Walter Licht, *Working for the Railroad: The Organization of Work in the Nineteenth Century* (Princeton University Press, 1983), p. 248.

100. For a thorough, depressing account, see Mark W. Summers, *The Plundering Generation: Corruption and the Crisis of the Union, 1849–1861* (Oxford University Press, 1987).

101. Gay Wilson Allen and Sculley Bradley, *The Collected Writings of Walt Whitman.* vol. 11 (New York University, 1964), p. 762.

102. John Stilgoe, *Metropolitan Corridor: Railroads and the American Scene* (Yale University Press, 1983), p. 3. This is the indispensable work on how railways shaped American imagination between 1880 and the 1930s.

103. Robert G. Athearn, *Union Pacific Country* (Rand McNally, 1971), p. 54.

104. *Complete Poetry and Selected Prose by Walt Whitman*, ed. Miller, p. 291. The idea that the transcontinental railroad would link the United States to Asia turned up in numerous speeches and newspaper articles. See, e. g., "Opening of the Atlantic and Pacific Railroad and Completion of South Pacific Railroad to Springfield" (proceedings, Springfield, Mo., May 3, 1870; copy in Smithsonian Institution), which includes (p. 13) the assertion that "the products of India and Japan will pass our very doors."

105. For a discussion of Russell's photograph see Novak, *Nature and Culture*, pp. 177–178. On how technology and art intertwined in western landscapes, see Nancy K. Anderson, "The Kiss of Enterprise: The Western Landscape as Symbol and Resource," in *The West as America: Reinterpreting Images of the Frontier, 1820–1920*, ed. W. Treuttner (Smithsonian Institution, 1991). Anderson (p. 244) notes that another Russell photograph showed that "the natural and technological sublime could coexist, that the landscape that had long claimed mythic status in the American imagination was not compromised by the tracks and trestles."

106. Andrew F. Rolle, *California: A History* (Crowell, 1969), pp. 340–344; Warren A. Beck and David A. Williams, *California: A History of the Golden State* (Doubleday, 1972), pp. 216–225. On the Union Pacific see Athearn, *Union Pacific Country*, pp. 97–102.

107. *New York Times*, May 8, 1869.

108. "Completion of the Union Pacific Railroad," *Frank Leslie's Illustrated Newspaper*, June 5, 1869.

109. "Celebrations in Chicago," *New York Times*, May 11, 1869.

110. Ibid.

111. He continued: "Of course, it may happen that, in abandoning myself to them unreservedly, I do not feel the pressure they exert upon me. . . . We are then victims of the illusion of having ourselves created that which actually forced itself from without. If the complacency with which we permit ourselves to be carried along conceals the pressure undergone, nevertheless it does not abolish it. Thus, the air is no less heavy because we do not detect its weight." (Émile Durkheim, *The Rules of Sociological Method* [Free Press, 1938], pp. 4–5)

112. *New York Times*, May 11, 1869.

113. Ibid.

114. Ralph Waldo Emerson, "Works and Days," in *Complete Works*, vol. 7 (Houghton Mifflin, 1903), pp. 160–161.

115. On painters and the transcontinental railroad see Albert Boime, *The Magisterial Gaze: Manifest Destiny and American Landscape Painting, c. 1830–1865* (Smithsonian Press, 1991), pp. 127–137.

116. On Russell's work for the railroad and its use in lecturing see Susan Danly, "Andrew Joseph Russell's The Great West Illustrated," in *The Railroad in American Art: Representations of Technological Change*, ed. S. Danly and L. Marx (MIT Press, 1988), p. 100.

117. Ibid., p. 10.

Chapter 4

1. Emerson, cited in Marx, *Machine in the Garden*, p. 231.

2. See James A. Ward, *Railroads and the Character of America, 1820–1887* (University of Tennessee Press, 1986), pp. 116–125.

3. David Plowden, *Bridges: The Spans of North America* (Viking, 1974), pp. 9–12, 29. Also very useful is David P. Billington, *The Tower and the Bridge: The New Art of Structural Engineering* (Basic Books, 1983), pp. 72–85.

4. Opened in 1883, this bridge remained in service for 79 years before undergoing alterations. See Plowden, *Bridges*, p. 31.

5. Many of the early American bridges were built by James Finley or based on his 1808 patent. In part based on study of these American developments, the British took the lead, notably in the work of Thomas Telford. They in turn were studied and surpassed by the French, whose mathematical theorization of bridge design, notably in the work of Navier, spread throughout Europe and back to the United States. See Emory L. Kemp, "Roebling, Ellet, and the Wire-Suspension Bridge," in *Bridge to the Future: A Centennial Celebration of the Brooklyn Bridge.*, ed. M. Latimer (New York Academy of Sciences, 1984), pp. 44–45.

6. Roebling, cited in Elizabeth McKinsey, *Niagara Falls* (Cambridge University Press, 1985), p. 253.

7. McKinsey (ibid., p. 254) notes: "Almost without exception, they assume a position below the bridge looking upriver, placing the bridge squarely in the center of the picture left to right." In this perspective, Niagara Falls "appears framed beneath the span, diminished by perspective and distance and thoroughly controlled by the bridge."

8. Plowden, *Bridges*, pp. 117–119.

9. Some of the material in this and the following paragraphs is drawn from Quinta Scott and Howard S. Miller, *The Eads Bridge* (University of Missouri Press, 1979).

10. Plowden, *Bridges*, p. 128. See also John H. McCarthy, *Historic American Engineering Record MO-12, Eads Bridge* (Sverdrup Corp., 1990).

11. John Bogart, "Feats of Railway Engineering," in *The American Railway* (Castle, 1988), pp. 93–94.

12. Louis Sullivan, *The Autobiography of an Idea* (Dover, 1954), p. 247.

13. Calvin M. Woodward, *History of the St. Louis Bridge* (Jones, 1881), pp. 196–197.

14. John Kouwenhoven, "Eads Bridge: The Celebration," *Bulletin of the Missouri Historical Society,* April 1974, pp. 166–168.

15. Woodward, *History of the St. Louis Bridge,* p. 198.

16. For more on the creation of the engineer as cultural hero see Cecelia Tichi, *Shifting Gears: Technology, Literature, Culture in Modernist America* (University of North Carolina Press, 1987), pp. 117–132.

17. Walt Whitman, *Specimen Days.* vol. 1, ed. F. Stovall (New York University Press, 1963), p. 229.

18. "St. Louis," *Harper's New Monthly Magazine,* March 1884, pp. 497–498.

19. David McCullough, *The Great Bridge* (Avon, 1972). The important role of Emily Roebling, wife of the chief engineer, has only recently begun to be appreciated. Because her husband was an invalid, unable to visit the site for virtually the whole period of its construction and often unable to receive guests or even to write, she functioned as his eyes and ears and transmitted all his instructions to the subordinate engineers. Her extensive mathematical training and such engineering knowledge as she obtained from her brother, a West Point-trained civil engineer, made it possible for her to perform these duties. Without her constant help, Washington Roebling could not have continued. See Alva T. Matthews, "Emily W. Roebling: One of the Builders of the Bridge," in *Bridge to the Future: A Centennial Celebration of the Brooklyn Bridge.,* ed. M. Latimer (*Annals of the New York Academy of Sciences,* vol. 424, 1984), pp. 63–70.

20. For photographs, see Ralph Greenhill, *Engineer's Witness* (Godine, 1985) and Peter Hales, *Silver Cities* (Temple University Press, 1985).

21. McCullough, *The Great Bridge,* p. 536.

22. Cited in Alan Trachtenberg. *Brooklyn Bridge: Fact and Symbol* (University of Chicago Press, 1965), p. 118.

23. Ibid., pp. 121–122.

24. For a selection of paintings and other art works see *The Great East River Bridge, 1883–1983* (Abrams, 1983).

25. Ibid., p. 155.

26. Joseph Stella, "The Brooklyn Bridge (A page of my life)," *transition* 16 (June 1929), pp. 87–88.

27. Burke, *Enquiry,* pp. 58, 70, 71, 75.

28. Mario Salvadori and Christos Tountas, "The Brooklyn Bridge as a Work of Art," in *Bridge to the Future,* ed M. Latimer (*Annals of the New York Academy of Sciences,* vol. 424, 1984), p. 75.

29. Trachtenberg, *Brooklyn Bridge,* p. 192. See also his important essay "Brooklyn Bridge as Culture Text" in *Bridge to the Future* (ibid.). Here Trachtenberg contrasts the rhetoric of the official dedication with Montgomery Schuyler's more critical views to show how contradiction creates a "compelling cultural text."

30. Mark Girouard notes that in London, Paris, and Berlin "a sense of hierarchy led to high buildings being prohibited by law: it was for long unthinkable

that cathedrals, palaces or public buildings should be overshadowed by commercial structures" (*Cities and People: A Social and Architectural History* [Yale University Press, 1985], p. 329).

31. Girouard, *Cities and People*, pp. 320–321.

32. Montgomery Schuyler, "The Evolution of the Sky-Scraper," *Scribner's Magazine* 46, September 1909; reprinted in Leland M. Roth, *America Builds: Source Documents in American Architecture and Planning* (Harper & Row, 1983). Likewise, the Revere House in Boston had an elevator in the 1870s.

33. Raymond H. Merritt, *Engineering in American Society, 1850–1875* (University of Kentucky Press, 1969), pp. 18–19. Steel was preferable to iron because in a fire iron lost its ability to bear weight.

34. John W. Root, "A Great Architectural Problem," *Inland Architect and News Record* 15, June 1890; reprinted in Roth, *America Builds*, p. 297. In the same article Root explains how to deal with the settling of a new building and its effect on nearby structures, by using jack screws and cantilevers. On the first Chicago School see Billington, *The Tower and the Bridge*, pp. 99 –111.

35. The first steel-frame skyscraper in New York was the Tower Building (1889). See Paul Goldberger, *The Skyscraper* (Knopf, 1981), p. 27.

36. Girouard, *Cities and People*, p. 322.

37. Mona Domosh, "The Symbolism of the Skyscraper: Case Studies of New York's First Tall Buildings," *Journal of Urban History* 14, May 1988, p. 334.

38. Edgar H. Hall, *The Hudson Fulton Celebration, 1909*, vol. 1 (State of New York, 1910), p. 252. The mayor's comments echoed those of many critics in *Scribner's*. See Annette Larsen Benert,"Reading the Walls: The Politics of Architecture," *Arizona Quarterly* 47 (1991), no. 1.

39. Cited in Alan Trachtenberg, *Reading American Photographs*, (Hill and Wang, 1989), p. 213.

40. William R. Taylor ("New York and the Origin of the Skyline," *Prospects* 13 [1988], p. 234) traces the first use of the term to May 3, 1896. For earlier uses of the term in England and America see Thomas A. P. van Leeuwen, *The Skyward Trend of Thought: The Metaphysics of the American Skyscraper* (MIT Press, 1988).

41. Montgomery Schuyler, "The Sky-line of New York, 1881–1897," *Harper's Weekly*, March 20, 1897, p. 295.

42. Arnold Bennett, *Those United States* (Martin Secker, 1912), pp. 45–46.

43. Robert Thurston, "The Scientific Basis of Belief," *North American Review* 153, August 1891, pp. 181–192.

44. Merritt, *Engineering in American Society*, p. 150. Billington notes in *The Tower and the Bridge* (p. 85) that Roebling "saw his works as promises for a new utopia" but that he also had "a deep reverence for changeless laws of wind and gravity."

45. See van Leeuwen, *Skyward Trend of Thought*, p. 128.

46. See Sullivan, "The Tall Office Building Artistically Considered," *Lippincott's Magazine* 57, March 1896, pp. 403–409. This often-reprinted essay presents the

problem of building the skyscraper as one of finding the proper organic form that grows out of the functions of the office building. The essay, which is Emersonian in tone, concludes with an exhortation to reject European traditions, turn to Nature, and create a democratic architecture.

47. Goldberger, *The Skyscraper*, p. 42.

48. On Woolworth see van Leeuwen, pp. 60–62, and Goldberger, pp. 41–46.

49. John P. Nichols, *Skyline Queen and the Merchant Prince: The Woolworth Story* (Trident, 1973), pp. 86–87, 93.

50. Girouard, p. 322. William R. Taylor, "The Evolution of Public Space in New York City," in *Consuming Visions: Accumulation and Display of Goods in America, 1880–1920*, ed. S. J. Bronner (Norton, 1989), p. 305.

51. "55 Story Building Opens on a Flash," *New York Times*, April 25, 1913.

52. *New York Tribune*, Illustrated Supplement, 1902, cited in Philip William Kreitler, *Flatiron: A photographic history of the world's first steel frame skyscraper, 1901–1990* (American Institute of Architects, 1990), p. 10.

53. Kreitler (ibid.), p. 18.

54. Dean MacCannell, *The Tourist: A New Theory of the Leisure Class* (Schocken, 1976), p. 122.

55. Roland Marchand, *Advertising the American Dream* (University of California Press, 1985), pp. 240, 242.

56. See also William R. Taylor, "Psyching Out the City" in *Uprooted Americans: Essays to Honor Oscar Handlin*, ed. Richard Bushman (Little, Brown, 1979).

57. Henry James, *The American Scene* (Chapman and Hall, 1907), pp. 81, 139, 228, 232. James (p. 185) much preferred the horizontality of the "great Palladian pile just erected by Messrs. Tiffany. . . . One is so thankful to it, I recognize, for not having twenty-five stories."

58. William Dean Howells, *The World of Chance* (Harpers., 1893), pp. 12–13.

59. On the conflict between the horizontal, European style and the vertical, vernacular style, see William R. Taylor and Thomas Bender, "Culture and Architecture: Some Aesthetic Tensions in the Shaping of Modern New York City," in *Visions of the Modern City*, ed. W. Sharpe and L. Wallock (Johns Hopkins University Press, 1987).

60. Girouard, *Cities and People*, p. 323.

61. Schuyler in Roth, *America Builds*.

62. Goldberger, *The Skyscraper* (Knopf, 1981), pp. 10–11.

63. Cited in Girouard, *Cities and People*, p. 324.

64. Merrill Schleier, *The Skyscraper in American Art, 1890–1931* (Da Capo, 1986), pp. 10–15; Edgar Saltus, "New York from the Flatiron," *Munsey's Magazine* 33, July 1905, p. 382.

65. "Chicago Poems," in *The Complete Works of Carl Sandburg* (Harcourt Brace Jovanovich, 1969), p. 31.

66. Schleier, *Skyscraper in American Art*, pp. 47–58.

67. Ibid., p. 77.

68. "Tallest Tower Built in Less Than a Year," *New York Times,* May 2, 1931. For a short popular history see Jonathan Goldman, *The Empire State Building Book* (St. Martin's, 1980).

69. Douglas Haskell, "A Temple of Jehu," *The Nation,* May 27, 1931, pp. 589–590.

70. Paul Starrett, *Changing the Skyline: An Autobiography* (McGraw-Hill, 1939), pp. 299, 308. One of the main architects on the project agreed that construction was "like an assembly line—the assembly of standard parts" (ibid., p. 296).

71. "Throngs Inspect Tallest Building," *New York Times,* May 30, 1931; Starrett, *Changing the Skyline,* p. 298.

72. "Tallest Building Opened by Hoover," *New York Times,* May 2, 1931.

73. "Building in Excelsis" [editorial], *New York Times,* May 1, 1931.

74. Edmund Wilson, "Progress and Poverty," *New Republic,* May 20, 1931.

75. "Big Sunday Crowd Sees Empire Tower," *New York Times,* May 4, 1931; "Throngs Inspect Tallest Building," *New York Times,* May 30, 1931.

76. *New Yorker,* May 16, 1931.

77. "Our Tallest Sky-Piercing Pinnacle," *Literary Digest,* May 16, 1931, p. 11.

78. Cited in ibid.

79. "Empire State Roof Pays $3,100 Daily," *New York Times,* October 11, 1931.

80. Thirty-five years after it opened the visitor total reached 50 million (Goldman, pp. 83–84).

81. "Other Skyscrapers Are Dwarfed," *New York Times,* May 2, 1931.

82. "Panorama Viewed from 85th Story," *New York Times,* May 2, 1931.

83. Saltus,"New York from the Flatiron," pp. 382–383. A widely distributed Underwood and Underwood stereograph (reproduced in Kreitler, *Flatiron*) made the same point.

84. Cited in "Sky Boys Who 'Rode the Ball' on Empire State," *Literary Digest,* May 23, 1931, p. 33.

85. Roland Barthes, *The Eiffel Tower and other Mythologies* (Hill and Wang, 1979), p. 8.

86. Ibid., p. 9.

87. James Oppenheim, *The Olympian* (Harper, 1912), pp. 416–417.

Chapter 5

1. Louis C. Hunter, *A History of Industrial Power in the United States, 1780–1930.* vol. 2, *Steam Power* (University Press of Virginia, 1985), p. 74. Not only did the Americans have relatively few steam engines, but those in operation were seldom large. While the English built a great number of "walking beam" engines in the form perfected by James Watt, Americans built smaller, less efficient, but cheaper engines on the model provided by Oliver Evans.

2. Francis Klingender, *Art and the Industrial Revolution* (Augustus M. Kelley, 1968), pp. 87–90.

3. Hunter, *History of Industrial Power*, vol. 2, p. 1.

4. Edward A. Kendall, *Travels through the Northern Parts of the United States in the Years 1808 and 1809*, vol. 3 (New York, 1809), pp. 33–34.

5. From a report of the American Society for the Encouragement of Domestic Manufactures, cited in Thomas Bender, *Toward an Urban Vision: Ideas and Institutions in Nineteenth-Century America* (University Press of Kentucky, 1975), p. 28.

6. Hunter, *History of Industrial Power*, p. 181.

7. "The Excursion" (eighth book, pp. 269, 171), in *The Poetical Works of William Wordsworth*, vol. 5 (Oxford University Press, 1949).

8. See Bender, *Toward an Urban Vision*, pp. 31–38, 79–80.

9. "Lowell Cotton and Woollen Goods," *Niles' Register*, July 6, 1833, p. 315.

10. John Kasson, *Civilizing the Machine* (Penguin, 1977), pp. 64–65.

11. "Excursion to Lowell," *Niles' Register*, July 6, 1833, pp. 313–315 [reprinted from *The Salem Register*].

12. Michel Chevalier, *Society, Manners, and Politics in the United States* (Doubleday Anchor, 1961), p. 130.

13. Gayle Graham Yates, ed., *Harriet Martineau on Women* (Rutgers University Press, 1985), p. 161.

14. Dickens, *American Notes*, pp. 66–67.

15. Marvin Fisher, *Workshops in the Wilderness: The European Response to American Industrialization*, 1830–1860 (Oxford University Press, 1967), pp. 90–102.

16. Burke, *Enquiry*, p. 129.

17. Cited in Tamera K. Hareven, *Family Time and Industrial Time* (Cambridge University Press, 1982), p. 9. The oral history of these mills is detailed in Tamara K. Hareven and Randolph Langenbach, *Amoskeag: Life and Work in an American Factory-City* (Pantheon, 1978).

18. Fisher, *Workshops in the Wilderness*, p. 95.

19. Cited in Russell B. Nye, *Society and Culture in America, 1830–1860* (Harper & Row, 1974), p. 278.

20. "Lowell Cotton and Woollen Goods," *Niles' Register*, July 6, 1833, p. 315. See Hareven, *Family Time and Industrial Time*, p. 12.

21. "President at Lowell," *Niles' Weekly Register*, July 6, 1833, p. 316; Robert V. Remini, *Andrew Jackson and the Course of American Democracy, 1833–1845* (Harper & Row, 1984), pp. 80–82.

22. Soam, *The Popular Mood of Pre-Civil War America*, pp. 161–162.

23. *Lowell Offering*, December 1840; reprinted in *The Factory Girls.*, ed. P. S. Foner (University of Illinois, 1977), p. 36.

24. Anthony F. C. Wallace, *Rockdale* (Norton, 1978), p. 181.

25. Quotation from Foner, *Factory Girls*, p. 27. On mill conditions see Norman Ware, *The Industrial Worker, 1840–1860* (Elephant, 1990), pp. 98–100, 120–124.

26. Cited in Wallace, *Rockdale*, p. 181.

27. Cited in ibid., pp. 182–183.

28. Quotation in Foner, *Factory Girls*, p. 45.

29. Workers were not the only ones who refused to see mills as a harmonious merging of man and nature. A number of critics voiced their disapproval, none more tellingly than Herman Melville in his short piece "A Tartarus of Maids."

30. Thomas Dublin, ed. *Farm to Factory: Women's Letters, 1830–1860* (Columbia University Press, 1981), p. 21.

31. Census statistics from David R. Goldfield and Blaine A. Brownell, *Urban America: From Downtown to No Town* (Houghton Mifflin, 1979), pp. 16–17.

32. On working-class holidays and revels see Davis, *Parades and Power.*

33. Diana Karter Appelbaum, *The Glorious Fourth: An American Holiday, an American History* (Facts on File, 1989), pp. 77–79.

34. Roy Rosenzweig, *Eight Hours For What We Will* (Cambridge University Press, 1983), p. 153.

35. The spectacular effects of the amusement park presaged a new form of the sublime, which would be far more effective than the "Safe and Sane" movement, as it used electricity to embellish technology, a topic to be taken up in chapters seven and eight. How a genteel elite transformed Shakespeare from a popular to an elite playwright in the later nineteenth century is the subject of Lawrence W. Levine's essay "William Shakespeare and the American People: A Study of Cultural Transformation," in *The Unpredictable Past: Explorations in American Cultural History* (Oxford University Press, 1993). On the eight-hour movement and the strikes of May 1886 see Jeremy Breecher, *Strike!* (Straight Arrow, 1972), pp. 37–50.

36. H. Beale, *Picturesque Sketches of American Progress, comprising Official description of Great American Cities, Showing their origin, development, present condition, commerce and manufacturers* (New York: Empire Co-operative Association, 1889), p. 423.

37. Edward Bellamy, *Looking Backward* (Modern Library, 1982). See Kenneth Roemer, *The Obsolete Necessity: America in Utopian Writings, 1888–1900* (Kent State University Press, 1981).

38. Hunter, *History of Industrial Power,* vol. 2, pp. 251–254.

39. Corliss typically guaranteed his engine's performance by agreeing to a penalty for every day it did not operate. He was equally innovative in sales tactics, offering customers the choice between a fixed price or a price based on the amount of fuel saved by installing his engines. See Hunter, p. 297, and Horace Greeley, et al., *The Great Industries of the United States* (Hartford, 1872).

40. On the engine and its critics see Hunter, vol. 2, pp. 297–299 and 453–462. The definitive article is Eugene Ferguson's "Power and Influence: The Corliss Steam Engine in the Centennial Era," in *Bridge to the Future*, ed. M. Latimer (*Annals of the New York Academy of Sciences* 424, 1984). Ferguson points out that, although parts had to be replaced at times, Corliss engines virtually never wore out, routinely lasted more than 50 years, and usually were replaced not because of mechanical failure but for other reasons.

41. Ferguson, "Power and Influence," p. 228.

42. *Scientific America Supplement,* 1 (1876), no. 24, p. 370.

43. William Dean Howells, "A Sennight of the Centennial," *Atlantic Monthly,* July 1976, p. 96.

44. Cited in Kasson, *Civilizing the Machine,* p. 164.

45. *Scientific American,* May 27, 1876.

46. Philip S. Klein and Ari Hoogenboom, *A History of Pennsylvania* (McGraw-Hill, 1973), pp. 334–336.

47. Martha Banta, *Imaging American Women: Idea and Ideals in Cultural History* (Columbia University Press, 1987), p. 525.

48. Cited in Robert W. Rydell, *World of Fairs: The Century of Progress Expositions* (University of Chicago Press, 1993), pp. 15–16.

49. Kasson, *Civilizing the Machine,* p. 162.

50. See ibid., pp. 167–172.

51. John Stilgoe, *Metropolitan Corridor: Railroads and the American Scene* (Yale University Press, 1983), p. 78.

52. John T. Holdsworth, "The Smoky City," *Report of the Economic Survey of Pittsburgh.* Pittsburgh, 1912, pp. 35–37.

53. Margaret Byington, *Homestead: The Households of a Mill Town* (Pittsburgh: University Center for International Studies, 1974; reprint of Russell Sage Foundation edition of 1910), p. 3.

54. Willard Glazier, "The Great Furnace of America," in *Peculiarities of American Cities* (Hubbard Brothers, 1883), pp. 332–334.

55. Quoted in Stilgoe, *Metropolitan Corridor,* pp. 79–81.

56. Nicholas Taylor, "The Awful Sublimity of the Victorian City: Its Aesthetic and Architectural Origins," in *The Victorian City, Images and Realities,* vol. 2, ed. H. J. Dyos and Michael Wolff (Routledge & Kegan Paul, 1973), p. 436.

57. Banta, *Imaging American Women,* p. 674. Portions of the mural are reproduced on pp. 675–677.

58. Ibid., p. 674.

59. Theodore Dreiser, *A Hoosier Holiday* (Oxford University Press, 1932), pp. 178–180.

60. Julian Ralph, "The Industrial Region of Northern Alabama, Tennessee and Georgia," *Harper's,* February 1895, pp. 607–617.

61. Karen Tsujimoto, *Images of America: Precisionist Painting and Modern Photography* (University of Washington Press, 1982), p. 189.

62. James Oppenheim, *The Olympian* (Harper, 1912), p. 2.

63. Burke, *Enquiry,* p. 73.

64. Dreiser, *Hoosier Holiday,* p. 180.

65. Susan Strasser, *Satisfaction Guaranteed: The Rise of the American Mass Market* (Pantheon, 1989), p. 113.

66. Roland Marchand's forthcoming book *Creating the Corporate Soul* will be a detailed history of public relations. I am indebted to him for pointing out several of the examples mentioned here.

67. See David E. Nye, *Image Worlds: Corporate Identities at General Electric* (MIT Press, 1985), pp. 71–92.

68. Marshall Everett, *The Book of the Fair: The Greatest Exposition the World has Ever Seen, A Panorama of the St. Louis Exposition* (Henry Neil, 1904), pp. 172–173, 197.

69. Cited by Roland Marchand, "Corporate Imagery and Popular Education: World's Fairs and Expositions in the United States, 1893–1940," in *Consumption and American Culture*, ed. D. E. Nye and C. Pedersen (Amsterdam: Free University Press, 1991).

70. See Nye, *Electrifying America*, pp. 221–235.

71. See James J. Flink, *The Automobile Age* (MIT Press, 1988), p. 30.

72. David L. Lewis, *The Public Image of Henry Ford* (Wayne State University Press, 1976), p. 54. For another view of Ford see David E. Nye, *Henry Ford: Ignorant Idealist* (Kennikat. 1979).

73. Henry Ford, *My Life and Work* (Heinemann, 1922), p. 280.

74. William Stidger, *Henry Ford: The Man and His Motives* (Doran, 1923), pp. 113–118.

75. See ibid., pp. 113, 118.

76. These rhythmic patterns of control became the subject of a famous series of photographs by Charles Sheeler, commissioned by the Ford Company. Elsie Driggs also responded to the organization of the Rouge: "There seemed to be a relation between function and felicitous proportion. . . . The Greeks had the law of human perfection, but the well-functioning machine and the well-functioning factory had its own law of inevitable rightness." (Tsujimoto, *Images of America*, p. 69)

77. Oppenheim, *The Olympian*, p. 358.

78. Ibid., pp. 358–359.

79. Conceiving of nature as female was not new; it could be traced back to ancient times, when the extraction of ores from the earth and the work of smelting was often compared to procreation. See Mircea Eliade, *The Forge and the Crucible* (Harper, 1962).

80. For discussion of the development summarized here see Nye, *Electrifying America*, pp. 187–237.

81. Arnold Bennett, *Those United States* (Martin Secker, 1912), pp. 101–102.

82. John Stilgoe, "Central Stations and the Electric Vision, 1890–1930," in *Essays from the Lowell Conference on Industrial History*, ed. R. Weible (1985), p. 55.

83. Donald Des Granges, "The Designing of Power Stations," *Architectural Forum*, 51, September 1929, p. 362.

84. This was a major early demonstration of the practicality of alternating current for long-distance transmission of electricity. See Louis C. Hunter and Lynwood Bryant, *A History of Industrial Power in the United States, 1790–1930*, vol.

3 (MIT Press, 1991), p. 271. A standard history from an engineering perspective is Edward D. Adams, *Niagara Power: History of the Niagara Power Company, 1896–1918* (Niagara Falls Power Co., 1927). See also Robert Belfield, The Niagara Frontier, Ph. D. dissertation, University of Pennsylvania, 1981.

85. Martha Moore Trescott, *The Rise of the American Electrochemical Industry, 1880–1910* (Greenwood, 1981). See also Harold Passer, *The Electrical Manufacturers. 1875–1900* (Harvard University Press, 1953), pp. 291–293. A useful source is *The Niagara Falls Electrical Handbook* (American Institute of Electrical Engineers, 1904), published as a guide for foreign engineers.

86. Peter Conrad, *Imagining America* (Oxford University Press, 1980), pp. 25–26.

87. Alton D. Adams, "The Destruction of Niagara Falls," *Cassier's Magazine*, 1905, pp. 413–417. See also John M. Clarke, "The Menace to Niagara," *Popular Science Monthly*, October 1905, pp. 489–504; Orrin E. Dunlap, "Is Niagara Doomed?" *Technical Word.* July, 1905, pp. 557–568. By the following year the controversy had reached all levels of the popular press. See, for example, "The Desecration of Niagara," *Ladies Home Journal,* June 1906.

88. On shrinkage see J. W. Spence, "The Spoilation of the Falls of Niagara," *Popular Science Monthly*, October 1908. This article was summarized in *Nature* (November 5, 1908). See also letters in *Scientific American*, March 31, 1906, p. 271. (An earlier letter in *Scientific American* (February 24) had opposed saving the falls for "sentimental reasons."

89. Roderick Nash, *Wilderness and the American Mind* (Yale University Press, 1982), pp. 167–168.

90. Robert S. Lanier, "International Aid for Niagara," *Review of Reviews*, April 1906, pp. 432–439.

91. "Niagara Falls from a New Point of View," *Scientific American* 105, September 9, 1911, p. 227.

92. See Nicholas B. Wainwright, *History of the Philadelphia Electric Company* (Philadelphia Electric Co., 1961), pp. 166–185; "Roosevelt Calls for Abundant Life in this Power Age," *New York Times*, September 12, 1936, p. 1 (text of speech: p. 3).

93. On the TVA see William U. Chandler, *The Myth of TVA* (Ballinger, 1984); *TVA: Fifty Years of Grassroots Bureaucracy*, ed. P. K. Conklin (University of Illinois, 1983); Thomas K. McCraw, *TVA and the Power Fight, 1933–1939* (Lippincott, 1971).

94. "Roosevelt Calls for Abundant Life in This Power Age," *New York Times*, September 12, 1936.

95. Frank Waters, *The Colorado* (Rinehart, 1946), p. 337.

96. U.S. Department of the Interior and Bureau of Reclamation, *Boulder Canyon Project, Final Reports, General History and Description of Project.* Boulder City, Nevada, 1948, pp. 72–74.

97. "Giant Boulder Dam Begins Operation," *New York Times*, September 12, 1936. At this time it was called Boulder Dam, but later the original name was restored. To avoid confusion, I call it Hoover Dam throughout.

98. Richard Guy Wilson, Dianne H. Pilgrim, and Dickran Tashjian, *The Machine Age in America, 1918–1941* (Abrams, 1986), p. 117.

99. John E. Stevens, *Hoover Dam: An American Adventure* (University of Oklahoma Press, 1988), pp. 266–267.

100. Martin Friedman, *Charles Sheeler* (Watson-Guptill, 1975), p. 129.

101. "Power, A Portfolio by Charles Sheeler," *Fortune*, December, 1940.

102. Ibid. Sheeler's final painting on electricity depicted the interior of the "world's largest steam power plant" in Brooklyn.

Chapter 6

1. Burton Benedict, *The Anthropology of World's Fairs* (Scholar Press, 1985), p. 5.

2. See Nye, *Electrifying America*, chapter 5. An earlier version of the present chapter appeared as "Republicanism and the Electrical Sublime" in *American Transcendental Quarterly* (New Series 4 [1990], no. 3).

3. Burke, pp. 80, 81, 143.

4. *The Railroad Jubilee, An Account of the Celebration Commemorative of the Opening of Railroad Communication between Boston and Canada* (Boston: J. H. Eastburn, 1852), p. 101.

5. "Death of Luther Stieringer," *Electrical World and Engineer* 42, no. 4, p. 132. Stieringer was engineer for the first exposition in America to use electric lights at Cincinnati in 1883. He also helped design the lighting for the fairs at Chicago, Omaha, and Buffalo. Biographical material in Hammer Papers, Museum of American History, Smithsonian Institution.

6. "Spectacular Lighting in the Past," *The Illuminating Engineer* 2 (1908), no. 12, pp. 847–850. In the same issue also see Raymond T. Vredenburgh, "Spectacular Lighting" (pp. 857–858).

7. *Louisville Courier-Journal*, July 4, 1883.

8. Jane Mark Gibson, The International Electrical Exhibition of 1884 and the National Conference of Electricians, M. A. thesis, University of Pennsylvania, 1984, pp. 44, 85.

9. *Journal of the Franklin Institute*, May 1886, p. 121.

10. *Cincinnati Commercial Gazette*, June 10, 1888, p. 4.

11. From the caption accompanying a full-page reproduction in *The Columbian Gallery: A Portfolio of Photographs from the World's Fair* (Chicago: Werner, 1894), p. 86.

12. Letter to Elizabeth Cameron, October 8, 1893, in *The Letters of Henry Adams*, vol. 4, ed. J. C. Levinson et al. (Harvard University Press, 1988), p. 132.

13. Theodore Dreiser, *Newspaper Days*, (University of Pennsylvania Press, 1991), pp. 309–310.

14. On Hassam see Robert Muccigrosso, *Celebrating the New World: Chicago's Columbian Exposition of 1893* (Dee, 1993), p. 90.

15. For a more detailed account see Nye, *Electrifying America*, pp. 43–47.

16. Marshall Everett, *The Book of the Fair, The Greatest Exposition the World has Ever Seen. A Panorama of the St. Louis Exposition* (Henry Neil, 1904), pp. 201–203.

17. Ibid., p. 202.

18. Burke, *Enquiry,* p. 73.

19. On the "electrical sublime" see also David E. Nye, "Social Class and the Electrical Sublime, 1880–1915," in *High Brow Meets Low Brow,* ed. R. Kroes (Amsterdam: Free University Press, 1988), pp. 1–20.

20. See Nye, *Electrifying America,* pp. 122–132.

21. F. Scott Fitzgerald, "Basil: A Night at the Fair," in Malcolm Cowley, ed. *Scott Fitzgerald,* vol. 6 (Bodley Head, 1963), p. 49. It is quite significant that Fitzgerald compares the fairgrounds to a woman, for between Basil's two visits to the fairgrounds he acquires long pants for the first time, and on the night in question he must decide with whom to spend his evening.

22. Edmund Pendleton Tompkins and J. Lee Davis, *The Natural Bridge and its Historical Surroundings* (Natural Bridge of Virginia, Inc., 1939), p. 54.

23. Also among the 300 appointed members were General Stewart L. Woodford (the president), Joseph H. Choate, Seth Low, Levi P. Morton, Alton B. Parker, Oscar S. Straus, W. B. Van Rensselaer, General James Grant Wilson, Isaac N. Seligman (the treasurer), and Henry Sackett (the secretary). For a brief summary see Office of the Commission of Education, *Hudson–Fulton Celebration 1609–1807–1909* (Albany: State Education Department, 1909).

24. "The Hudson–Fulton Celebration in New York City," *Electrical World* 54, (1909), no. 14, p. 770.

25. Forbin, *L'illustration,* October 9, 1909, pp. 254–255.

26. *New York Times,* October 3, 1909.

27. *New York Times,* September 19, 1909.

28. *World's Work,* November 1909.

29. David Glassberg, *American Historical Pageantry* (University of North Carolina Press, 1990), pp. 16–18.

30. *New York Times,* October 2, 1909.

31. Edgar H. Hall, *The Hudson Fulton Celebration, 1909,* vol. 1 (State of New York, 1910, pp. 282–283.

32. Ibid., p. 287.

33. Ibid., p. 362.

34. Ibid., p. 292.

35. Ibid., p. 303.

36. Ibid., p. 292.

37. In September and October of 1909 Cook was generally believed, on the basis of his own statements, to have been the first man to reach the North Pole. Peary disputed his claims, but at first this was thought to be sour grapes. Cook was feted and given the keys to the city; later his claims of discovery were found to be false and Peary was recognized as having the better claim.

38. Wright was the first to fly over New York.

39. *Electrical World* 54 (1909), no. 14, p. 776.

40. "Spectacular Electric Illumination During the Hudson-Fulton Celebration," *Electrical World* 54 (1909), no. 6, p. 290. For photographs see "The Hudson Fulton Celebration," *Edison Monthly*, October 1909, pp. 120–133.

41. *Electrical World* 54 (1909), no. 6, p. 290.

42. *Electrical World* 54 (1909), no. 14, p. 771.

43. L. Elliott, "Plain Talks on Illuminating Engineering, XV. Spectacular Lighting," *Illuminating Engineering*, February 1908. p. 878.

44. Ryan's system had first been tested on the North Shore of Massachusetts in 1906.

45. Raymond T. Vrendenburgh, "Spectacular Lighting," *Illuminating Engineering*, January 1908, p. 858.

46. Elliott, "Plain Talks on Illuminating Engineering," p. 876.

47. For more on Ryan and the uses of the electrical sublime see Nye, *Electrifying America*, pp. 54–66.

48. "Feast of Lights Ends Pageant," *New York Times*, October 3, 1909.

49. See Raymond O'Brien, *American Sublime: Landscape and Scenery of the Lower Hudson Valley* (Columbia University Press, 1981), pp. 261–264. O'Brien points out (p. 328) that writers often appreciated the stone quarries and other industrial establishments that defaced the Palisades as being attractive in their own right. In short, they admired the technological and the natural sublime together.

50. *London Times*, October 4, 1909.

51. Glassberg, *American Historical Pageantry*, p. 213.

52. Ibid., p. 213.

53. Ibid., p. 215.

54. Ibid., p. 223.

55. Ibid, pp. 226–227.

56. F. Dickerson, "Spectacular Lighting," in Proceedings of the 47th Convention of the National Electric Lighting Association, 1924, pp. 493–494.

57. *New York Times*, July 5, 1919, p. 8. On the sometimes-strained relations between the War Camp Community Service and the pageant organizers see Glassberg, *American Historical Pageantry*, pp. 214–224.

58. *Washington Star*, July 4, 1919, p. 1.

59. For a brief account of this event, see George William Douglas and Helen Douglas Compton, *The American Book of Days* (Wilson, 1948), pp. 373–376.

60. Description drawn from *Washington Evening Star*, July 4, 1919.

61. Glassberg, *American Historical Pageantry*, pp. 135–136.

62. *Washington Star*, July 4, 1919.

63. Ibid.

64. *Washington Star*, July 5, 1919.

65. The dedication of the Golden Gate Bridge in 1937 shows that the pageant tradition did not disappear. A pageant with a cast of 3000, called "The Span of

Gold," traced California's history from Spanish discovery to statehood and culminated in spectacular fireworks and the first lighting of the bridge."From barges in the Bay, from planes flying overhead, fireworks were hurled toward the bridge, and the red paint on the towers and cables and truss spans reflected the light of the bursting rockets and star shells." (Allen Brown, *Golden Gate* [Doubleday, 1965], p. 123]

66. William Zinsser, *American Places* (HarperCollins, 1992), p. 16.

Chapter 7

1. Frank Presbury, *The History and Development of Advertising* (Doubleday, Doran, 1929), p. 506.

2. Richard Harding Davis, "Broadway," *Scribner's Magazine*, 10 (1891), pp. 598, 603, 604. A decade before electric signs became prominent, Mark Twain complained that "government has snatched out all the snags and lit up the shores like Broadway" (*Life on the Mississippi* [Harper and Row, 1951], p. 237).

3. Davis, "Broadway."

4. Ibid.

5. On the development of various kinds of electric lights, see Arthur A. Bright, *The Electric-Lamp Industry: Technological Change and Economic Development from 1800 to 1947* (Macmillan, 1949).

6. See Nye, *Electrifying America*, pp. 60–61.

7. Neon tubes were first used in 1925 in a sign advertising the movie *Ben-Hur*. By 1938 about 60 percent of midtown Manhattan's advertising lights were neon, according to an estimate made by the General Outdoor Advertising Company and reported in the *New York Times Magazine* of November 13, 1938.

8. For example, electric signs are scarcely mentioned in Stephen Fox's *The Mirror Makers* (William Morrow, 1984) or in Roland Marchand's *Advertising the American Dream* (University of California Press, 1985).

9. See Nicholas B. Wainwright, *History of the Philadelphia Electric Company, 1881–1961* (Philadelphia Electric Company, 1961), p. 13.

10. John Winthrop Hammond, *Men and Volts* (Lippincott, 1941), pp. 28–29.

11. Chicago's first arc light was set up in April 1878 by John P. Barrett. See Harold L. Platt, *The Electric City* (University of Chicago Press, 1991), p. 3.

12. "Bright Lights in Boston: A Theatre Milestone," Boston Edison Company, December 11, 1982.

13. G. Rogers and Mildred Weston, *Carnival Crossroads, The Story of Times Square* (Doubleday, 1960), p. 92.

14. All forms of night lighting were particularly profitable to the utilities, because the overall demand for electricity dropped at night, leaving them with unused capacity. In Wichita, for example, the contract called for the lights to come on at 6:15 P. M., 20 minutes after the load usually peaked for the day.

15. Platt, *The Electric City*, pp. 100–101.

16. See Frank B. Rae, "Creating Demands for Electricity," in Proceedings of the 31rst Convention of the National Electric Light Association, Chicago, 1908.

17. Platt, p. 101.

18. Local utilities used the same technique to sell the idea of the electrified home, setting up model demonstration homes under the auspices of retailers associations, usually with the agreement that no overt advertising would be permitted at the site. See Nye, *Electrifying America*, pp. 265–267 and 356–359.

19. Note the following: "Since the installation of memorial lamp standards around the City Hall, nearly two years ago, Philadelphia has been constantly agitating the question of more and better public lighting." (*The Illuminating Engineer*, August, 1910, p. 315) On developments in Philadelphia, Minneapolis, and other cities, see *The Illuminating Engineer*, October 1910, pp. 425–428.

20. F. Dickerson, "Spectacular Lighting," in Proceedings of the 47th Convention of the National Electric Lighting Association, 1924, p. 485.

21. *Chicago Tribune, Chicago American*, and *Chicago Post*, October 15, 16, 1926.

22. *Edison Light*, March 1903, p. 7.

23. *Edison Light*, November 1907, p. 4.

24. Information (courtesy of Barbara R. Kelley, Library Services, Commonwealth Edison) from Bernadine Skeels, A History of Commonwealth Edison's Merchandising (company publication), 1972.

25. Information courtesy of Lisa McDonough, Boston Edison Company, from *Edison Life*, May-June 1938, pp. 6–7.

26. Before floodlighting, buildings were often decorated with strings of lights that emphasized cornices and other architectural details. For more information see chapter 2 of *Electrifying America*. Early electrical displays were not permanently available. They had to be searched out, and when seen they could evoke a sense of unearthly wonder. In F. Scott Fitzgerald's story "Absolution" (in *The Bodley Head Scott Fitzgerald*, vol. 5, ed. M. Cowley [Bodley Head, 1963]) a priest in a small town finds that a boy has never been to an amusement park and urges him to go and see one: "It's a thing like a fair, only much more glittering. Go to one at night and stand a little way off from it in a dark place— under dark trees. You'll see a big wheel made of lights turning in the air, and a long slide shooting boats down into the water. A band playing somewhere, and . . . everything will twinkle. But it won't remind you of anything, you see. It will all just hang out there in the night like a colored balloon—like a big lantern on a pole. But don't get up close because if you do you'll only feel the heat and the sweat and the life." The boy is frightened by this strange talk, but he intuits the basic premise of the manufactured sublime: ". . . underneath his terror he felt that his own inner convictions were confirmed. There was something ineffably gorgeous somewhere that had nothing to do with God."

27. See Nye, *Electrifying America*, p. 66.

28. Paul Goldberger, *The Skyscraper* (Knopf, 1981), pp. 38–40, 59–61.

29. Electrical Advertising: Its Forms, Characteristics, and Design,. Bulletin 50, Engineering Department, National Lamp Works of General Electric Company, 1925.

30. Ibid.

31. On "reason why" advertising before 1910 see T. J. Jackson Lears, "Some Versions of Fantasy: A Cultural History of Advertising," *Prospects* 9 (1984), pp. 349–406.

32. Accounts of the wording of the sign vary somewhat, because the message was changed from time to time. This version is cited in Theodore Dreiser, *The Color of a Great City* (Boni & Liveright, 1923), p. 119.

33. "Electricity in Advertising," *Edison Light* 1 (1903), no. 3, p. 4.

34. D. Ziegler,"The Living Electric Sign," *The Illuminating Engineer*, April 1910, p. 74.

35. Electrical Advertising (n. 29 above), p. 11.

36. General Electric Company, pocket-size handbook for salesmen, p. 122.

37. Ziegler,"Living Electric Sign," pp. 74–75.

38. Ibid.

39. Presbury, *History and Development of Advertising*, p. 508.

40. "Broadway Aquarium," *Architectural Forum*, May 1936, pp. 56–57.

41. Eric J. Sandeen, "The Value of Place: The Redevelopment Debate over New York's Times Square," in *American Studies and Consumption*, ed. D. Nye and C. Pedersen (Amsterdam: Free University Press, 1991), pp. 157–158.

42. "New York's Avenue of Light," *New York Times*, October 10, 1926.

43. The parties reached a settlement in 1929, in the form of a consent decree. Under its provisions, the contract between GOAC and the Bureau was canceled; the Bureau became an independent agent for display business, and it was specifically enjoined from discriminating against independent operators (*Harvard Law Review*, April 1933, pp. 931–932). Similarly, a consent decree signed in *United States vs. Foster & Kleiser Co.*, (S. D. Cal. March 13, 1931) required the defendant, which had a near monopoly of the sign business on the West Coast, to sell the business and assets of the La Fon System, Inc., a competitive display plant in Los Angeles that it had acquired.

44. "The Savagery of Electric Sky-Signs: A Conversation with George Moore," *World Today*, 1924, pp. 222–223. This article provoked a reply ("Signs of the Times," *World Today*, 1924, pp. 357–358) in which Charles Higham states: "I entirely disagree with Mr. Moore's statement that we become as savages. On the contrary we are infinitely better off in every way than we were, and this is largely due to advertisement." See also L. W. Diggs, "Refinement of Outdoor Advertising," *National Municipal Review* 14, October 1925, pp. 613–617.

45. Cited in Rogers and Weston, *Carnival Crossroads*, pp. 105–106.

46. Charles Mulford Robinson, *The Improvement of Towns and Cities* (Putnam, 1902), pp. 87–88.

47. John DeWitt Warner, "Advertising Run Mad," *Municipal Affairs* 4 (1900), no. 2, p. 276.

48. Comment by Mr. Gille, Proceedings of 31st Convention, National Electric Light Association, Chicago, 1908, p. 763.

49. Cited in Raymond C. Miller, *Kilowatts at Work: A History of the Detroit Edison Company* (Wayne State University Press, 1957), p. 153.

50. Chester H. Liebs, *Main Street to Miracle Mile* (Little, Brown, 1985), pp. 41–42.

51. *New York Times*, February 3, 1924.

52. "Broadway Association Will Fight Petition," December 10, 1922; "Sign Committee," December 16, 1923.

53. See *New York Times,* May 1, May 6, May 7, May 8, May 9, May 18, and May 23, 1929.

54. "Fight Hanging Signs in Retail District," *New York Times,* April 30, 1929.

55. "Paris Bans American Type Advertising Signs," *New York Times,* October 30, 1926. A year later an editorial appeared about this decision in Paris (November 22, 1927). Also see June 18, 1928.

56. Bill Blackbeard and Martin Williams. *Smithsonian Collection of Newspaper Comics* (Smithsonian Institution Press, 1977), p. 156.

57. "The Sky Is His Blackboard," *The American Magazine,* March, 1931, p. 78.

58. James Oppenheim, *The Olympian* (Harper, 1912), p. 22.

59. Ibid.

60. More recently Piri Thomas has described that landscape's continuing appeal in *Down These Mean Streets* (Random House, 1967): "In the daytime Harlem looks kinda dirty and the people a little drab and down. But at night, man, it's a swinging place, especially Spanish Harlem. The lights transform everything into life and movement and blend the different colors into a magic cover-all that makes the drabness and garbage, wailing kids and tired people invisible. Shoes and clothes that by day look beat and worn out, at night take on a reflected splendor that the blazing multi-colored lights burn on them. Everyone seems to develop a sense of urgent rhythm and you get the impression that you have to walk with a sense of timing."

61. Lewis Mumford, *Sketches from Life: The Autobiography of Lewis Mumford: The Early Years* (Dial, 1982), pp. 129–130.

62. Pound, cited in Hugh Kenner, *A Homemade World: The American Modernist Writers* (William Morrow, 1975), p. 5.

63. Edward Hungerford,"The Night-Glow of the City," *Harper's* 54 (1910), no. 11, p. 3.

64. The electrified city became a central subject for American painters and photographers. The night cityscape was a new visual reality, signifying a break in the continuity of lived experience. It challenged artists with a new subject, and most important painters of the early twentieth century attempted to portray the city at night, including Alfred Stieglitz, John Sloan, Edward Hopper, Georgia O'Keeffe, Charles Burchfield, and Joseph Stella. See chapter 2 of Nye, *Electrifying America.*

65. Arnold Bennett, *Those United States* (Martin Secker, 1912), pp. 25–26.

66. Vladimir Mayakovsky, *My Discovery of America,* quoted in Wiktor Woroszylski, *The Life of Mayakovsky* (London, 1972).

67. Ibid.

68. Claude McKay, "On Broadway," c. 1920. In the early 1960s another "On Broadway," written by Barry Mann, Cynthia Weil, Jerry Leiber, and Mike Stoller and performed by the Drifters, offered a more pointed critique: "They say the

neon lights are bright / on Broadway / They say there's always magic in the air / But when you're walking down that street / and you ain't had enough to eat / the glitter rubs right off, and you're nowhere. . . ." (Screen Gems-EMI Music, 1962)

69. Cited in Cervin Robinson and Rosemari Haag Bletter, *Skyscraper Style: Art Deco New York* (Oxford University Press, 1975), p. 67.

70. Kenneth Burke, letter dated January 18, 1923, in *The Selected Correspondence of Kenneth Burke and Malcolm Cowley*, ed. P. Jay (Viking, 1988), p. 131. Cowley, then in Europe, replied: "A single electric light on Broadway has the same value as a single lower-case 'e' in Hamlet. As for America being the concentration point for all the vices and vulgarities of the world—shit. New York is refinement itself beside Berlin." (p. 135)

71. H. G. Wells, *The Future in America* (Bernhard Tauchnitz, 1907), p. 48.

72. Jean Baudrillard, *For a Critique of the Political Economy of the Sign* (Telos, 1981). For a related discussion in an American context, see Lears, "Some Versions of Fantasy" (n. 31 above).

73. "History of a Decade Told in a Ribbon of Light," *New York Times Magazine*, November 13, 1938.

Chapter 8

1. I am indebted to Roland Marchand for his work on corporate display techniques. See Roland Marchand, "Corporate Imagery and Popular Education: World's Fairs and Expositions," in *Consumption and American Culture*, ed. D. E. Nye and C. Pedersen (Amsterdam: Free University Press, 1991), pp. 21–23.

2. Ibid.

3. "Proposed Basis of Illumination at the New York World's Fair," from *Edison Life* (n. d.) (courtesy of Boston Edison Company).

4. Joseph Corn, *America's Romance with Aviation, 1900–1950* (Oxford University Press, 1983), pp. 6–8; Elsbeth E. Freudenthal, *Flight Into History* (University of Oklahoma Press, 1949), pp. 85–101. Experts doubted that two unknown mechanics had really achieved human flight, and those members of the general public who heard about it tended to confuse their airplane with lighter-than-air craft (airships), which had been familiar since the 1880s.

5. Corn, *America's Romance*, p. 4.

6. Corn, chapter 5, passim; Robert J. Serling, *Eagle: The Story of American Airlines* (St. Martins, 1985), pp. 108–109. I thank Robert Baehr for sharing his knowledge of early aviation history with me.

7. *Amsterdam News*, April 29, 1939. There were also many stories about the pressure to train blacks as pilots in the *Washington Afro-American*—see, for example, the issues of March 18 and March 4. Edgar Brown, president of the United Government Employees, appealed to the senate to increase President Roosevelt's request from 5500 planes to 6000, with the extra 500 earmarked for blacks.

8. *Chicago Defender*, May 6, 1939; follow-up stories, May 13 and 20.

9. Two secondary works are devoted entirely to the New York Fair: Helen A. Harrison, ed., *Dawn of a New Day: The New York World's Fair 1939/40* (New York University Press, 1980); Larry Zim, Mel Lerner, and Herbert Rolfes, *The World of Tomorrow, 1939: New York World's Fair* (Harper and Row, 1988). The latter also reprints several interesting contemporary accounts. Warren Susman later published a revised version of an essay in Harrison's volume as chapter 11 of *Culture as History* (Pantheon, 1985). An important visual source is Stanley Appelbaum's book *The New York World's Fair* (Dover, 1979). See also Folke Kihlstedt, "Utopia Realized: The World's Fairs of the 1930s," in *Imagining Tomorrow.*, ed. J. Corn (MIT Press, 1986); Alice Goldfarb Marquis, *Hopes and Ashes: The Birth of Modern Times* (Free Press, 1986); Michael Robertson, "Cultural Hegemony Goes to the Fair," *American Studies*, spring, 1992; David E. Nye, "Yesterday's Ritual Tomorrows: The New York World's Fair of 1939," *Anthropology and History*, spring 1992. The New York World's Fair gets only a passing reference in each of the following standard surveys of the 1930s: Robert S. McElvaine: *The Great Depression, America 1929–1941* (Time Books, 1984); Robert Goldstone, *The Great Depression* (Bobbs-Merrill, 1969); Donald R. McCoy: *Coming of Age* (Penguin, 1973). Richard Pells does not mention the fair in *Radical Visions and American Dreams: Culture and Social Thought in the Depression Years* (Harper and Row, 1973). Of survey books, only Dixon Wecter's *Age of the Great Depression* (Macmillan, 1948) contains as much as a paragraph about the fair. Popular histories of the 1930s give it more space; see, e. g., J. C. Furnas, *Stormy Weather: Crosslights on the 1930s* (Publisher, 1977), pp. 589–593.

10. Burton Benedict, *The Anthropology of World's Fairs* (Scholar Press, 1985), p. 10.

11. On the Japanese exhibit see *Official Guide Book, New York World's Fair, 1939,* pp. 107–108. Plans for a German cultural pavilion emphasizing pre-Nazi culture, announced January 13, were abandoned in February (*New York Times*, January 13 and February 2).

12. *New York Times,* July 16, 1939.

13. See Kasson, *Civilizing the Machine*, pp. 23–28. Some rides were popular from the start, particularly a 250- foot parachute jump. The most popular attraction was Billy Rose's Aquacade, a kind of mermaid ballet in the style of the Radio City Music Hall. Exotic acts and exhibits, long a staple of expositions, also fared badly. Much of the midway struggled to break even over the fair's two seasons, and the concessions pressured the fair's management to extend hours and reduce entrance fees (*New York Times*, May 3, 1939). For a description of the amusement park see the special supplement to the *Times* of April 30; on costs see July 12; on the bankruptcy of Children's World see August 4; on attendance problems see July 23.

14. On the development of the museum see Neil Harris. "The Gilded Age Revisited: Boston and the Museum Movement," *American Quarterly*, winter 1962; Stephen Bann, "Historical Text and Historical Object: The Poetics of the Musée de Cluny," *History and Theory* 17 (1978), no. 3.

15. On the Washington memorials see the *New York Times'* special supplement on the World's Fair, April 30, 1939; see also "Inauguration Reenacted," May 1.

16. Paul Greenhalgh, *Ephemeral Vistas* (University of Manchester Press, 1988), pp. 27–41.

17. Bond subscribers are listed in the *New York Times* of May 1, 1939.

18. Source of costs: *New York Times*, April 30, 1939.

19. Cited in Jeffery Hart, "Yesterday's America of Tomorrow," *Commentary*, July 1985, p. 62.

20. Map, *New York Times*, April 30.

21. Ibid.

22. Robert D. Kohn, "Social Ideals in a World's Fair," reprinted in *Culture and Commitment, 1929–1945*, ed. W. Susman (Braziller, 1973), p. 298.

23. David Noble (*America by Design: Science, Technology and the Rise of Corporate Capitalism* [Oxford University Press, 1977]) describes how corporations shaped the engineering profession in the United States. See particularly the last four chapters.

24. Donald Bush (*The Streamlined Decade* [Braziller, 1975], pp. 17–21) outlines the backgrounds of these designers. See also Jeffrey Meikle, *Twentieth Century Limited: Industrial Design in America, 1925–1939* (Temple University Press, 1979).

25. "Jitterbugs Break Ground for Fair," *Washington Afro-American*, March 25, 1939.

26. The fair had a poor record of minority employment. Mayor La Guardia officially invited a number of African-American groups to the fair, and he gave his word that the civil rights laws would be upheld. There was to be no segregation anywhere on the grounds. In general this rule was upheld, although the Standard Brands exhibit maintained separate toilets for black and white employees. Yet blacks were offered only a few, menial jobs. A month before the opening, several hundred African-Americans picketed fair president Grover Whalen's office at the Empire State Building, demanding more employment. On opening day, 300 African-Americans protested in Times Square and 500 stood outside the fair's entrance. (Inside, about 25,000 African-Americans attended the fair, largely to hear President Roosevelt speak.) Because of the protests, 150 more black people were hired. Nevertheless, on May 13 a community meeting in Harlem condemned "the Jim Crow Job Situation" and protested "the position of the Negro in 'the world of tomorrow.'" This led to additional hiring of blacks and to the creation of a "Negro exhibit" in the Federal Building. African-Americans were not hostile to the fair's themes; they wanted to be included. This meant more than getting jobs at the fair; it also meant allowing white tourists into Harlem, where a flood of out-of-towners, both black and white, bolstered the local economy. The *Amsterdam News* reported the following: "Noting the alacrity with which white visitors to the Fair scurried to Harlem no sooner than they had peeked into the exhibit out in Queens the authorities were said to have thrown up their hands in disgust and decided to abandon the hard and shut policy of 'no whites in Harlem' and to have agreed to let Harlem 'live' insofar as getting a share of the world's fair money is concerned." ("Harlem is Wide Open During Fair," *Amsterdam News*, May 13) The story went on to say that "big money gamblers and playboys were already on the train heading for Harlem" and that prostitutes were being allowed to ply their trade without much interference from police." "As long as they don't rob,

cut, or otherwise injure guests of the fair . . . they can have their fun," said one official. For a detailed look at the treatment of African-Americans at the fair see Robert Rydell, *World of Fairs: The Century of Progress Expositions* (University of Chicago Press, 1993), pp. 157–190. In the second year of the fair, more than 130 black leaders, including W. E. B. Dubois, sponsored a Negro Week celebration.

27. "125 D. C. Students Visit New York fair; Cruise Hudson," *Washington Afro-American*, May 27, 1939.

28. *Chicago Defender*, April 29.

29. See "Race Achievement to Get Spot at Exposition," *Amsterdam News*, May 22.

30. This was an important gesture, because Anderson had recently been denied the use of a concert hall in Washington by the Daughters of the American Revolution. After this incident, First Lady Eleanor Roosevelt resigned from the DAR and invited her to perform at the White House. See *Amsterdam News*, April 15, 1939.

31. See Zim et al., *The World of Tomorrow, 1939*, p. 174.

32. See "Sissle Certain to Play World's Fair," *Chicago Defender*, May 13; "Trip to World's Fair Contest Starts Next Week," June 17; "You'll Enjoy N. Y. Trip, Says Defender Winner," June 17; photographs of candidates on front page, June 24. Winners received a free trip that included Niagara Falls, Pittsburgh, Atlantic City, and New York. The contest was designed as a circulation-booster. Thirteen social clubs in the black community were asked to nominate one candidate each. The "votes" came in the form of coupons cut out of the paper, which were countered periodically during the campaign. The contest closed with a "mammoth dance" at which the five winners were announced.

33. Market Analysts, Inc., interviewed 1020 people in August 1939 and presented the results to fair's managers in a report entitled "Attendance and Amusement Area Survey of the New York World's Fair 1939, Incorporated" (New York Public Library World's Fair Collection, 1934–40, p. 11).

34. Dean MacCannell provides a theory for the ways tourists identify and experience sites in *The Tourist: A New Theory of the Leisure Class* (Schocken, 1976). See particularly chapter 6,"A Semiotic of Attraction."

35. *Official Guide*, pp. 27–29.

36. Kohn, "Social Ideals in a World's Fair," p. 298–299.

37. *New York Times*, May 1.

38. Ibid.

39. On the reproduction of disasters at fairs see Benedict, *Anthropology of World's Fairs*, p. 56.

40. *New York Times*, May 7. On the Perisphere exhibit see Zim et al., *The World of Tomorrow, 1939*, pp. 39 and 49–55.

41. Quotations from The City of Light (Consolidated Edison, 1939), pp. 1, 16.

42. *New York Times*, May 7, April 8, April 30.

43. *Official Guide*, pp. 151, 166; *New York Times*, April 30.

44. Victor Turner, *The Ritual Process* (Cornell University Press, 1969), p. 91.

45. See Zim et al., *The World of Tomorrow, 1939*, pp. 43, 84–86.

46. I discuss the General Electric exhibit in more detail in *Electrifying America* (pp. 373–375).

47. Roland Marchand and Michael Smith, "Corporate Science on Display," in *Science and Reform in Industrial America*, ed. R. Walters (Johns Hopkins University Press, 1992).

48. On the Westinghouse exhibit see Zim et al., *The World of Tomorrow, 1939*, pp. 78–80.

49. *New York Times*, April 30, 1939.

50. *Official Guide*, p. 161.

51. Bush, *The Streamlined Decade*, p. 167.

52. *New York Times*, July 5.

53. General Motors Company, Highways and Horizons (brochure). See also Bush, *The Streamlined Decade*, p. 161.

54. Meikle, *Twentieth Century Limited*, p. 202.

55. *Life*, June 5, 1939; reprinted in *This Fabulous Century, 1930–1940* (American Heritage, 1975), pp. 279–280.

56. Ibid.

57. *New York Times*, April 30.

58. Alan Trachtenberg, *The Incorporation of America*, pp. 221–222.

59. For a discussion of sacred time and sacred space see Mircea Eliade, *The Sacred and the Profane* (Harcourt, Brace, 1968).

60. Turner, *The Ritual Process*, pp. 108–111.

Chapter 9

1. On early ideas of space travel see Brian Horrigan, "Popular Culture and Visions of the Future in Space, 1901–2001," in *New Perspectives on Technology and American Culture*, ed. B. Sinclair (American Philosophical Society, 1986). See also Sam Moskowitz, "The Growth of Science Fiction from 1900 to the Early 1950s," Fred L. Whipple,"Recollections of the Pre-Sputnik Days," and Randy Liebermann,"The Collier's and Disney Series," all in *Blueprint for Space: Science Fiction to Science Fact*, ed. F. Ordway and R. Lieberman (Smithsonian Institution, 1992).

2. Rip Bulkeley, *The Sputnik Crisis and Early United States Space Policy* (Macmillan, 1991), p. 5.

3. Cited in Walter A. McDougall, *The Heavens and the Earth: A Political History of the Space Age* (Basic Books, 1985), p. 145. The most militant black publications often took satisfaction in reporting Soviet successes. On July 4, 1969, shortly before Apollo XI, *Muhammad Speaks* noted: "Obviously Russia is ahead of the United States in the one important aspect of the space race. This area is the unmanned interplanetary exploration. . . . Some suggestion of the magnitude and thrust of the Soviets came this May, when the Russians, only 52 years after the monumental revolution, zoomed an unmanned flight to Venus."

4. See McDougall, *The Heavens and the Earth*, pp. 149–176.

5. Gen. Leslie Groves, "Top Secret Memorandum for the Secretary of War," in H. Feis, *Between War and Peace: The Potsdam Conference* (Princeton University Press, 1960). On the unexpected size of the explosion see *History of the Atomic Energy Commission,* vol. 1, *The New World: 1939–1946* (Pennsylvania State University Press, 1962), p. 380.

6. Groves, "Top Secret Memorandum," p. 381.

7. Cited in Richard Rhodes, *The Making of the Atom Bomb* (Simon & Schuster, 1986), p. 672.

8. Responses cited in ibid., p. 675.

9. Burke, *Enquiry,* p. 79.

10. See Edith Wyschogrod, *Spirit in Ashes* (Yale University Press, 1985), p. 16.

11. Ibid, p. 15. Wyschogrod's analysis is directed primarily at the death camps and slave labor camps of totalitarian governments, but the concept is clearly applicable to atomic weapons as well, as she confirmed in private conversation. Such an awareness of the death-world, of life under threat, is fundamentally inconsistent with the Lockean idea of freely uniting individuals, who form a social contract. It likewise undermines the whole concept of a "state of nature" from which society evolves, and laissez-faire notions of free competition. Free nuclear competition can only lead to the apocalypse. Instead, the relation of the individual to the state in a nuclear age looks more like a Hobbesian life without the state: nasty, brutish, and short.

12. For an overview of the enormous and complex response to nuclear weapons in literature and the arts see James R. Bennett and Karen Clark, "Hiroshima, Nagasaki, and the Bomb: A Bibliography of Literature and the Arts," *Arizona Quarterly,* fall 1990.

13. *History of the Atomic Energy Commission,* p. 379.

14. Brian Easlea, *Fathering the Unthinkable* (Pluto, 1983), p. 130.

15. Easlea makes an important argument, but it is not my point here. He is concerned with the internal dynamic that leads to the creation of weapons of mass destruction. In contrast, throughout this volume I am concerned with the general public and the meaning of new technologies for them. For too long it has been assumed that the social meaning of a new machine was defined by the inventor.

16. Thomas P. Hughes, *American Genesis: A Century of Invention and Technological Enthusiasm* (Penguin, 1989), p. 383.

17. Ibid., p. 393, 401. Many of the Manhattan Project scientists were unhappy with the increasing military-industrial control, but Du Pont's conservative design of the plutonium production plant at Hanford was a success precisely because its capacity exceeded what seemed necessary according to theory and small-scale experiments.

18. Cited in Werner Meihofer, "The Ethos of the Republic and the Reality of Politics," in *Machiavelli and Republicanism,* ed. G. Bok et al. (Cambridge University Press, 1990).

19. Lewis Mumford, "Gentlemen, You Are Mad!" *Saturday Review of Literature,* March 2, 1946.

20. Everett Mendelsohn, "Prophet of Our Discontent: Lewis Mumford Confronts the Bomb," in *Lewis Mumford: Public Intellectual* (Oxford University Press, 1990), p. 343.

21. Spencer R. Weart, *Nuclear Fear: A History of Images* (Harvard University Press, 1988), p. 107.

22. Carole Gallagher, *American Ground Zero: The Secret Nuclear War* (MIT Press, 1993).

23. Thomas H. Saffer and Orville Kelly, *Countdown Zero: GI Victims of U.S. Atomic Testing* (Penguin, 1983), pp. 35–47; quotation from p. 43.

24. Catherine Caufield, *Multiple Exposures* (University of Chicago Press, 1990), p. 106.

25. John M. Findlay, *People of Chance: Gambling in American Society from Jamestown to Las Vegas* (Oxford University Press, 1986), p. 154.

26. Costandina Titus, *Bombs in the Backyard* (University of Nevada Press, 1986), pp. 93–95.

27. Lawrence Wright, *In the New World: Growing Up with America, 1960–1984* (Knopf, 1988), p. 35.

28. On early reactions to the atom bomb see Paul Boyer, *By the Bomb's Early Light: American Thought and Culture at the Dawn of the Atomic Age* (Pantheon, 1985).

29. Quoted in Weart, *Nuclear Fear*, pp. 12–13.

30. Cited in Daniel Ford, *The Cult of the Atom*, (Simon and Schuster, 1984). For more on early predictions of atomic abundance see Stephen Hilgartner, Richard C. Bell, and Rory O'Connor, *Nukespeak: The Selling of Nuclear Technology in America* (Penguin, 1983), p. 19; Stephen L. Del Sesto, "Wasn't the Future of Nuclear Energy Wonderful?" in *Imagining Tomorrow*, ed. Corn, p. 61. See also John O'Neill, "Enter Atomic Power," *Harper's*, June 1940; William L. Laurence, "The Atom Gives Up," *Saturday Evening Post*, September 7, 1940. Laurence was later the only journalist in residence at Los Alamos.

31. Michael Smith, "Advertising the Atom," in *Government and Environmental Politics*. ed. M. J. Lacy (Wilson Center Press, 1989), pp. 245–247.

32. Ibid., p. 247.

33. See J. Samuel Walker, "Nuclear Power and the Environment: The Atomic Energy Commission and Thermal Pollution, 1965–1971," *Technology and Culture* 30 (1989), pp. 964–992.

34. Ford, *Cult of the Atom*, pp. 179–180.

35. Glenn T. Seaborg and William R. Corliss, *Man and Atom: Building a New World Through Nuclear Technology* (Dutton, 1971), p. 13.

36. Ibid., pp. 165–167.

37. Smith, "Advertising the Atom," pp. 248–252.

38. Seaborg and Corliss, *Man and Atom*, pp. 186–187.

39. On the government coverup see Titus, *Bombs in the Backyard*, Howard Ball, *Justice Downwind* (Oxford University Press, 1986), and Philip Fradkin, *Fallout:*

An American Nuclear Tragedy (University of Arizona Press, 1989). Ball analyzes changing community responses test by test; Fradkin emphasizes the official coverup and the devastating effects on particular victims.

40. On incompetence within the Pennsylvania reactor and in the NRC, see Daniel E. Ford, *Three Mile Island* (Penguin, 1983), pp. 42–44, 55–59, 246–250.

41. Sharon O'Brien, "Domesticating Technology: Structuring the Tourist Experience at Three Mile Island," delivered at American Studies Association National Convention, Toronto, 1989.

42. Daniel J. Boorstin, *The Image: A Guide to Pseudo-Events in America* (Harper Colophon, 1961), p. 116.

43. Dale Carter, *The Final Frontier: The Rise and Fall of the Rocket State* (Verso, 1988), pp. 187–188.

44. Robert Baehr, "The Moon, How Far Is It . . . to Charles Lindbergh?" in *American Studies in Transition.*, ed. D. Nye and C. Thomsen (Odense University Press, 1985), p. 162.

45. See Carter, *The Final Frontier,* passim.

46. Thomas Weiskel, *The Romantic Sublime* (Johns Hopkins University Press, 1976), p. 24.

47. "'Yeeeow!' and 'Doggone!' Are Shouted on Beaches As Crowds Watch Liftoff," *New York Times,* April 13, 1981.

48. Christian Williams, "Beach Blanket Liftoff, The Shuttle-Watchers Spaced-Out Bon Voyage Party," *Washington Post,* April 10, 1981.

49. Victor and Edith Turner, *Image and Pilgrimage in Christian Culture* (Columbia University Press, 1978), p. 7.

50. Ibid., p. 9.

51. *New York Times,* April 13, 1981.

52. Norman Mailer, *Of a Fire on the Moon* (Weidenfeld and Nicolson, 1970), p. 48.

53. Here and in the following paragraphs I rely for background upon *Facts on File,* vol. 29, no. 15 (July 24–30, 1969).

54. *New York Times,* July 16, 1969.

55. *Philadelphia Inquirer,* February 17, 1969. Those on the East and West Coasts were more supportive than those in the Midwest and the South. Support also correlated positively with education—65% of those with an eighth-grade education or less opposed the moon landing, while 62% of those with a college education supported it. Nevertheless, many college students favored cutting the space program (*San Francisco Examiner,* July 29, 1969). On youth reaction see "Teen-Agers Show Cynicism on America's Space Race," *New Orleans Times-Picayune,* February 8, 1968.

56. *Washington Post,* July 14, 1969.

57. Mailer, *Of a Fire on the Moon,* pp. 48–49.

58. Diana Ackerman, *A Natural History of the Senses* (Random House, 1990), pp. 279–280.

59. Mailer, pp. 79–81, 180.

60. Burke, *Enquiry*, p. 75.

61. *Changing Times*, October 28, 1983, p. 28.

62. Advertisement in *Space World*. February 1985.

63. Mailer, p. 180.

64. Courtney G. Brooks et al., *Chariots for Apollo: A History of Manned Lunar Spacecraft* (NASA, 1979), pp. 327–329.

65. *New York Times,* July 25, 1969.

66. "Americans Still Question Space Budget," *Washington Post,* August 25, 1969.

67. *USA Today* poll taken July 14, 1989, cited in Ordway and Liebermann, *Blueprint for Space,* p. 196.

68. *America,* November 28, 1981, p. 331.

69. "Watching the Space Shuttle," *Sunset,* March 1982, p. 80.

70. "News Media Swarm to Shuttle Site, Giving Launching a Carnival Air," *New York Times,* April 8, 1981.

71. Ibid.

72. Bruce Franklin, *War Stars: The Superweapon and the American Imagination* (Oxford University Press, 1988). On the technical weaknesses of SDI, particularly the problems of tactical decision-making it posed, see Sanford Lakoff and Herbert F. York, *A Shield in Space? Technology, Politics, and the Strategic Defense Initiative* (University of California Press, 1989).

73. Franklin, *War Stars,* p. 13.

74. Ibid., p. 198.

75. See Leslie R. Groves, *Now It Can Be Told: The Inside Story of the Development of the Atomic Bomb* (Andre Deutsch, 1963), p. 187.

76. Rob Wilson, *American Sublime: Genealogy of a Poetic Genre* (University of Wisconsin Press, 1991), p. 201. (For Los Alamos read Alamogordo.)

77. Cited in Kasson, *Civilizing the Machine,* p. 144.

Chapter 10

1. Anne Cannon Palumbo and Ann Uhry Abrams, "Proliferation of the Image," in *Liberty: The French American Statue in Art and History,* prepared by New York Public Library (Harper and Row, 1986), pp. 234–236, 244, 247.

2. Burke, *Enquiry,* pp. 164–165.

3. John J. Garnett, *The Statue of Liberty: Its Conception, Its Construction, Its Inauguration* (Dinsmore, 1886), p. 25.

4. Marina Warner, *Monuments and Maidens: The Allegory of the Female Form* (Picador, 1985), p. 8. Although I take issue with Warner on this point, her book is a valuable introduction to the problem.

5. Marvin Trachtenberg, *The Statue of Liberty* (Viking, 1976), p. 116–117.

6. Ibid., pp. 63–69.

7. Pierre Provoyeur, "Artistic Problems," in Palumbo and Abrams, *Liberty*, p. 182. Provoyeur's article covers much the same ground as one chapter in Trachtenberg's book (n. 5 above).

8. Martha Banta gives considerable attention to this theme in *Imaging American Women* (Columbia University Press, 1967). Her point is that women were sculpted and painted on an enormous scale, particularly at the Columbian Exposition and after, to represent militant democracy. These warrior queens provide another example of feminization that parallels the way machines were invested with sexuality and thus made more acceptable.

9. Cited in Provoyeur, "Artistic Problems," p. 82.

10. Ibid., p. 99.

11. Ibid.

12. Gösta M. Bergman, *Lighting in the Theater* (Rowman and Littlefield, 1977), pp. 278–286.

13. Trachtenberg, *The Statue of Liberty*, p. 129.

14. For details see June Hargrove's two articles, "The American Committee," and "The American Fund-Raising Campaign," in Palumbo and Abrams, *Liberty*.

15. Hertha Pauli and E. B. Ashton, *I Lift My Lamp: The Way of a Symbol* (Appleton-Century-Crofts, 1948), pp. 281–284.

16. John Higham, *Send These To Me*, revised edition (Johns Hopkins University Press, 1986), p. 74.

17. Pauli and Ashton, *I Lift My Lamp*, p. 302.

18. See Werner Sollors,"Of Plymouth Rock and Jamestown and Ellis Island—Or Ethnic Literature and Some Redefinitions of 'America'," in Hans Bak, ed., *Multiculturalism and the Canon of American Culture*, ed. H. Bak (Amsterdam: Free University Press, 1993), pp. 279–281.

19. See Anne Cannon Palumbo and Ann Ury Abrams, "Proliferation of the Image," in *Liberty*, p. 236.

20. Higham, *Send These to Me*, p. 76.

21. H. G. Wells, *The Future in America* (Bernhard Tauchnitz, 1907), p. 43.

22. Ibid., p. 234

23. I am indebted to Niels Thorsen, who drew my attention to the political significance of the Statue of Liberty ceremony as we incorporated it into our common work of producing educational television programs for Danish National Television. His essay "Political Rituals in America" appears in *Inventing Modern America* (Danmarks Radio, 1989).

24. Richard Seth Hayden and Thierry W. Despont describe the restoration work briefly in "Restoration of the Statue of Liberty, 1984–1986," in Palumbo and Abrams, *Liberty*. Also see "Cables, the Man Behind Liberty Gala Spectacular" (*Amsterdam News*, July 5, 1986), which is about the black Park Service administrator whose office issued the contracts for the restoration work.

25. Hayden and Despont, "Restoration," p. 223.

26. *New York Times,* July 4.

27. John Corry, "Networks and the Statue's Lighting," *New York Times,* July 4, 1986.

28. Among the other performers were more than 75 Elvis Presley impersonators.

29. *New York Daily News,* July 4, 1986; *New York Times,* June 4 and 13.

30. *New York Daily News,* July 4; *Dayton Daily News,* editorial, July 4; *Pittsburgh Press,* July 3; *Lexington Herald Leader,* July 1.

31. Maureen Dowd, "Yachts and Names Crowd the Harbor," *New York Times,* July 3.

32. Thomas J. Lueck, "Windfall Is Eluding Merchants," *New York Times,* July 5.

33. "Image of Liberty May Seem Ubiquitous This Weekend," *New York Times,* July 1.

34. *New York Times,* June 18.

35. John Gross, "The Ideal vs. Rassmatazz," *New York Times,* July 5. An interesting meditation on the commercialism of the event, by "Sob Sister," appeared in the July 5 *Baltimore Afro-American.*

36. Robert D. McFadden, "On the Eve of Liberty Salute, A Whirl of Final Touches," *New York Times,* July 3.

37. The weekly *Chicago Muslim Journal* made no mention of the rededication in its issues dated July 4 and 11.

38. These ideas reappeared in the July 12 issue of the *Amsterdam News* in an article titled "Weekend for the Super Rich."

39. See Martin Gottlieb, "Folk Hero for the '80s" *New York Times,* July 3; "For Iacocca, A Day of Accolades" and "About New York," *New York Times,* July 4.

40. Across the country, about 25,000 others also became citizens at the same time via closed-circuit television. This was the spectacle that some judges had refused to endorse.

41. See Michael Rogin, *Ronald Reagan, The Movie, and Other Episodes in Political Demonology* (University of California Press, 1987), pp. 1–43.

42. *Public Papers of Ronald Reagan,* 1986, vol. 2, June 28–December 31 (Government Printing Office, 1989), p. 920. Reagan also recalled the rapprochement between Jefferson and Adams in their last years through their correspondence, and the coincidence that both men died on July 4, 1826, precisely 50 years after the Revolution. On the singer, Kenneth Mack, see *Amsterdam News,* July 12.

43. Palumbo and Abrams, *Liberty,* p. 212.

44. *New York Times,* July 4.

45. Ibid.

46. Elizabeth Kolbert, "A 2,000 Microphone Man," *New York Times,* July 3.

47. David E. Sanger, "Fireworks By Computer," *New York Times,* July 3.

48. *Public Papers of Ronald Reagan,* 1986, vol. 2, p. 925.

49. David W. Dunlop, "Downtown Is Awaiting the Horde," *New York Times,* July 3.

50. *New York Daily News*, July 4.

51. "Fireworks Fill Sky After Ships Honor the 4th and the Statue," *New York Times*, July 5.

52. Joseph Berger, "New York Relaxes as It Savors a Celebration," *New York Times*, July 5.

53. This preservation was a logical part of the politics of the Reagan era, which required the positive validation of technology as part of its revamped nineteenth-century ideology of individualism and laissez-faire.

Chapter 11

1. George Ticknor, *Life, Letters and Journals of George Ticknor*, cited in Charles Mason Dow, *Anthology and Bibliography of Niagara Falls*, vol. 1 (State of New York, 1921), p. 221.

2. On European intellectuals and their postmodern visions of America see Rob Kroes, "Flatness and Depth," in *Consumption and American Culture.*, ed. D. E. Nye and C. Pedersen (Amsterdam: Free University Press).

3. I believe Professor Marx would agree with this statement.

4. Martha Banta, *Imaging American Women: Idea and Ideals in Cultural History* (Columbia University Press, 1987), p. 761.

5. Don Ihde, *Technology and the Lifeworld: From Garden to Earth* (Indiana University Press, 1990), pp. 21–30.

6. Catherine Caufield, *Multiple Exposures* (University of Chicago Press, 1990), pp. 106–107.

7. Poems of Ralph Waldo Emerson (Collected Works, vol. 9) (Houghton Mifflin, 1904), p. 78.

8. Slavoj Zizek, *The Sublime Object of Ideology* (Verso, 1989), pp. 203, 204.

9. National Park Service, *Grand Canyon Natural and Cultural Resource Management Plan*, April 1984.

10. See U.S. Department of the Interior, *Aircraft Management Plan Environmental Assessment, Grand Canyon National Park*, May 1986.

11. These questions had been put so many times that the rangers at the information counters had a list of the twenty most common queries printed up and taped it to the counter in early December 1993. I interviewed five staff members, all of whom confirmed that these were frequently asked questions. It is also common for people to ask where the geysers are and to ask for the location of the "carved stone faces." Some people also want to know if it is true that someone jumped the canyon on a motorcycle.

12. For a history of gambling with a focus on Las Vegas see John M. Findlay, *People of Chance: Gambling in American Society from Jamestown to Las Vegas* (Oxford University Press, 1986).

13. For a survey and a history of Las Vegas's casinos see David Spanier, *Welcome to the Pleasure Dome* (University of Nevada Press, 1992).

14. Robert Venturi, *Learning from Las Vegas: The Forgotten Symbolism of Architectural Form*, revised edition (MIT Press, 1991).

15. Statistics from Las Vegas Chamber of Commerce, *Las Vegas Perspective.* Las Vegas Review Journal, 1993.

16. My descriptions of the hotels are based on visits, brochures, and advertising. See also Sergio Lalli, "Big Projects Boost Vegas," *Hotel and Motel Management* 206, November 4, 1991; "Wynn's World: White Tigers, Blackjack, and a Midas Touch," *Business Week*, March 30, 1992; David B. Rosenbaum, "Resorts: Precast as Big Winner in Vegas" *ERN* 229, July 13, 1992.

17. This fact has interesting financial implications. Surveys already show that Las Vegas attracts mostly middle-class consumers who gamble "voraciously and with less regard for income" than those who go to Atlantic City. In effect, those who visit Las Vegas pay a highly regressive tax for their pleasure. See Mary O. Borg, "An Economic Comparison of Gambling Behavior in Atlantic City and Las Vegas," *Public Finance Quarterly* 18 (1990), no. 3, pp. 291–312.

18. For background information on the new casinos I am indebted to the Las Vegas News Bureau.

19. Mike Weatherford, "Where Dreams Come True," *Nevada*, January-February 1993, pp. 10–14.

20. Though water is used lavishly at the new family resorts, water shortages present a potentially crippling barrier to further development. See John Jesitus, "Growing Gambling Mecca Reacts to Dwindling Supply of Water," *Hotel and Motel Management* 206, November 4, 1991.

Bibliography

For complete and detailed citations, see the notes (pp. 297–342).

Ackerman, Diana. *A Natural History of the Senses*. Random House, 1990.

Applebaum, Diana Karter. *The Glorious Fourth: An American Holiday, an American History*. Facts on File, 1989.

Arensberg, Mary, ed. *The American Sublime*. State University of New York Press, 1986.

Athearn, Robert G. *Union Pacific Country*. Rand McNally, 1971.

Ball, Howard. *Justice Downwind*. Oxford University Press, 1986.

Banta, Martha. *Imaging American Women: Idea and Ideals in Cultural History*. Columbia University Press, 1987.

Barthes, Roland. *The Eiffel Tower and other Mythologies*. Hill and Wang, 1979.

Baudrillard, Jean. *For a Critique of the Political Economy of the Sign*. Telos, 1981.

Beck, Warren A., and David A. Williams. *California: A History of the Golden State*. Doubleday, 1972.

Beckerman, Bernard. *Theatrical Presentation: Performer, Audience and Act*. Routledge, 1990.

Belfield, Robert. The Niagara Frontier. Ph.D. dissertation, University of Pennsylvania, 1981.

Bender, Thomas. *Toward an Urban Vision: Ideas and Institutions in Nineteenth-Century America*. University Press of Kentucky, 1975.

Benedict, Burton. *The Anthropology of World's Fairs*. Scholar, 1985.

Bennett, Arnold. *Those United States*. Martin Secker, 1912.

Bergman, Gösta M. *Lighting in the Theater*. Rowman & Littlefield, 1977.

Billington, David P. *The Tower and the Bridge*. Basic Books, 1983.

Bilstein, Roger. *Flight in America*. Johns Hopkins University Press, 1984.

Blackbeard, Bill, and Martin Williams. *Smithsonian Collection of Newspaper Comics*. Smithsonian Institution Press, 1977.

Bogart, John. *The American Railway: Its Construction, Development, Management, and Appliances*. Reprint of 1890 edition. Castle, 1988.

Boime, Albert. *The Magisterial Gaze: Manifest Destiny and American Landscape Painting, c. 1830–1865.* Smithsonian Institution, 1991.

Boorstin, Daniel. *The Image: A Guide to Pseudo-Events in America.* Harper Colophon, 1961.

Boorstin, Daniel. *The Americans: The National Experience.* Random House, 1965.

Boyer, Paul. *By the Bomb's Early Light. American Thought and Culture at the Dawn of the Atomic Age.* Pantheon, 1985.

Breecher, Jeremy. *Strike!* Straight Arrow, 1972.

Bright, Arthur A. *The Electric-Lamp Industry: Technological Change and Economic Development from 1800 to 1947.* Macmillan, 1949.

Bronner, Simon J., ed. *Consuming Visions: Accumulation and Display of Goods in America, 1880–1920.* Norton, 1989.

Brooklyn Museum. *The Great East River Bridge, 1883–1983.* Abrams, 1983.

Brooks, Courtney G., et al. *Chariots for Apollo: A History of Manned Lunar Spacecraft.* NASA, 1979.

Brown, Allen. *Golden Gate: Biography of a Bridge.* Doubleday, 1965.

Bulkeley, Rip. *The Sputnik Crisis and Early United States Space Policy.* Macmillan, 1991.

Burke, Edmund. *A Philosophical Enquiry into the Origin of Our Ideas of the Sublime and Beautiful.* Oxford University Press, 1990.

Bush, Donald J. *The Streamlined Decade.* Braziller, 1975.

Byington, Margaret. *Homestead: The Households of a Mill Town.* Reprint of 1910 edition. Pittsburgh: University Center for International Studies, 1974.

Carter, Dale. *The Final Frontier: The Rise and Fall of the Rocket State.* Verso, 1988.

Caufield, Catherine. *Multiple Exposures.* University of Chicago Press, 1990.

Chandler, Alfred D. *The Visible Hand.* Harvard University Press, 1977.

Chandler, William U. *The Myth of TVA.* Ballinger, 1984.

Cochran, Thomas C. *Pennsylvania: A Bicentennial History.* Norton, 1978.

Columbian Gallery: A Portfolio of Photographs from the World's Fair. Werner, 1894.

Conkin, Paul K. ed. *TVA: Fifty Years of Grassroots Bureaucracy.* University of Illinois Press, 1983.

Conrad, Peter. *Imagining America.* Oxford University Press, 1980.

Corn, Joseph. *America's Romance with Aviation, 1900–1950.* Oxford University Press, 1983.

Corn, Joseph, ed. *Imagining Tomorrow: History, Technology, and the American Future.* MIT Press, 1986.

Czitrom, Daniel. *Media and the American Mind.* University of North Carolina Press, 1982.

Danly, Susan, and Leo Marx. *The Railroad in American Art: Representations of Technological Change.* MIT Press, 1988.

Davis, Susan G. *Parades and Power: Street Theatre in Nineteenth-Century Philadelphia.* University of California Press, 1988.

Dorsh, T. S., ed. *Aristotle, Horace, Longinus: Classical Literary Criticism.* Penguin, 1965.

Douglas, George William, and Helen Douglas Compton. *The American Book of Days.* H. W. Wilson, 1948.

Dow, Charles Mason. *Anthology and Bibliography of Niagara Falls.* Two volumes. State of New York, 1921.

Dreiser, Theodore. *The Color of a Great City.* Boni & Liveright, 1923.

Dreiser, Theodore. *A Hoosier Holiday.* Oxford University Press, 1932.

Dublin, Thomas, ed. *Farm to Factory: Women's Letters, 1830–1860.* Columbia University Press, 1981.

Durkheim, Émile. *The Rules of Sociological Method,* Free Press, 1938.

Dutton, Clarence E. *Tertiary History of the Grand Cañon District.* Reprint of 1882 edition. Peregrine Smith, 1977.

Easlea, Brian. *Fathering the Unthinkable.* Pluto, 1983.

Electrical Advertising: Its Forms, Characteristics, and Design. Bulletin 50, Engineering Department, National Lamp Works, General Electric Company, 1925.

Eliade, Mircea. 1954. *The Myth of the Eternal Return.* Princeton University Press, 1954.

Eliade, Mircea. *The Sacred and the Profane.* Harcourt, Brace, 1959.

Eliade, Mircea. *The Forge and the Crucible.* Harper, 1962.

Emerson, Ralph Waldo. *Complete Works.* Houghton Mifflin, 1903.

Emerson, Ralph Waldo. *Selected Writings of Emerson,* ed. B. Atkinson. Modern Library, 1950.

Everett, Marshall. *The Book of the Fair; The Greatest Exposition the World Has Ever Seen, A Panorama of the St. Louis Exposition.* Henry Neil, 1904.

Ferraro, William M. "The Baltimore Railway Celebration of 1857: Transportation, Economics, and the Question of Union." Presented at American Studies National Convention, Baltimore, 1991.

Findlay, John M. *People of Chance: Gambling in American Society from Jamestown to Las Vegas.* Oxford University Press, 1986.

Fisher, Marvin. *Workshops in the Wilderness: The European Response to American Industrialization, 1830–1860.* Oxford University Press, 1967.

Fletcher, Colin. *The Man Who Walked Through Time.* Knopf, 1968.

Flink, James J. *The Automobile Age.* MIT Press, 1988.

Foner, Philip S., ed. *The Factory Girls.* University of Illinois, 1977.

Ford, Daniel E. *Three Mile Island.* Penguin, 1983.

Ford, Daniel E. *The Cult of the Atom.* Simon & Schuster, 1984.

Ford, Ford Madox. *New York Is Not America.* A. and C. Boni, 1927.

Ford, Henry, and Samuel Crowther. *My Life and Work.* Heinemann, 1922.

Fox, Stephen. *The Mirror Makers.* Morrow, 1984.

Fradkin, Philip. *Fallout: An American Nuclear Tragedy*. University of Arizona Press, 1989.

Franklin, H. Bruce. *War Stars: The Superweapon and the American Imagination*. Oxford University Press, 1988.

Freudenthal, Elsbeth E. *Flight into History*. University of Oklahoma Press, 1949.

Friedman, Martin. *Charles Sheeler*. Watson-Guptil, 1975.

Fuller-Ossoli, Sarah Margaret. *Summer on the Lakes, in 1843*. Little, Brown, 1844.

Gallagher, Carole. *American Ground Zero: The Secret Nuclear War*. MIT Press, 1993.

Garnett, John J. *The Statue of Liberty: Its Conception, Its Construction, Its Inauguration*. Dinsmore, 1886.

Gibson, Jane Mark. The International Electrical Exhibition of 1884 and the National Conference of Electricians. M.A. thesis, University of Pennsylvania, 1984.

Girouard, Mark. *Cities and People: A Social and Architectural History*. Yale University Press, 1985.

Glassberg, David. *American Historical Pageantry*. University of North Carolina Press, 1990.

Goldberger, Paul. *The Skyscraper*. Knopf, 1981.

Goldfield, David R., and Blaine A. Brownell. *Urban America: From Downtown to No Town*. Houghton Mifflin, 1979.

Goldman, Jonathan. *The Empire State Building Book*. St. Martin's, 1980.

Goldthwait, John T. Translator's Introduction to Immanuel Kant, *Observations on the Feeling of the Beautiful and Sublime*. University of California Press, 1960.

Grabes, Herbert. "Pleasure and the Praxis of Processes: A Transcendentalist Aesthetic of the Strange." Delivered at Würzburg University, 1991.

Greeley, Horace, et. al. 1974. *The Great Industries of the United States*, two vols. Reprint. Garland, 1974.

Greenhalgh, Paul. *Ephemeral Vistas*. University of Manchester Press, 1988.

Greenhill, Ralph. *Engineer's Witness*. Godine, 1985.

Groves, Leslie R. *Now It Can Be Told: The Inside Story of the Development of the Atomic Bomb*. Deutsch, 1963.

Hales, Peter. *Silver Cities*. Temple University Press, 1985.

Hall, Edgar H. *The Hudson Fulton Celebration, 1909*, two volumes. State of New York, 1910.

Hammond, John Winthrop. *Men and Volts*. Lippincott, 1941.

Hance, John. *Personal Impressions of the Grand Canyon of the Colorado River*. Whitaker and Ray, 1899.

Hareven, Tamara K. *Family Time and Industrial Time*. Cambridge University Press, 1982.

Hareven, Tamara K., and Randolph Langenbach. *Amoskeag: Life and Work in an American Factory-City*. Pantheon, 1978.

Henry, R. S. 1947. *Trains.* Indianapolis: Bobbs-Merrill, 1947.

Higham, John. 1986. *Send These To Me.* Revised edition. Johns Hopkins University Press, 1986.

Hilgartner, Stephen, Richard C. Bell, and Rory O'Connor. *Nukespeak: The Selling of Nuclear Technology in America.* Penguin, 1983.

History of the Atomic Energy Commission. vol. 1. Pennsylvania State University Press, 1962.

Holbrook, Stewart H. *The Story of American Railroads.* Crown, 1947.

Horrigan, Brian. "Popular Culture and Visions of the Future in Space, 1901–2001." In *New Perspectives on Technology and American Culture,* ed. B. Sinclair. American Philosophical Society, 1986.

Hughes, Thomas Parke, ed. *Changing Attitudes Toward American Technology.* Harper & Row, 1975.

Hughes, Thomas Parke. *American Genesis: A Century of Invention and Technological Enthusiasm.* Penguin, 1989.

Hungerford, Edward. *Story of the Baltimore and Ohio Railroad, 1827–1927.* Putnam, 1928.

Hunter, Louis C. *A History of Industrial Power in the United States, 1780–1930,* vol. 2. University Press of Virginia, 1985.

Hunter, Louis C., and Lynwood Bryant. *A History of Industrial Power in the United States, 1790–1930.* vol. 3. MIT Press, 1991.

Jackson, J. B. *Landscapes.* University of Massachusetts Press, 1970.

James, Henry. *The American Scene.* Chapman and Hall, 1907.

Jay, Paul, ed. *The Selected Correspondence of Kenneth Burke and Malcolm Cowley.* Viking, 1988.

Jefferson, Thomas. *Notes on the State of Virginia.* ed. William Peden. University of North Carolina Press, 1955.

Jensen, Oliver. *Railroads in America.* American Heritage, 1975.

Johnson, Clifton. *Highways and Byways of the South.* Macmillan, 1904.

Jones, Howard Mumford. *O Strange New World.* Viking, 1967.

Kant, Immanuel. *Critique of Judgement* [1790], tr. J. C. Meredith. Oxford University Press, 1952.

Kant, Immanuel. *Observations on the Feeling of the Beautiful and Sublime.* University of California Press, 1960.

Kasson, John. *Civilizing the Machine: Technology and Republican Values in America, 1776–1900.* Penguin, 1977.

Kenner, Hugh. *A Homemade World: The American Modernist Writers.* Morrow, 1975.

Kertzer, David I. *Ritual, Politics, and Power.* Yale University Press, 1988.

Kihlstedt, Folke. "Utopia Realized: The World's Fairs of the 1930s." In *Imagining Tomorrow,* ed. Corn.

Klein, Philip S., and Ari Hoogenboom. *A History of Pennsylvania.* McGraw-Hill, 1973.

Klingender, Francis. *Art and the Industrial Revolution*, ed. A. Elton. Kelley, 1968.

Knapp, Steven. *Personification and the Sublime: Milton to Coleridge*. Harvard University Press, 1985.

Kolodny, Annette. *The Land Before Her: Fantasy and Experience of the American Frontiers, 1630–1860*. University of North Carolina Press, 1984.

Kramnick, Issac. *Republicanism and Bourgeois Radicalism: Political Ideology in Late Eighteenth-Century England and America*. Cornell University Press, 1990.

Kreitler, William. *Flatiron: A photographic history of the world's first steel frame skyscraper, 1901–1990*. American Institute of Archtects, 1990.

Kroes, Rob, Robert Rydell, and Doeko Bosscher, eds. *Cultural Transmissions and Receptions: American Mass Culture in Europe*. Free University of Amsterdam Press, 1993.

Krutch, Joseph Wood. *The Grand Canyon*. Sloane Associates, 1958.

Kulik, Gary, Roger Parks, and Theodore Penn, eds. *The New England Mill Village, 1790–1860*. MIT Press, 1982.

Lakoff, Sanford, and Herbert F. York. *A Shield in Space? Technology, Politics, and the Strategic Defense Initiative*. University of California Press, 1989.

Lears, T. J. Jackson. *No Place of Grace: Anti-Modernism and the Transformation of American Culture, 1880–1920*. Pantheon, 1981.

Levine, Lawrence W. *The Unpredictable Past: Explorations in American Cultural History*. Oxford University Press, 1993.

Levinson, J. C., et al., eds. *The Letters of Henry Adams*. vol. 4. Harvard University Press, 1988.

Lewis, David L. *The Public Image of Henry Ford*. Wayne State University Press, 1976.

Licht, Walter. *Working for the Railroad: The Organization of Work in the Nineteenth Century*. Princeton University Press, 1983.

Liebs, Chester H. *Main Street to Miracle Mile*. Little, Brown, 1985.

Lyotard, Jean-François. *The Postmodern Condition*. University of Minnesota Press, 1984.

MacCannell, Dean. *The Tourist: A New Theory of the Leisure Class*. Schocken, 1976.

Mailer, Norman. *Of a Fire on the Moon*. Weidenfeld and Nicolson, 1970.

Marchand, Roland, and Michael Smith. "Corporate Science on Display." In *Science and Reform in Industrial America*, ed. R. Walters. Johns Hopkins University Press, 1992.

Marchand, Roland. *Advertising the American Dream*. University of California Press, 1985.

Marchand, Roland. "Corporate Imagery and Popular Education: World's Fairs and Expositions in the United States, 1893–1940." In *Consumption and American Culture*, ed. Nye and Pedersen.

Marquis, Alice Goldfarb. *Hopes and Ashes: The Birth of Modern Times*. Free Press, 1986.

Marvin, Carolyn. *When Old Technologies Were New.* Oxford University Press, 1988.

Marx, Leo. *The Machine in the Garden: Technology and the Pastoral Ideal in America.* Oxford University Press, 1965.

Marx, Leo. *The Pilot and the Passenger: Essays on Literature, Technology, and Culture in the United States.* Oxford University Press, 1988.

McCraw, Thomas K. *TVA and the Power Fight, 1933–1939.* Lippincott, 1971.

McCullough, David. *The Great Bridge.* Avon, 1972.

McDougall, Walter A. *The Heavens and the Earth: A Political History of the Space Age.* Basic Books, 1985.

McKay, Claude. *Selected Poems of Claude McKay.* Harcourt, Brace and World, 1953.

McKinsey, Elizabeth. *Niagara Falls: Icon of the American Sublime.* Cambridge University Press, 1985.

Meihofer, Werner. "The Ethos of the Republic and the Reality of Politics." In *Machiavelli and Republicanism,* ed. G. Bock et al. Cambridge University Press, 1990.

Meikle, Jeffrey. *Twentieth Century Limited: Industrial Design in America, 1925–1939.* Temple University Press, 1979.

Melder, Keith. "Mighty Masses of Freemen: The Young Whigs of Baltimore, 1844." Presented at American Studies National Convention, Baltimore, 1991.

Merritt, Raymond H. *Engineering in American Society, 1850–1875.* University of Kentucky Press, 1969.

Miller, Howard S. *The Eads Bridge.* University of Missouri Press, 1979.

Miller, Perry. *The Life of the Mind in America from the Revolution to the Civil War.* Harcourt, Brace and World, 1965.

Miller, Raymond C. *Kilowatts at Work: A History of the Detroit Edison Company.* Wayne State University Press, 1957.

Monk, Samuel Holt. *The Sublime: A Study of Critical Theories in Eighteenth Century England.* University of Michigan Press, 1960.

Morris, David B. *The Religious Sublime: Christian Poetry and Critical Tradition.* University of Kentucky Press, 1972.

Muccigrosso, Robert. *Celebrating the New World: Chicago's Columbian Exposition of 1893.* Dee, 1993.

Mumford, Lewis. *Sketches from Life: The Autobiography of Lewis Mumford: The Early Years.* Dial, 1982.

Nash, Roderick. *Wilderness and the American Mind,* third edition. Yale University Press, 1982.

Niagara Falls Electrical Handbook. American Institute of Electrical Engineers, 1904.

Nichols, John P. *Skyline Queen and the Merchant Prince: The Woolworth Story.* Trident, 1973.

Nicolson, Marjorie Hope. *Mountain Gloom and Mountain Glory: The Development of the Aesthetics of the Infinite.* Cornell University Press, 1959.

Nisbet, Robert A. *Émile Durkheim.* Greenwood, 1965.

Novak, Barbara. *Nature and Culture.* Oxford University Press, 1980.

Nye, David E. *Henry Ford: Ignorant Idealist.* Kennikat, 1979.

Nye, David E. *The Invented Self: An Anti-biography from Documents of Thomas A. Edison.* Odense University Press, 1983.

Nye, David E. *Image Worlds: Corporate Identities at General Electric.* MIT Press, 1985.

Nye, David E. "Social Class and the Electrical Sublime, 1880–1915." In *High Brow Meets Low Brow*, ed. R. Kroes. Free University of Amsterdam Press, 1988.

Nye, David E. *Electrifying America: Social Meanings of New Technology, 1880–1940.* MIT Press, 1990.

Nye, David E. "European Self-Representations at the New York World's Fair of 1939. In *Cultural Transmissions and Receptions: American Mass Culture in Europe*, ed. R. Kroes et al. Free University of Amsterdam Press, 1993.

Nye, David E., and Carl Pedersen, eds. *Consumption and American Culture.* Free University of Amsterdam Press, 1991.

Nye, David E., and Christen Kold Thomsen, eds. *American Studies in Transition.* Odense University Press, 1983.

Nye, Russell B. *Society and Culture in America, 1830–1860.* Harper & Row, 1974.

O'Brien, Raymond. *American Sublime: Landscape and Scenery of the Lower Hudson Valley.* Columbia University Press, 1981.

Office of the Commission of Education. *Hudson-Fulton Celebration 1609–1807–1909.* (New York) State Education Department, 1909.

Oppenheim, James. *The Olympian.* Harper, 1912

Ordway, Frederick I., and Randy Liebermann. *Blueprint for Space: Science Fiction to Science Fact.* Smithsonian Institution, 1992.

Padover, Saul K., ed. *The Complete Jefferson.* Books for Libraries Press, 1969.

Pangle, Thomas L. *The Spirit of Modern Republicanism: The Moral Vision of the American Founders and the Philosophy of Locke.* University of Chicago Press, 1988.

Passer, Harold. *The Electrical Manufacturers. 1875–1900.* Harvard University Press, 1953.

Pauli, Hertha, and E. B. Ashton. *I Lift My Lamp: The Way of a Symbol.* Appleton-Century-Crofts, 1948.

Pavlovskis, Zoja. *Man in an Artificial Landscape: The Marvels of Civilization in Imperial Roman Literature.* Mnemosyne, Biblioteca Classica Batava, 1973.

Pease, Donald. "Sublime Politics." In *The American Sublime*, ed. M. Arensberg. State University of New York Press, 1986.

Pells, Richard. *Radical Visions and American Dreams: Culture and Social Thought in the Depression Years.* Harper & Row, 1973.

Pickering, W. S. G., ed. *Durkheim on Religion.* Routledge & Kegan Paul, 1975.

Platt, Harold L. *The Electric City.* University of Chicago Press, 1991.

Plowden, David. *Bridges: The Spans of North America.* Viking, 1974.

Pocock, J. G. A. *Politics, Language, and Time.* Atheneum, 1973.

Pomery, Earl. *In Search of the Golden West.* Knopf, 1957.

Presbury, Frank. *The History and Development of Advertising.* Doubleday, Doran, 1929.

Railroad Jubilee, An Account of the Celebration Commemorative of the Opening of Railroad Communication between Boston and Canada. J. H. Eastburn, 1852.

Remini, Robert V. *Andrew Jackson and the Course of American Democracy. 1833–1845.* Harper & Row, 1984.

Rhodes, Richard. *The Making of the Atom Bomb.* Simon & Schuster, 1986.

Robinson, Cerwin and Rosemari Haag Bletter. *Skyscraper Style: Art Deco New York.* Oxford University Press, 1975.

Robinson, Charles Mulford. 1902. *The Improvement of Towns and Cities.* Putnam, 1902.

Robinson, Douglas. 1985. *American Apocalypses: The Image of the End of the World in American Literature.* Johns Hopkins University Press, 1985.

Roemer, Kenneth. *The Obsolete Necessity: America in Utopian Writings, 1888–1900.* Kent State University Press, 1981.

Rogers, W. G., and Mildred Weston. *Carnival Crossroads, the Story of Times Square.* Doubleday, 1960.

Rogin, Michael. *Ronald Reagan, The Movie, and Other Episodes in Political Demonology.* University of California Press, 1987.

Rolle, Andrew F. *California: A History.* Crowell, 1969.

Rosenzweig, Roy. *Eight Hours for What We Will.* Cambridge University Press, 1983.

Roth, Leland M. *America Builds: Source Documents in American Architecture and Planning.* Harper & Row, 1983.

Runte, Alfred. *National Parks: The American Experience.* University of Nebraska Press, 1979.

Rydell, Robert W. *All the World's a Fair.* University of Chicago Press, 1984.

Rydell, Robert W. *World of Fairs: The Century of Progress Expositions.* University of Chicago Press, 1993.

Saffer, Thomas H., and Orville Kelly. *Countdown Zero: GI Victims of U.S. Atomic Testing.* Penguin, 1983.

Sandeen, Eric J. "The Value of Place: The Redevelopment Debate over New York's Times Square." In *American Studies and Consumption,* ed. Nye and Pedersen.

Sanford, Charles L., ed. *Quest for America, 1810–1824.* New York University Press, 1964.

Sanford, Charles L. *The Quest for Paradise: Europe and the American Moral Imagination.* University of Illinois Press, 1961.

Saum, Lewis O. *The Popular Mood of Pre-Civil War America.* Greenwood, 1980.

Scharf, Col. J. Thomas. *The Chronicles of Baltimore; Being a Complete History of Baltimore Town and Baltimore City from the Earliest Period to the Present Time.* Turnbull Brothers, 1874.

Schleier, Merrill. *The Skyscraper in American Art, 1890–1931.* Da Capo, 1986.

Schull, William J. *Song Among The Ruins.* Harvard University Press, 1990.

Seaborg, Glenn T., and William R. Corliss. *Man and Atom: Building a New World Through Nuclear Technology.* Dutton, 1971.

Sears, John F. *Sacred Places: American Tourist Attractions in the Nineteenth Century.* Oxford University Press, 1989.

Serling, Robert J. *Eagle: The Story of American Airlines.* St. Martin's, 1985.

Shaw, Ronald E. *Canals for a Nation: The Canal Era in the United States, 1790–1860.* University of Kentucky Press, 1990.

Shaw, Ronald E. *Erie Water West: A History of the Erie Canal, 1792–1854.* University of Kentucky Press, 1966.

Shields, Rob. *Places on the Margin.* Routledge, 1991.

Sibley, Mulford Q. *Political Ideas and Ideologies: A History of Political Thought.* Harper, 1970.

Silverman, Hugh J., and Gary E. Aylesworth, eds. *The Textual Sublime. Deconstruction and its Differences.* State University of New York Press, 1990.

Singer, Charles, et al. *A History of Technology,* vol 5. Oxford University Press, 1958.

Smith, Henry Nash. *Virgin Land.* Harvard University Press, 1950.

Smith, Michael L. "Back to the Future: EPCOT, Camelot, and the History of Technology." In *New Perspectives on Technology and American Culture,* ed. B. Sinclair. American Philosophical Society, 1986.

Smith, Michael L. *Pacific Visions: California Scientists and the Environment, 1850–1915.* Yale University Press, 1987.

Smith, Michael L. "Advertising the Atom." In *Government and Environmental Politics,* ed. M. Lacey. Wilson Center Press, 1989.

Sollors, Werner. "Of Plymouth Rock and Jamestown and Ellis Island—or Ethnic Literature and Some Redefinitions of 'America'." In *Multiculturalism and the Canon of American Culture,* ed. H. Bak. Free University of Amsterdam Press, 1993.

Sower, John F. *History of the Baltimore and Ohio Railroad.* Purdue University Press, 1987.

Spanier David. *Welcome to the Pleasure Dome.* University of Nevada Press, 1992.

Starrett, Paul. *Changing the Skyline: An Autobiography.* McGraw-Hill, 1939.

Stidger, William. *Henry Ford: The Man and His Motives.* Doran, 1923.

Stilgoe, John. *Metropolitan Corridor: Railroads and the American Scene.* Yale University Press, 1983.

Stover, John F. *American Railroads.* University of Chicago Press, 1961.

Stover, John F. *The Life and Decline of the American Railroad.* Oxford University Press, 1970.

Strasser, Susan. *Satisfaction Guaranteed: The Rise of the American Mass Market.* Pantheon, 1989.

Sullivan, Louis. *The Autobiography of an Idea*. Dover, 1954.

Summers, Mark W. *The Plundering Generation: Corruption and the Crisis of the Union, 1849–1861*. Oxford University Press, 1987.

Taylor, Nicholas. "The Awful Sublimity of the Victorian City: Its aesthetic and architectural origins." In *The Victorian City, Images and Realities*, vol. 2, ed. H. J. Dyos and M. Wolff. Routledge & Kegan Paul, 1973.

Taylor, William R., and Thomas Bender. "Culture and Architecture: Some Aesthetic Tensions in the Shaping of Modern New York City." In *Visions of the Modern City*, ed. W. Sharpe and L. Wallock. Johns Hopkins University Press, 1987.

Taylor, William R. "The Evolution of Public Space in New York City." In *Consuming Visions*, ed. Bronner.

Thorsen, Niels. "Political Rituals in America." In *Inventing Modern America*, ed. D. Nye, C. Pedersen, and N. Thorsen. Copenhagen: Danmarks Radio, 1989.

Tichi, Cecelia. *Shifting Gears: Technology, Literature, Culture in Modernist America*. University of North Carolina Press, 1987.

Titus, A. Costandina. *Bombs in the Backyard*. University of Nevada Press, 1986.

Tompkins, Edmund Pendleton, and J. Lee Davis. *The Natural Bridge and its Historical Surroundings*. Natural Bridge of Virginia, Inc., 1939.

Trachtenberg, Alan. *Brooklyn Bridge: Fact and Symbol*. University of Chicago Press, 1979.

Trachtenberg, Alan. *The Incorporation of America*. Hill & Wang, 1982.

Trachtenberg, Alan. *Reading American Photographs*. Hill and Wang, 1989.

Trachtenberg, Marvin. *The Statue of Liberty*. Viking, 1976.

Trescott, Martha Moore. *The Rise of the American Electrochemical Industry, 1880–1910*. Greenwood, 1981.

Truettner, William H., ed. *The West as America: Reinterpreting Images of the Frontier, 1820–1920*. Smithsonian Institution, 1991.

Tsujimoto, Karen. *Images of America: Precisionist Painting and Modern Photography*. University of Washington Press, 1982.

Turner, Victor. *The Ritual Process*. Cornell University Press, 1969.

Turner, Victor and Edith. *Image and Pilgrimage in Christian Culture*. Columbia University Press, 1978.

Tuveson, Ernest Lee. *Redeemer Nation*. University of Chicago Press, 1968.

van Leeuwen, Thomas A. P. *The Skyward Trend of Thought: The Metaphysics of the American Skyscraper*. MIT Press, 1988.

Wallace, Anthony F. C. *Rockdale*. Norton, 1978.

Ward, James A. *Railroads and the Character of America, 1820–1887*. University of Tennessee Press, 1986.

Ward, John William. *Andrew Jackson: Symbol for an Age*. Oxford University Press, 1955.

Ware, Norman. *The Industrial Worker, 1840–1860*. Reprint of 1924 edition. Elephant, 1990.

Warner, Marina. *Monuments and Maidens: The Allegory of the Female Form.* Picador, 1985.

Waters, Frank. *The Colorado.* Rinehart, 1946.

Watson, John. *The Philosophy of Kant Explained.* Facsimile reprint of 1908 edition. Garland, 1976.

Weart, Spencer R. *Nuclear Fear: A History of Images.* Harvard University Press, 1988.

Weiskel, Thomas. *The Romantic Sublime; Studies in the Structure and Psychology of Transcendence.* Johns Hopkins University Press, 1976.

Wilentz, Sean. *Chants Democratic: New York City and the Rise of the American Working Class, 1788–1850.* Oxford University Press, 1984.

Wills, Gary. *Inventing America: Jefferson's Declaration of Independence.* Doubleday, 1978.

Wilson, Richard Guy, Dianne H. Pilgrim, and Dickran Tashjian. *The Machine Age in America, 1918–1941.* Brooklyn Museum/Abrams, 1986.

Wilson, Rob. *American Sublime: Genealogy of a Poetic Genre.* University of Wisconsin Press, 1991.

Wolf, Bryan. "When is a Painting Most Like a Whale?: Ishmael, *Moby Dick*, and the Sublime." In *New Essays on Moby Dick*, ed. R. H. Brodhead. Cambridge University Press, 1986.

Wright, Lawrence. *In the New World: Growing Up with America, 1960–1984.* Knopf, 1988.

Wyschogrod, Edith. *Spirit in Ashes.* Yale University Press, 1985.

Yates, Gayle Graham, ed. *Harriet Martineau on Women.* Rutgers University Press, 1985.

Zelinsky, Wilbur. *Nation Into State: The Shifting Symbolic Foundations of American Nationalism.* University of North Carolina Press, 1988.

Zim, Larry, Mel Lerner, and Herbert Rolfes. *The World of Tomorrow, 1939: New York World's Fair.* Harper & Row, 1988.

Zinsser, William. *American Places.* HarperCollins, 1992.

Zizek, Slavoj. *The Sublime Object of Ideology.* Verso, 1989.

Index